Functional Metal-Organic Frameworks

Scrivener Publishing
100 Cummings Center, Suite 541J
Beverly, MA 01915-6106

Publishers at Scrivener
Martin Scrivener (martin@scrivenerpublishing.com)
Phillip Carmical (pcarmical@scrivenerpublishing.com)

Functional Metal-Organic Frameworks

Structure, Properties and Applications

Ali Morsali and
Sayed Ali Akbar Razavi
Department of Chemistry, Faculty of Sciences,
Tarbiat Modares University, Tehran, Iran

Scrivener
Publishing

WILEY

This edition first published 2021 by John Wiley & Sons, Inc., 111 River Street, Hoboken, NJ 07030, USA and Scrivener Publishing LLC, 100 Cummings Center, Suite 541J, Beverly, MA 01915, USA

For more information about Scrivener publications please visit www.scrivenerpublishing.com.

Wiley Global Headquarters
111 River Street, Hoboken, NJ 07030, USA

For details of our global editorial offices, customer services, and more information about Wiley products visit us at www.wiley.com.

Library of Congress Cataloging-in-Publication Data

ISBN 9781119640431

Cover image: Pixabay.Com
Cover design by Russell Richardson

Set in size of 11pt and Minion Pro by Manila Typesetting Company, Makati, Philippines

Printed in the USA

Contents

Preface

As a subclass of coordination polymers and porous materials, metal–organic frameworks (MOFs) are composed of a dual organic–inorganic structure based on organic (organic linkers) and inorganic (metal ions/clusters) building blocks. In structural view, a unique kind of connection between organic linkers and inorganic nodes leads to construction of a three-dimensional framework with vacant spaces between building blocks. Owing to unlimited possibility in selection of organic ligands and metal ions/clusters, theoretically it is feasible to synthesis an unlimited number of frameworks.

In recent decades, MOFs received lots of attention in the world of material science and chemistry. Such tremendous attention is owing to their unique chemical characters such as hybrid organic–inorganic nature, high porosity and surface area, tunability in chemical functionality, highly ordered and crystalline structure and moderate-to-high chemical and thermal stabilities. Each one of these chemical properties enable MOFs to apply for specific purpose, but the ability to functionalize MOFs is a specific character to improve the capability of MOFs in different field of applications.

There are three ways for functionalization of MOFs including: (I) using functional organic linker, (II) pore functionalization through immobilization of other functional materials and (III) functionalization of inorganic nodes of the framework. Owing to versatile kind of organic functional groups, linker functionalization is recognized as a favorite strategy to tailor the chemical properties and enrich the host-guest chemistry of functional metal–organic frameworks (FMOFs).

In this book, we tried to review the literature to gain deep insight about the effects of linker functionalization on structure and host-guest chemistry of FMOFs. The content of this book is useful for gaining better understanding of the structural and chemical properties of FMOFs. Considering our strategy in this book, we believe that this book is interesting for diverse group of scientists like chemists, material engineers and anyone who is working on supramolecular chemistry of MOFs and designing functional materials.

The authors
October 2020

1

Introduction to Functional Metal–Organic Frameworks

Abstract
In this chapter, we discuss about the advantages of porous materials and crystalline materials and explain that what kinds of benefits are attainable if these advantages combine together in the structure of functional materials like metal-organic frameworks. Then, functional metal-organic frameworks are discussed and classified based on the roles of organic functions in the structure and application of MOFs.

Keywords: Porous materials, crystalline materials, functional metal-organic frameworks, coordination polymers, host-guest chemistry, function-application properties, function-structure properties

1.1 Coordination Polymers

Solid materials are generally classified in amorphous and crystalline (single-crystalline or poly-crystalline) solids in chemistry and material science. Crystalline solids are constructed based on periodic symmetrical arrays of constituents giving rise to definite, regular and repeating pattern of the solid in three dimensions over a large distance. Such long-range structural order rises in the beneficial fact that crystalline solids represent specific and repeatable chemical properties. This is a very pivotal advantage which is not observed in amorphous solids. For example, crystalline solids are of sharp melting point and definite heat of fusion while amorphous solids have not a characteristic heat of adsorption and sharp melting point. As a result, crystalline solids benefit from repeatable structure and chemical properties which are fitting characters in application of novel materials.

Ali Morsali and Sayed Ali Akbar Razavi. Functional Metal-Organic Frameworks: Structure, Properties and Applications, (1–14) © 2021 Scrivener Publishing LLC

Another classification of materials is based on their porosity. Porosity, which also is called void fraction, is defined as the ratio of vacant space (void) in material to the total volume that the materials occupy. This fraction is always between 0 and 1. Porous materials encompass vacant spaces in their structure based on accessible pore volume (vacant cavities or channels) for guest molecules. This is a unique advantage of porous material in which not only can they interact with guest molecules on their surface, but also they can adsorb and interact with guests within their pores inside the bulk material. The characteristics of a porous material define by their surface area ($m^2 \cdot g^{-1}$), accessible pore volume ($m^3 \cdot g^{-1}$), shape, size and distribution of pores. Based on pore size, porous materials are classified in three major groups including microporous (in the range of 2 nm and below), mesoporous (in the range of 2 to 50 nm) and macroporous (above 50 nm) [1]. Another way to classify porous materials is pursuant to uniformity in the pore size, volume and distribution [2]. In this approach, porous materials are classified as ordered (uniform) and non-ordered groups. Uniform porous materials are developed based on same pore size, shape and distribution. To observe such uniformity in porosity, a porous material must be founded on uniform and repeatable structural patterns. This uniformity in the structure and porosity is essential for some of superior applications like size selective separation of a small molecule from a mixture containing large molecules. In size selective applications, guest molecules with smaller size (or kinetic diameter) than pore aperture of the host are able to diffuse into the pores of ordered porous material while molecules with larger size cannot. Definitely, porous materials without uniformity in their pore size and distribution could not be applied in size-selective applications because they cannot differentiate guest molecules with different sizes. These contents indicate that crystalline porous solids with regular and repeatable structure and porosity are very efficacious in molecular-sieving and also other kinds of applications.

Crystalline porous solids can be extended by different types of interactions (ionic and hydrogen bonds, covalent interactions and coordination interactions) between their individual molecular building blocks [3]. Especially, crystalline porous materials which are developed by coordination interactions are coordination polymers (CPs). In structural view, CPs could be extended in different dimensions, so they could be 1-dimensional (1D), 2D or 3D. Also, they are synthesized based on linkers and metal ion/clusters when a polydentate linker is able to associate multiple metal centers through coordination bonds in self-assembly process (Figure 1.1) [4]. As a subclass of CPs, metal–organic frameworks (MOFs) are porous,

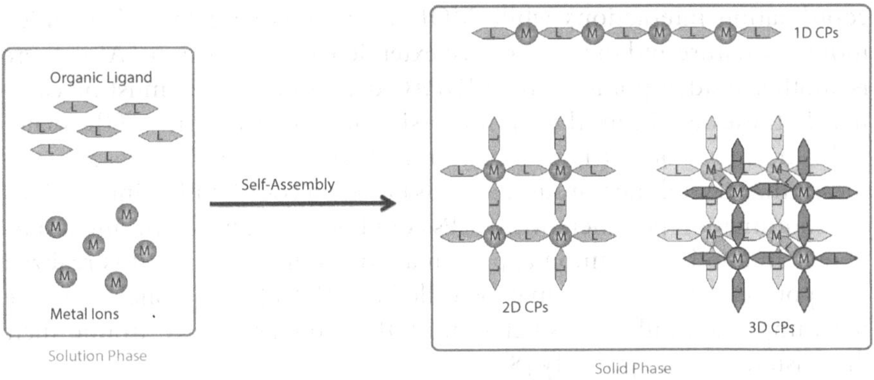

Figure 1.1 Representation of CPs building blocks, synthesis and dimensionality.

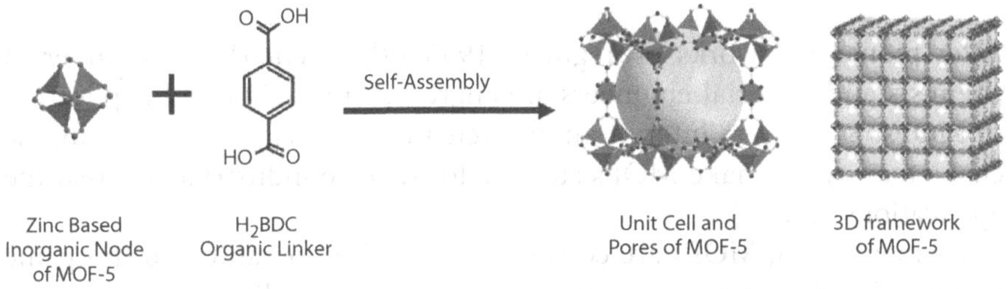

Figure 1.2 Depiction of MOF-5 ($Zn_4O(BDC)_3$) as one of the most well-known MOFs. MOF-5 is synthesized using zinc nitrate and 1,4-benzenedicarboxyxlic acid (H_2-BDC). In self-assembly process Zn(II) ions and deprotonated 1,4-benzenedicarboxyxlic acid bond together through coordination interactions between Zn(II) ions and BDC^{2-}linkers to develop MOF-5 in solid phase [5].

three-dimensional and developed based on polydentate organic ligands and metal ion/clusters (Figure 1.2). Although these terms, MOFs and CPs, are widely applied interchangeably in the literature, there are some similarities and differences between both terms.

In 2013, IUAC recommended that CP could be defined as a "coordination compound with repeating coordination entities extending in 1, 2, or 3 dimensions" while MOF is defined as "coordination network with organic ligands containing potential voids" [6]. It is necessary to mention that IUPAC defines coordination networks as a "coordination compound extending through repeating coordination entities in 2 or 3 dimensions" [6]. Based on these recommendations it could be concluded that CPs could be 1D, 2D or 3D with repetitive building blocks connected with coordination interactions while MOF are porous with hybrid organic–inorganic nature and they are at least extended in 2 dimensions. Also, there is another hidden point in these IUPAC definitions which must be clarified. In these definitions there is no persistence on crystallinity of the materials. In other word, it is not mentioned that CPs must be crystalline. It means that crystalline porous materials based on coordination interactions are coordination polymers while CPs could be both crystalline and amorphous. In another common approach about MOFs and CPs, it is realized that porous 3D structures may be called as MOFs [7]. In conclusion, the most important differences between MOFs and CPs revolve around their dimensionality and porosity [8].

1.2 Metal–Organic Frameworks

After pioneering works of Yaghi in 1995 [9], a remarkable number of chemists and material engineers are engrossed in design and application of MOFs. Such great interest among scientists is because of their unique characters which make MOFs suitable for diverse industrial and real-life applications [10, 11].

As mentioned, MOFs are constructed based on organic and inorganic building blocks. In most cases, organic ligands are ditopic or polytopic O-donor ligands based on carboxylates linkers or N-donor ligands based on pyridine pillar spacers. Metal-containing units founded on different kind of metal ions, mostly based on lanthanide cations (like Ln(III), Tb(III), Eu(III), Dy(III) and Sm(III)), transition metals (like 3d cations like Zn(II), Cu(II), Ni(II), Co(II), Fe(III) or Fe(II), Mn(II) and Cr(III) or Cr(II) and heavy transition metals like Cd(II), Zr(IV), Hf(IV)) and main metal ions (Al(III), or some of alkaline or alkaline-earth cations). As a result of such diversity in selection of building blocks, unlimited number of MOFs with different structural and practical properties can be developed by changing metal ion/clusters, using various combinations of these inorganic building blocks and infinite types of organic linkers with different lengths, functionalities and geometries. Additionally, since MOFs are developed based on organic and inorganic building blocks, their hybrid organic–inorganic nature is suitable for tuning the structure and application of MOFs.

MOFs are synthesized by coordination of organic linker to inorganic units by strong bonds [12]. These connections between organic and inorganic building blocks through coordination interactions are such that they create vacant spaces (pores) between the each individual building block.

As a result of such porosity, MOFs provide accessible pore volume in the bulk of the materials moreover than accessible area at the surface of material. Right selection of the building blocks makes it possible to vary some parameters, such as the pore size (to increase pore diameter to 98 Å), density (to decrease to 0.126 $g{\cdot}cm^{-3}$) and surface area (typically in range from 1,000 to 10,000 $m^2{\cdot}g^{-1}$) which are exceeding those of traditional porous materials such as zeolites and carbons [13]. In summary, porosity and surface area of MOFs could be tuned through right selection of organic ligands with correct size, flexibility and appropriate inorganic nodes.

The nature, strength and the number of coordination interactions between organic and inorganic building blocks of MOFs are the main reasons for evaluation of their stability. On one hand, selection of building blocks based on Hard-Soft acid-based theory is very beneficial for synthesis of highly stable MOFs [14]. For example hard metal ions such as Al(III) and Zr(IV) [15] could develop stable MOFs through connection with carboxylate-donor organic linker because these building blocks are hard Lewis acid and base, respectively. Another group of stable MOFs are based on selection of soft metal ions like late 3D metal ions and soft N-donor organic ligands like pyrazolate based linkers. On the other hand, the number of coordination bonds between inorganic nodes and organic linkers is another critical factor on the stability of MOFs. The higher number of coordination bonds, the higher stability of the MOF.

Another desirable character of MOFs is their crystalline structure. Selection of well-defined individual molecular building blocks could develop regular structure and periodic frameworks. As a result of their regular crystalline structure, we can tune their chemical and physical properties through logical designing of the framework and right selection of building blocks.

One of the most important advantages of MOFs from other conventional polymers or porous materials is the fact that we can tailor their chemical properties through rational choice of functional groups. Tunability in chemical functionality of MOFs could be provided via three functionalization strategies including selection of functional organic ligand, functionalization of inorganic nodes and functionalization of vacant spaces (pores) inside the framework of MOFs (Figure 1.3) [14, 16]. These functions could be introduced prior to the self-assembly process by selection of desirable functional building blocks (pre-synthesis functionalization strategy) or after synthesis of the frameworks (post-synthesis functionalization strategy) [17, 18]. Unlimited ways for functionalization of MOFs is key factor to control their chemical properties and host–guest

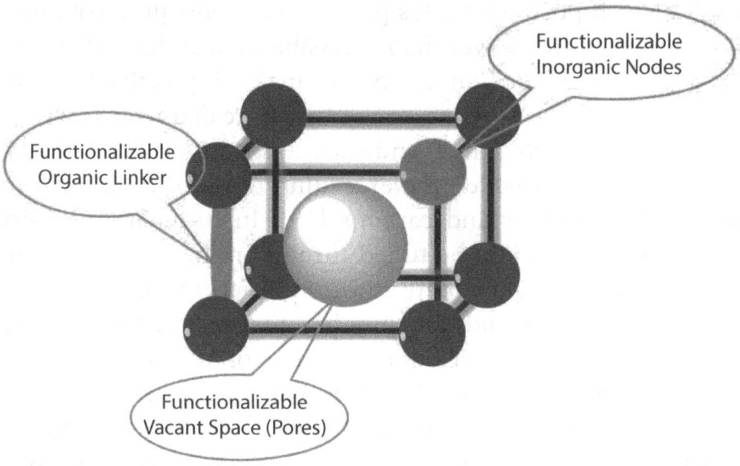

Figure 1.3 Three strategies for functionalization of MOFs.

interactions. Definitely, we can state that tunability in chemical functionality and consequent ability in control over host–guest chemistry of MOFs has very essential roles in absorbing the attraction of scientist toward MOFs.

Considering the diverse and useful characteristics of MOFs including regular crystalline structure, high porosity and surface area, hybrid inorganic–organic nature, satisfactory structural stability and tunability in chemical functionality, they are employed for different types of applications such as gas storage and separation [19–23], heterogeneous catalysis [24, 25] and photocatalysis [26–28], sensing [29–33], removal and separation of hazardous chemicals [34–36], drug delivery [37], bio sensing [38–41] and other medical applications [42] electrical conductivity [43] and electrochemical applications [44], ion storage and conductivity [45, 46], and designing MOF-based energetic materials [47]. Such diversity in chemical properties and application of MOFs show that intensive studies are necessary to identify the maximum capability of MOFs for each type of applications.

1.3 Functional Metal–Organic Frameworks

We mentioned that there are three different strategies for functionalization of MOFs. Among these methods, functionalization of organic ligand absorbed massive attention thanks to unlimited diversity in synthesis of organic ligands with different length and shape as well as rich host–guest chemistry of organic functional groups. So, functionalized organic linkers are extensively applied for construction of functional metal-organic frameworks (FMOFs) [14, 48].

The most important reason for introduction of organic functional groups into the framework of MOFs is to control the host–guest chemistry of FMOFs and gain selectivity to special guest molecules in presence of other analytes or tailoring the photoelectrochemical properties of FMOFs. However, moreover than domination in the host–guest chemistry and chemical properties of FMOFs, introduction of organic functional groups influence on the structural properties of FMOFs through induced structural changes and different types of secondary interactions (Figure 1.4).

Although linker functionalization strategy gained lots of attention to control the chemical properties of FMOFs, FMOFs could be developed through other methods. For example a special kind of linker can be stabilized on inorganic nodes or special kind of functional materials like carbonitride, nanoparticles, dyes, graphene or graphene oxide or polyoxometalates can be incorporated into the pores of MOFs to fabricate a multi-functional MOF-based hybrid material (Figure 1.5) [49]. In addition, synergic cooperation of functions in the structure of FMOFs is useful for improvement of practical application of FMOFs.

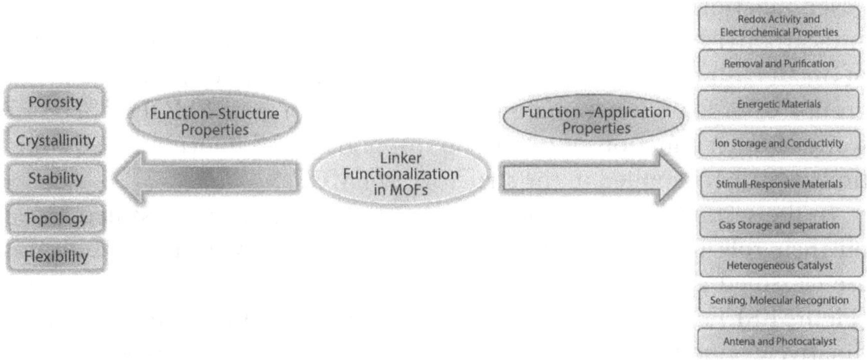

Figure 1.4 Function–structure and function–application properties of functional MOFs. Function–structure properties include relationship between structural features of MOFs and functional groups while function–application properties encompass the effects of functional groups on different type of applications.

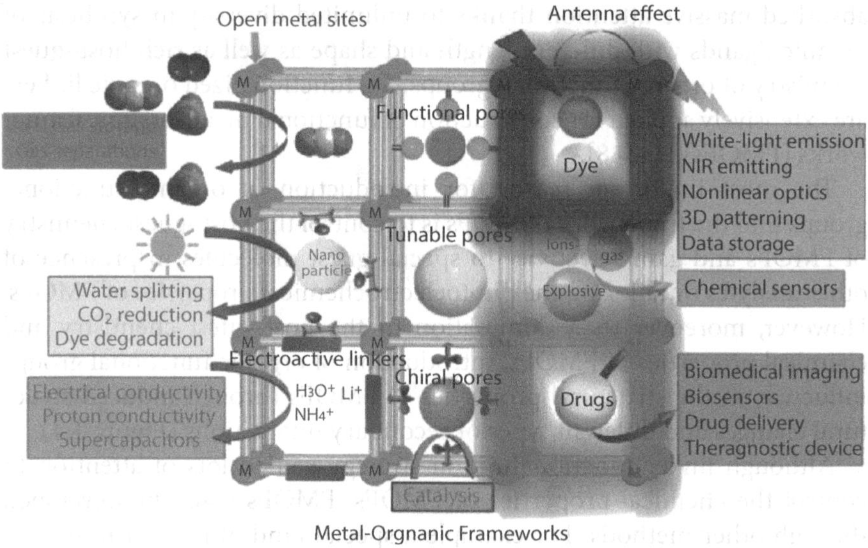

Figure 1.5 Different application of functional or multi-functional MOF-based materials [16].

For a coherent and purposeful study, organic functional groups are sorted in two different methods. Based on first approach, organic functional groups are classified owing to their chemical properties and structural similarities. In this view, applied functional groups into the structure of FMOFs are categorized in four major groups including (I) nitrogen-based functions, (II) oxygen-based functions, (III) sulfur-based functions and carbonyl-based functions. All these four major groups entail a number of organic functional groups (Figure 1.6). In addition to these four major groups of organic functional groups into the pores of FMOFs, there are some of other functions especially phosphonate and fluorine-based functions which are applied in the structure of FMOFs.

In the second approach, organic functional groups are classified based on their role in the structure of FMOFs. In this view functional groups are classified as coordinating sites (Figure 1.7) and gust-interactive sites (Figure 1.8). Functional groups as coordinating site are those which can coordinatively bond to metal ions during the synthesis of MOFs. Coordinating functional groups must be strong coordination bond donor with high chelation ability to develop stable FMOFs. There are two common types of

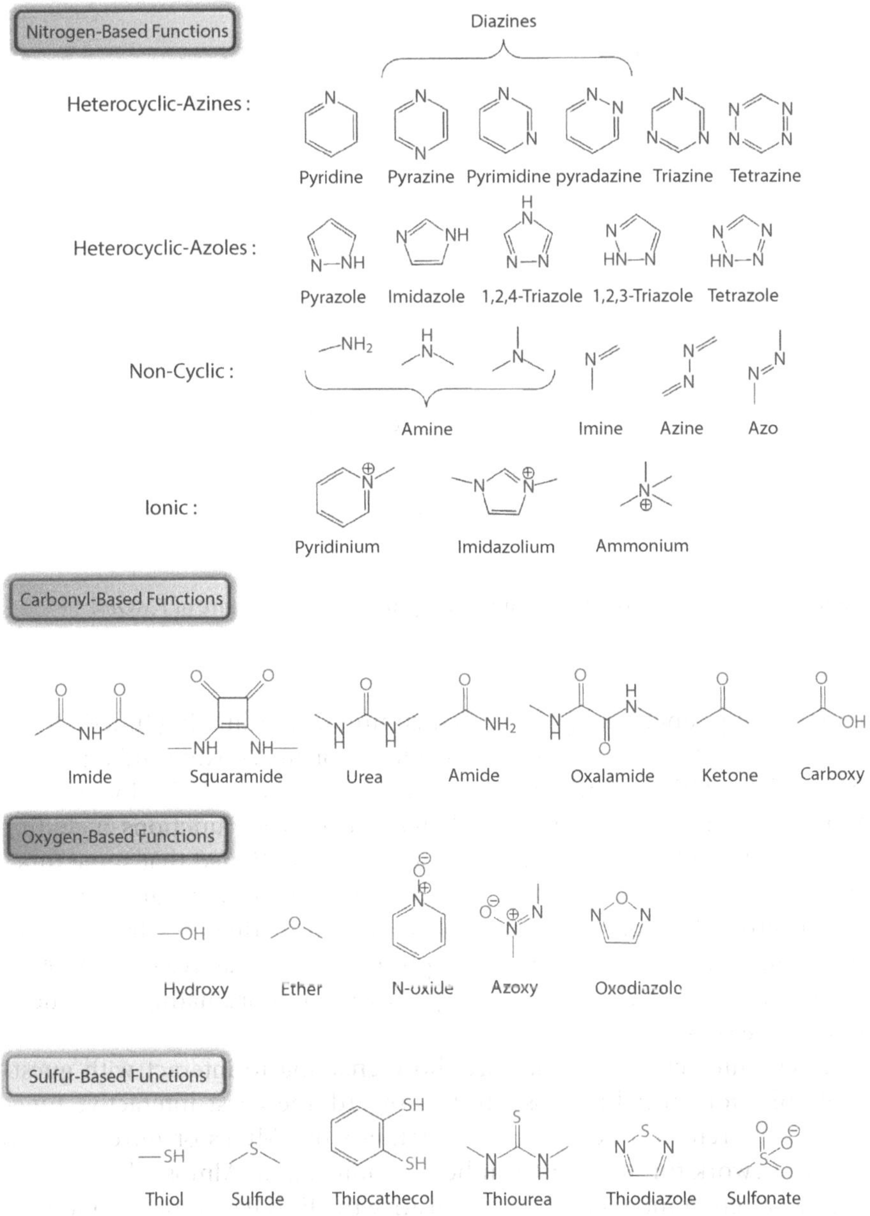

Figure 1.6 Classification of organic functional groups which are applied in synthesis of FMOFs based on their chemical characteristics and structural properties.

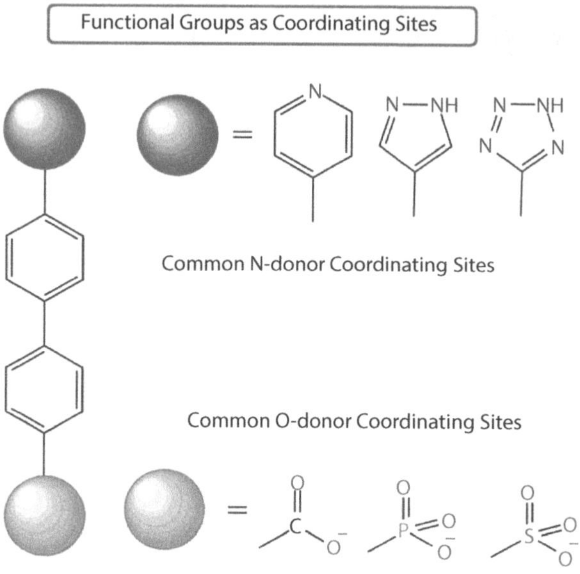

Figure 1.7 Common coordinating functional groups in the structure of FMOFs.

coordinating functional groups. First common coordinating functions are O-donor functions such as sulfonate, phosphonate, enoxide and especially carboxylate. Other common coordinating functions are N-donor functions like pyridine, pyrazine and heterocyclic azole functions including pyrazole, triazole and tetrazole. Although, we mentioned that some functions like carboxylate, sulfonate, phosphonate, enoxide, pyridine, pyrazine and heterocyclic azole functions applied as coordinating sites, this does not mean that these functions did not apply as guest interactive sites. In other word, mentioned functions can apply as both coordinating and guest-interactive sites.

Guest-interactive functions are those enabling to interact with guest-molecules for special purpose. In this regard, the guest-interactive functions must remain free during the synthesis of FMOFs or introduced in the framework through post-synthesis modification. Almost all functions applied as guest interactive site to sensitize the FMOFs to special guest.

The next chapters of this book are structured based on these two approaches and effects of any function on the structure and application of FMOFs are discussed in details.

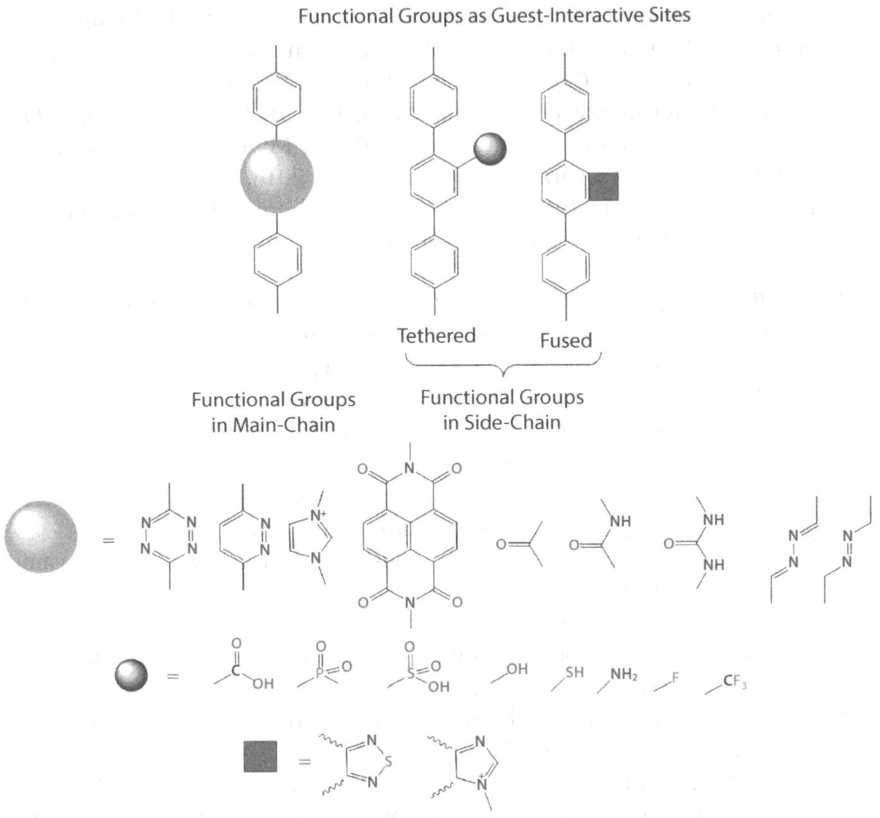

Figure 1.8 Functional groups as guest-interactive sites inside the structure of FMOFs and position of some of common functional groups.

References

1. Zdravkov, B., Čermák, J., Šefara, M., Janků, J., Pore classification in the characterization of porous materials: A perspective. *Open Chem.*, 5, 385–395, 2007.
2. Davis, M.E., Ordered porous materials for emerging applications. *Nature*, 417, 813–821, 2002.
3. Parnham, E.R. and Morris, R.E., Ionothermal synthesis of zeolites, metal–organic frameworks, and inorganic–organic hybrids. *Acc. Chem. Res.*, 40, 1005–1013, 2007.

4. Tranchemontagne, D.J., Mendoza-Cortés, J.L., O'Keeffe, M., Yaghi, O.M., Secondary building units, nets and bonding in the chemistry of metal-organic frameworks. *Chem. Soc. Rev.*, 38, 1257–1283, 2009.
5. Perez, E.V., Karunaweera, C., Musselman, I.H., Balkus, K.J., Ferraris, J.P., Origins and evolution of inorganic-based and MOF-based mixed-matrix membranes for gas separations. *Processes*, 4, 32, 2016.
6. Bennett, T.D. and Horike, S., Liquid, glass and amorphous solid states of coordination polymers and metal–organic frameworks. *Nat. Rev. Mater.*, 3, 431–440, 2018.
7. Biradha, K., Ramanan, A., Vittal, J.J., Coordination polymers versus metal-organic frameworks. *Cryst. Growth Des.*, 9, 2969–2970, 2009.
8. Batten, S.R., Champness, N.R., Chen, X.-M., Garcia-Martinez, J., Kitagawa, S., Öhrström, L., O'Keeffe, M., Suh, M.P., Reedijk, J., Coordination polymers, metal–organic frameworks and the need for terminology guidelines. *CrystEngComm*, 14, 3001–3004, 2012.
9. Yaghi, O.M., Li, G., Li, H., Selective binding and removal of guests in a microporous metal–organic framework. *Nature*, 378, 703, 1995.
10. Czaja, A.U., Trukhan, N., Müller, U., Industrial applications of metal–organic frameworks. *Chem. Soc. Rev.*, 38, 1284–1293, 2009.
11. Kirchon, A., Feng, L., Drake, H.F., Joseph, E.A., Zhou, H.-C., From fundamentals to applications: A toolbox for robust and multifunctional MOF materials. *Chem. Soc. Rev.*, 47, 23, 8611–8638, 2018.
12. Long, J.R. and Yaghi, O.M., The pervasive chemistry of metal–organic frameworks. *Chem. Soc. Rev.*, 38, 1213–1214, 2009.
13. Butova, V.V.E., Soldatov, M.A., Guda, A.A., Lomachenko, K.A., Lamberti, C., Metal–organic frameworks: Structure, properties, methods of synthesis and characterization. *Russ. Chem. Rev.*, 85, 280, 2016.
14. Ali Akbar Razavi, S. and Morsali, A., Linker functionalized metal–organic frameworks. *Coord. Chem. Rev.*, 399, 213023, 2019.
15. Bai, Y., Dou, Y., Xie, L.-H., Rutledge, W., Li, J.-R., Zhou, H.-C., Zr-based metal–organic frameworks: Design, synthesis, structure, and applications. *Chem. Soc. Rev.*, 45, 2327–2367, 2016.
16. Li, B., Wen, H.M., Cui, Y., Zhou, W., Qian, G., Chen, B., Emerging multifunctional metal–organic framework materials. *Adv. Mater.*, 28, 40, 8819–8860 2016.
17. Cohen, S.M., Postsynthetic Methods for the Functionalization of Metal-Organic Frameworks. *Chem. Rev.*, 112, 970–1000, 2012.
18. Deria, P., Mondloch, J.E., Karagiaridi, O., Bury, W., Hupp, J.T., Farha, O.K., Beyond post-synthesis modification: Evolution of metal–organic frameworks via building block replacement. *Chem. Soc. Rev.*, 43, 5896–5912, 2014.
19. Sumida, K., Rogow, D.L., Mason, J.A., McDonald, T.M., Bloch, E.D., Herm, Z.R., Bae, T.-H., Long, J.R., Carbon dioxide capture in metal–organic frameworks. *Chem. Rev.*, 112, 724–781, 2011.

20. Getman, R.B., Bae, Y.-S., Wilmer, C.E., Snurr, R.Q., Review and analysis of molecular simulations of methane, hydrogen, and acetylene storage in metal–organic frameworks. *Chem. Rev.*, 112, 703–723, 2011.
21. Mason, J.A., Veenstra, M., Long, J.R., Evaluating metal–organic frameworks for natural gas storage. *Chem. Sci.*, 5, 32–51, 2014.
22. He, Y., Zhou, W., Qian, G., Chen, B., Methane storage in metal–organic frameworks. *Chem. Soc. Rev.*, 43, 5657–5678, 2014.
23. Murray, L.J., Dincă, M., Long, J.R., Hydrogen storage in metal–organic frameworks. *Chem. Soc. Rev.*, 38, 1294–1314, 2009.
24. Corma, A., García, H., Llabrés i Xamena, F., Engineering metal organic frameworks for heterogeneous catalysis. *Chem. Rev.*, 110, 4606–4655, 2010.
25. Huang, Y.-B., Liang, J., Wang, X.-S., Cao, R., Multifunctional metal–organic framework catalysts: Synergistic catalysis and tandem reactions. *Chem. Soc. Rev.*, 46, 126–157, 2017.
26. Dolgopolova, E.A., Rice, A.M., Martin, C.R., Shustova, N.B., Photochemistry and photophysics of MOFs: Steps towards MOF-based sensing enhancements. *Chem. Soc. Rev.*, 47, 4710–4728, 2018.
27. Zhang, T. and Lin, W., Metal–organic frameworks for artificial photosynthesis and photocatalysis. *Chem. Soc. Rev.*, 43, 5982–5993, 2014.
28. Wang, C.-C., Li, J.-R., Lv, X.-L., Zhang, Y.-Q., Guo, G., Photocatalytic organic pollutants degradation in metal–organic frameworks. *Energy Environ. Sci.*, 7, 2831–2867, 2014.
29. Cui, Y., Yue, Y., Qian, G., Chen, B., Luminescent functional metal–organic frameworks. *Chem. Rev.*, 112, 1126–1162, 2011.
30. Kreno, L.E., Leong, K., Farha, O.K., Allendorf, M., Van Duyne, R.P., Hupp, J.T., Metal–organic framework materials as chemical sensors. *Chem. Rev.*, 112, 1105–1125, 2011.
31. Razavi, S.A.A. and Morsali, A., Metal ion detection using luminescent-MOFs: Principles, strategies and roadmap. *Coord. Chem. Rev.*, 415, 213299, 2020.
32. Hu, M.-L., Razavi, S.A.A., Piroozzadeh, M., Morsali, A., Sensing organic analytes by metal–organic frameworks: A new way of considering the topic. *Inorg. Chem. Front.*, 7, 1598–1632, 2020.
33. Lustig, W.P., Mukherjee, S., Rudd, N.D., Desai, A.V., Li, J., Ghosh, S.K., Metal–organic frameworks: functional luminescent and photonic materials for sensing applications. *Chem. Soc. Rev.*, 46, 3242–3285, 2017.
34. DeCoste, J.B. and Peterson, G.W., Metal–organic frameworks for air purification of toxic chemicals. *Chem. Rev.*, 114, 5695–5727, 2014.
35. Li, J., Wang, X., Zhao, G., Chen, C., Chai, Z., Alsaedi, A., Hayat, T., Wang, X., Metal–organic framework-based materials: superior adsorbents for the capture of toxic and radioactive metal ions. *Chem. Soc. Rev.*, 47, 2322–2356, 2018.
36. Kobielska, P.A., Howarth, A.J., Farha, O.K., Nayak, S., Metal–organic frameworks for heavy metal removal from water. *Coord. Chem. Rev.*, 358, 92–107, 2018.

37. Horcajada, P., Serre, C., Vallet-Regí, M., Sebban, M., Taulelle, F., Férey, G., Metal-organic frameworks as efficient materials for drug delivery. *Angew. Chem.*, 118, 6120–6124, 2006.
38. Zhang, J.-W., Zhang, H.-T., Du, Z.-Y., Wang, X., Yu, S.-H., Jiang, H.-L., Water-stable metal-organic frameworks with intrinsic peroxidase-like catalytic activity as a colorimetric biosensing platform. *Chem. Commun.*, 50, 1092–1094, 2014.
39. Ai, L., Li, L., Zhang, C., Fu, J., Jiang, J., MIL-53 (Fe): A Metal-Organic Framework with Intrinsic Peroxidase-Like Catalytic Activity for Colorimetric Biosensing. *Chem.-Eur. J.*, 19, 15105–15108, 2013.
40. Hu, P.-P., Liu, N., Wu, K.-Y., Zhai, L.-Y., Xie, B.-P., Sun, B., Duan, W.-J., Zhang, W.-H., Chen, J.-X., Successive and Specific Detection of Hg2+ and I-by a DNA@ MOF Biosensor: Experimental and Simulation Studies. *Inorg. Chem.*, 57, 8382–8389, 2018.
41. Qiu, G.-H., Lu, W.-Z., Hu, P.-P., Jiang, Z.-H., Bai, L.-P., Wang, T.-R., Li, M.-M., Chen, J.-X., A metal-organic framework based PCR-free biosensor for the detection of gastric cancer associated microRNAs. *J. Inorg. Biochem.*, 177, 138–142, 2017.
42. Horcajada, P., Gref, R., Baati, T., Allan, P.K., Maurin, G., Couvreur, P., Férey, G., Morris, R.E., Serre, C., Metal-Organic Frameworks in Biomedicine. *Chem. Rev.*, 112, 1232–1268, 2012.
43. Campbell, M.G., Sheberla, D., Liu, S.F., Swager, T.M., Dincă, M., Cu3(hexaiminotriphenylene)2: An Electrically Conductive 2D Metal-Organic Framework for Chemiresistive Sensing. *Angew. Chem. Int. Ed.*, 54, 4349–4352, 2015.
44. Morozan, A. and Jaouen, F., Metal organic frameworks for electrochemical applications. *Energy Environ. Sci.*, 5, 9269–9290, 2012.
45. Ramaswamy, P., Wong, N.E., Shimizu, G.K., MOFs as proton conductors–challenges and opportunities. *Chem. Soc. Rev.*, 43, 5913–5932, 2014.
46. Meng, X., Wang, H.-N., Song, S.-Y., Zhang, H.-J., Proton-conducting crystalline porous materials. *Chem. Soc. Rev.*, 46, 464–480, 2017.
47. Wang, S., Wang, Q., Feng, X., Wang, B., Yang, L., Explosives in the cage: Metal-organic frameworks for high-energy materials sensing and desensitization. *Adv. Mater.*, 29, 1701898, 2017.
48. Lu, W., Wei, Z., Gu, Z.-Y., Liu, T.-F., Park, J., Park, J., Tian, J., Zhang, M., Zhang, Q., Gentle, T.J.C.S.R., III, Tuning the structure and function of metal-organic frameworks via linker design. *Chem. Soc. Rev.*, 43, 5561–5593, 2014.
49. Li, B., Chrzanowski, M., Zhang, Y., Ma, S., Applications of metal-organic frameworks featuring multi-functional sites. *Coord. Chem. Rev.*, 307, 106–129, 2016.

2
Amine Decorated Metal–Organic Frameworks

Abstract
In this chapter, chemical properties of amine functions discussed and then these chemical features have been extended in the synthesis of amine decorated metal-organic frameworks. Application of amine decorated metal-organic frameworks reveals that, they applied successfully in deferent applications especially CO_2 post-combustion capture and release, metal ion and picric acid detection and removal of some pollutants like indole and quinoline from oil.

Keywords: Amine, functional metal-organic frameworks, CO_2 capture, picric acid detection, oil denitrogenation, metal ion detection, electrophilic substitution reactions, Lewis basic catalysis

2.1 General Chemical Properties of Amine Function

Amine is among the frequently applied organic functional groups in the structure of functional MOFs (FMOFs). Although, the chemistry of amine is simple, it is useful and effective. The chemistry of amine is dominated by the non-bonding electron pair on nitrogen atom and positively charged hydrogen atom.

The electrostatic potential map for the van der Waals surface of amine function reveals localization of negative charge on N atom due to high electronegativity of N atom engaged in N-H bond and its non-bonding electrons. In this view, amine can (as Lewis basic site) interact with Lewis acid species. Also, it can accept one proton from Brønsted acid to form ammonium cations. Moreover, through nitrogen atom, amine group can act as hydrogen bond acceptor site. In addition, accumulation of negative charge on nitrogen atom enables it to engage with polar and quadruple molecules.

Another characteristic of amine function is revolved around H atom(s). Since these H atoms are positively charged, they can participate in a

Ali Morsali and Sayed Ali Akbar Razavi. Functional Metal-Organic Frameworks: Structure, Properties and Applications, (15–36)

hydrogen bonding. Overally, amine function can interact as both hydrogen bond donor and hydrogen bond acceptor sites.

2.2 Function–Application Properties

Such simple but useful chemistry of amine is practical for development of FMOFs for certain applications. For example, they can apply as a polarizing site for interaction with quadruple gas molecules like carbon dioxide [1–8] and H-bond donor/acceptor gases [9], in sensing of metal ions as Lewis acid sites [10, 11], removal or sensing of hydrogen bond donor or hydrogen bond acceptor chemicals [12–17] or in heterogeneous catalysis as Lewis basic sites [18–20]. In this chapter, we deeply discuss about application of amine FMOFs in different fields [21].

Gas adsorption is a field that amine FMOFs extensively applied owing to delicate host–guest chemistry of amine function with polar or quadrupolar gas molecules [22]. Owing to environmental issues, selective CO_2 capture is a concern that extensively studied by scientists [23]. In this field, amine FMOFs show very high efficiency. Possible (which is proved experimentally and theoretically) CO_2(C)·(N)amine and CO_2(O)·(H)amine interactions give rise to high affinity between quadrupole carbon dioxide molecules and polar amine site. However, a mark difference exists between interactions of CO_2 molecule with aliphatic or aromatic amines. This observation is being caused by different basicity of arylamine and alkylamine functional groups. Due to delocalization of non-bonding lone pairs of N atom within aromatic ring in arylamine groups, they carry lower amount of negative charge on N atoms rather aliphatic amines. So, there is a stronger interaction between N atom of aliphatic amine with partially positive carbon of carbon dioxide molecule. The use of arylamines could favor strong physisorption (30–50 $kJ \cdot mol^{-1}$) with CO_2 while in case of alkylamines host–guest interaction is based on chemisorption process. Although, stronger interaction in case of CO_2 capture by alkylamine could lead to higher selectivity, but it is necessary to mention that release of carbon dioxide molecules are not energy conservative in this case while strong physisorption between CO_2 and arylamine decorated FMOFs is very favorable in case of carbon dioxide release. So, to attain maximum level of CO_2 release and optimized CO_2–amine interaction, it is absolutely essential to engineer the structural features of MOFs as well as their Lewis basicity.

2-aminoterphthalic acid is favorite amine functionalized ligand for synthesis of amine decorated FMOFs for different purposes especially CO_2 separation. UiO-66-NH_2 and CAU-1 are two amine decorated based on

2-aminoterphthalate ligand. Since UiO-66-NH_2 and CAU-1 are decorated with arylamine functions, they represent higher affinity to CO_2 molecules rather non-functional parent frameworks through strong physisorption. UiO-66-NH_2 shows higher adsorption capacity (CO_2 uptake (mmol·g^{-1}) = ≈ 8.5 vs ≈ 7), isosteric heat of CO_2 adsorption at low loading (−32 vs −25.5 kJ·mol^{-1}) and CO_2/N_2 15:85 selectivity (66.5 vs 37.5) rather UiO-66. CAU-1 has improved zero coverage enthalpy (−48 vs −32 kJ.mol^{-1}) and CO_2/N_2 15:85 selectivity (101 vs 66.5) rather UiO-66-NH_2 [24, 25]. In these FMOFs, both CAU-1 and Uio-66-NH_2 synthesized using 2-aminoterphthalic acid ligand and the difference between CO_2·(−NH_2) interaction (which is understood through zero coverage enthalpy) is attributed to the structural differences of these FMOFs.

One applied strategy for stabilization of alkylamine in the structure of MOFs is grafting alkylamine ligands on open metal sites (OMSs) through post-synthesis procedure. Different types of alkylamine ligands like tetraethylenepentamine, N,N'-dimethylethylenediamine, 1-methylethylenediamine, 1,1-dimethylethylenediamine, ethylenediamine, piperazine, 3 and 4-picolylamine, N,N'-dimethylethylenediamine were grafted through this strategy to prove higher Q_{st} and selectivity compared to arylamine functionalized MOFs [5–8, 26–33].

Immobilization of alkylamine ligands on open metal sites of MOFs is an ideal strategy to improve the affinity of the frameworks to CO_2 molecules through altering the Lewis basic open metal sites to the Lewis basic alkylamine functionalized nodes. Based on this approach, the favorable CO_2 physisorption by open metal site·(O)carbon dioxide interaction change into strong CO_2 chemisorption by alkylamine (N)·(C)carbon dioxide interaction which leads to high CO_2 adsorption enthalpy and selectivity. Homodiamine N,N'-dimethylethylenediamine (mmen) ligand applied in the structure of CuBTTri and Mg(dobpdc)$_2$ to develop post-synthetically modified materials with high affinity to carbon dioxide molecules (Q_{st}^{0}(kJ·mol^{-1}) = −24, −96, −47 and −71 for Cu-BTTri, mmen-Cu-BTTri, Mg(dobpdc)$_2$ and mmen-Mg(dobpdc)$_2$ respectively) [6, 34].

Originally, Mg-MOF-74 could adsorb exceptional amount of carbon dioxide (20.6 wt%) under relevant post-combustion flue gas conditions while it display only about 16% recovery of its initial amount of CO_2 in breakthrough experiment with 70% humidity [35]. To remove this barrier and improvement of CO_2 capacities at lower CO_2 partial pressures and presence of humidity, Mg(dobpdc)$_2$ post-synthetically modified with homodiamine ligands like ethylenediamine and dimethylethylenediamine [31]. Anyway, the working capacity of these alkylamine grafted FMOFs is not high at low temperatures. This is another issue that must be eliminated for efficient release of CO_2 and increase in carbon dioxide working capacity.

Chang Seop Hong and coworkers synthesized a diamine-grafted FMOF (with the name of dmen-Mg_2(dobpdc) where dmen = N,N' dimethylethylenediamine and H_4dobpdc = 4,4'-dihydroxy-(1,10-biphenyl)-3,3'-dicarboxylic acid) and applied it for carbon dioxide capture and evaluation of working capacity at post-combustion conditions (Figure 2.1) [8]. dmen is a heterodiamine with both primary and tertiary amines and incorporated into framework through post-synthesis modification. The results reveal that dmen-Mg_2(dobpdc) represent high CO_2/N_2 selectivity (S = 554 at 25 °C and p(CO_2) = 0.15 bar) which is higher that mmen-Mg_2(dobpdc) (S = 200) and en-Mg_2(dobpdc) (S = 230). At 25 °C and 0.15 bar, activated dmen-Mg_2(dobpdc) adsorb 3.77 mmol·g^{-1} of CO_2which is higher than en-Mg_2(dobpdc) (3.62 mmol·g^{-1}) and mmen-Mg_2(dobpdc) (3.13 mmol·g^{-1}). Efficiency of an adsorbent in post-combustion process is assessed by measuring the working capacity which is defined as difference between the adsorbed quantities at P_{ads} = 0.15 bar CO_2/T_{ads} = 40 °C and P_{des} = 1 bar CO_2/T_{des}. The higher working capacity at lower desorption temperatures, the lower energy consumption. dmen-Mg_2(dobpdc) could adsorb 18.8 wt% for pure CO_2, and 14.1 wt% for N_2/CO_2 (85/15) which is close to that at the corresponding CO_2 partial pressure in the isotherm at 40 °C, while no obvious adsorption was observed for pure N_2 at 40 °C. Regeneration of the material is evaluated via vacuum-swing adsorption (VSA) and temperature-swing adsorption (TSA) methods. In TSA method, CO_2 was adsorbed at 40 °C for 1 h and desorbed at 75 °C for 1 h under Ar. After 24 cycles, no capacity loss was observed, revealing that the dmen-Mg_2(dobpdc) is thermally stable under these experimental conditions as well as maintenance in its adsorption capacity. In VSA method, at 25 °C, material was saturated with CO_2 at 1.2 bar and then placed under high vacuum. The removal of adsorbed CO_2 from the solid was performed repeatedly by applying a vacuum to the adsorbent. Based on observed results, such a large amount of adsorbed CO_2 (4.5 mmol·g^{-1}) at 1.2 bar can be completely desorbed only under vacuum, without heating. The working capacities of dmen-Mg_2(dobpdc) at 130 to 90 °C desorption temperature is in the range of 11.7–13.5 wt%. Notably, experimental results reveal that at T_{des} = 75 °C working capacity is 11.6 wt% which is higher that top performing MOFs such as Mg-MOF-74 (3.7 wt%), Mg_2(dobpdc) (4.5 wt%), en-Mg_2(dobpdc) (2.9 wt%), mmen-Mg_2(dobpdc) (2.1 wt%), and tmen-Mg_2(dobpdc) (3.9 wt%) (tmen = N,N,N',N'-tetramethylethylenediamine). However, the working capacity of dmen-Mg_2(dobpdc) sharply reduced to almost zero at 70 °C. To evaluate the reusability of dmen-Mg_2(dobpdc) in the presence of water vapor, the material exposed to water vapor (100% RH for 10 min). Since the solid sample can

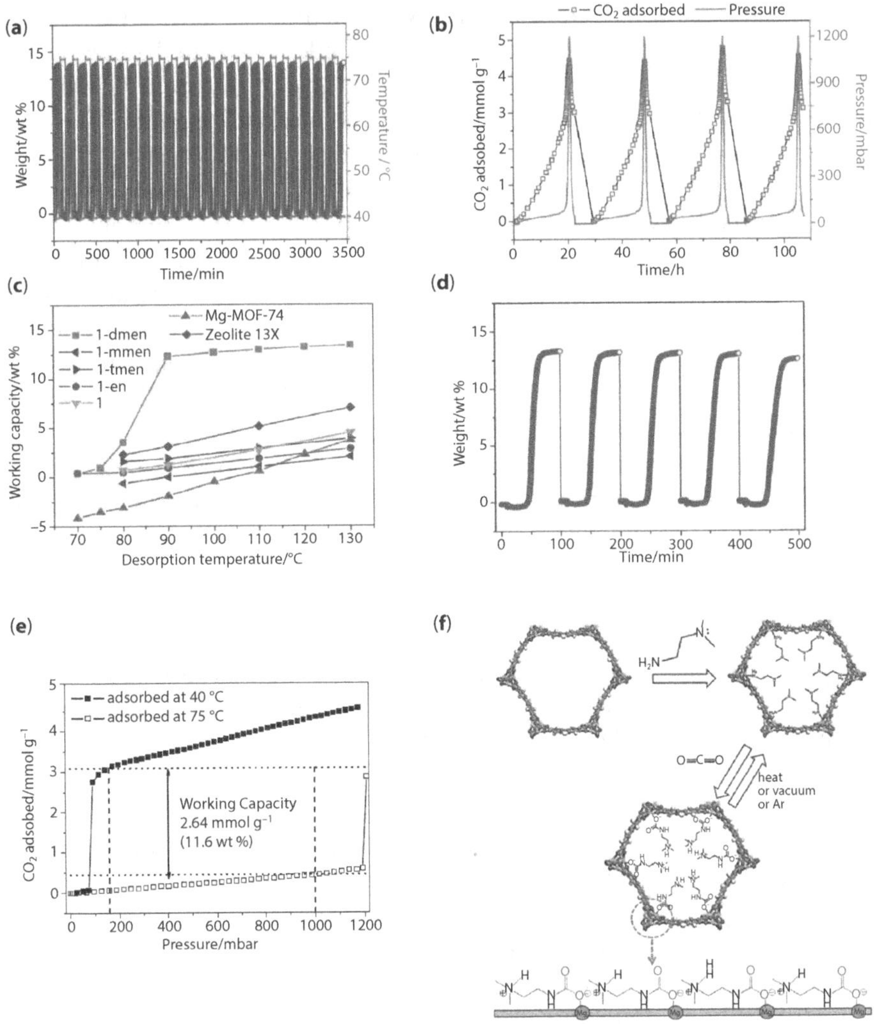

Figure 2.1 Application of dmen-Mg_2(dobpdc) in selective capture-release of carbon dioxide. (a) TSA process: Adsorption (40 °C)-desorption (75 °C) cycling of CO_2 under Ar. (b) Vacuum-swing adsorption at 25 °C. (c) The working capacities of 1-dmen and the other porous solids obtained under the same conditions. (d) CO_2 adsorption of 1-dmen in flue gas using the sequence (adsorption at 40 °C, desorption at 130 °C under pure CO_2-10 min exposure to 100% RH). (e) Estimated working capacity from q_{ads} (P_{ads} = 0.15 bar CO_2, T_{ads} = 40 °C)-q_{des} (P_{des} = 1 bar CO_2, T_{des} = 75). (f) Framework structure of 1 with open metal sites, grafting modes of dmen onto the open metal sites, and subsequent CO_2 adsorption. The schematic diagram (bottom) indicates the arrangement of ammonium carbamates running along the c-axis [8].

be fully saturated with CO_2 within the exposure time (10 min), it exposed to water vapors for 10 min. Then, dmen-Mg_2(dobpdc) reactivated under a pure CO_2 flow at 130 °C for 4 h, followed by CO_2 adsorption at 40 °C. After 5 cycles, a capacity loss of 5% is observed which is markedly higher than working capacities of the other MOFs during the first cycle (lower than 4.5 wt%). Clausius–Clapeyron equation applied for estimation of heat of adsorption ($-Q_{st}$) so that increases to 75 kJ·mol^{-1} at a loading of 0.25 mmol.g^{-1}, and remains almost invariant in the range of 71–75 kJ·mol^{-1} below loadings of 2.6 mmol.g^{-1}. Based on DFT calculations, the open metals site are mostly occupied by the primary amine end of dmen-Mg_2(dobpdc), although some tertiary amine ends are probably grafted onto the exposed metal site as well. Possible mechanism based on DFT calculations and *in-situ* FT-IR analysis possible CO_2 adsorption mechanism is illustrated in Figure 2.1f.

Although extensive studies conducted on application of amine decorated MOFs is carbon dioxide capture and release, but there is an urgent need to clarify the effective conditions of arylamine or alkylamine groups in practical CO_2 capture and release for real-life applications.

Christian Serre and coworkers applied amine decorated MIL-125(Ti) (MIL-125 formula is ($Ti_8O_8(OH)_4(BDC)_6$) where BDC is benzenedicarboxylate), denoted as NH_2-MIL-125(Ti) for separation of CO_2 and H_2S [9]. They mentioned this material could improve separation of these acid gases from biogas or natural gas markedly. Based on *in-situ* FT-IR analysis, they mentioned that $-NH_2$ function interact weakly with CO_2 through lone pair of relatively negative N-atom of amine and relatively positively charged C-atom of CO_2. Also, hydrogen bonding is the main reason for improved H_2S separation in a way that H_2S acts as hydrogen bond donor and amine, through its N atom, acts as hydrogen bond acceptor.

Amine function is an ideal gust-adsorptive site to interact with different type of guests. Since it contains positively charged H atoms, it can interact as hydrogen bond donor site. Also, it can interact as hydrogen acceptor or hydrogen bond acceptor site as well as Lewis basic site through its N atom. These multiple chemical features enable amine decorated MOFs to apply as a host for different type of analytes. Their Lewis basicity is potentially suitable to interact with metal ions and their ability to participate in hydrogen bonding is ideal to interact with small organic molecules.

It is reported that MOFs based on central d-metal ions (especially Zn(II) or Cd(II)) functionalized with Lewis basic sites are best sensors toward 2,4,6-trinitrophenol (TNP) with highest K_{SV} (Stern–Volmer constant) and best detection limits [36]. Especially adenine decorated MOFs are best candidate in this field because adenine molecule contain different kind of Lewis basic site especially free primary amine groups which is able to interact with TNP.

Mandal and coworker(s) synthesized [Cd(ATAIA)].4H_2O (where H_2ATAIA = 5-((4,6-diamino-1,3,5-triazin-2-yl)amino)isophthalic acid) for selective vapor sorption and nanomolar sensing of TNP (Figure 2.2) [14]. The pore walls of Cd-ATAIA are decorated with two primary amino groups and one secondary amino group. Activated Cd-ATAIA displays UV–Vis absorption peaks centered at 278 and 323 nm and fluorescence emission at 376 nm upon excitation at 310 nm. Application of Cd-ATAIA

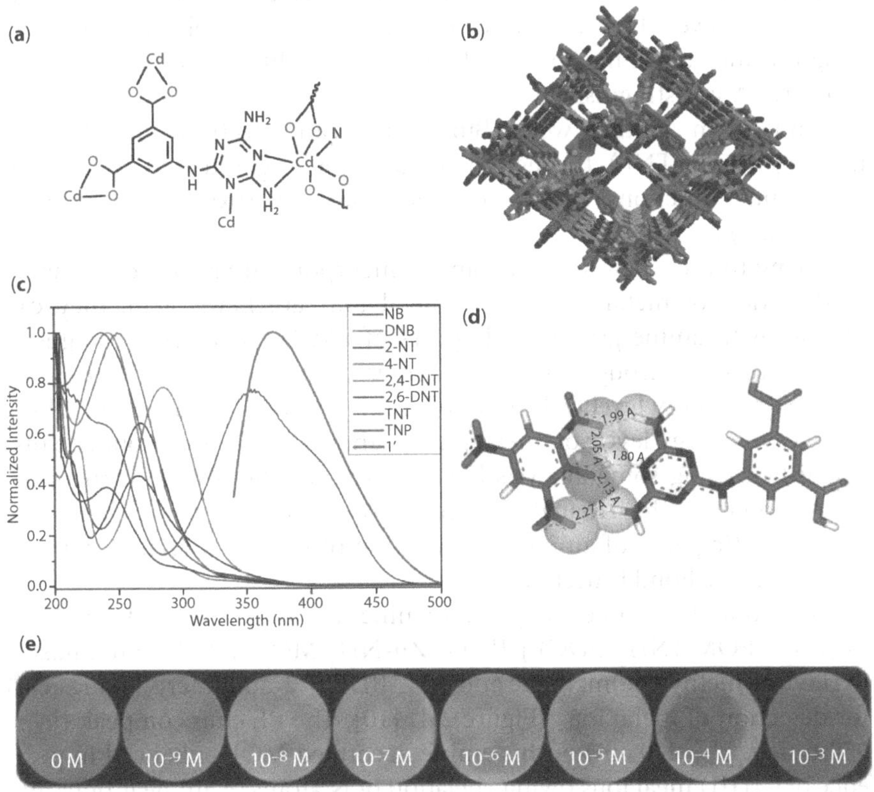

Figure 2.2 Application of [Cd(ATAIA)].4H_2O in detection of TNP. (a) Schematic representation view showing coordination environment around Cd(II) center. (b) Perspective view of packing along c-axis (free guests and hydrogen have been omitted for clarity) (Color code; carbon: grey, oxygen: red, nitrogen: blue, Cadmium: green). (c) spectral overlap of absorption spectra of different nitro-analytes and emission spectrum of activated Cd-ATAIA. (d) Energy minimized structures of H_2ATAIA and TNP complex. (e) Photograph of Whatman filter paper strips coated with activated Cd-ATAIA at different concentration of TNP [14].

in aqueous solution detection of nitroaromatic compounds (NACs) reveals that activated Cd-ATAIA represents the largest quenching efficiency (98%, 60 μl) in presence of TNP as well as a significant color change (K_{SV} = $1.59 \times 10^{+7}$ mol^{-1}, detection limit = 0.94 nM). DFT calculation performed to investigate about possible detection mechanism. The results show that there is strong electrostatic interactions and hydrogen bonding between two types of Lewis basic amine groups of the ATAIA ligand and the highly acidic hydroxyl group of TNP. Also UV–Vis experiments reveal that there is large overlap between ATAIA and TNP. As a result of such overlap and host-guest interactions, activated Cd-ATAIA could detect TNP in presence of other NACs. Also, Cd-ATAIA could detect TNP in vapor phase. The results reveal that a visual color change is observable after 30 while a significant quenching in photoluminescence (PL) emission spectra of CD-ATAIA is distinguishable.

Similarly in another work, Bio-MOF-1 can selectively detect 2,4,6-trinitrophenol (TNP) through hydrogen bonding interaction between free amine functions of adenine and acidic hydroxyl group of TNP (Figure 2.3) [37].

Owing to their Lewis basicity amine incorporated FMOFs could apply in detection of metal ions. Since metal ions are Lewis acid, they can interact with amine group inside pores of FMOFs as Lewis basic guest-interactive site. Through this kind of interaction, some changes could be induced in energy level of N atom of amine or even the metal ion. These changes in energy levels of metal ion·(N)amine interactions could be distinguished using some analysis like XPS (X-ray photoelectron spectroscopy). Sometime FT-IR analysis through observation of some changes in characteristic peaks of amine or generation of new peak related to metal ion ·(N)amine bond is useful.

Xin Liu and coworkers applied a amine decorated FMOF, formulated as $[Zn_2(TPOM)(NH_2\text{-}BDC)_2]\cdot 4H_2O$ (Zn-NH_2-MOF, TPOM = tetrakis(4-pyridyloxymethylene)methane and NH_2-BDC = 2-aminoterphthalic acid) for detection of metal ions (Figure 2.4) [10]. The PL emission peak (λ_{ex} = 353 nm, λ_{em}^{max} = 420 nm) of this amine decorated MOF enhanced in presence of Cr(III) meal ions owing chelation of N atoms of amine groups and O atoms of free carboxylate groups. Application of isostructure framework without -NH_2 group, Zn-MOF, reveal that the PL emission of this MOF did not change significantly in presence of Cr(III) ions. This blank experiment confirm the vital role of amine group in detection of Cr(III) ions (detection limit = 4.9 μM) [11]. In other works, Lili and coworkers applied a amine decorated MOF for detection of Hg(II) metal ions (detection limit = 4.2×10^{-8} M, K_{SV} = 4550 mol^{-1}). XPS analysis of this materials after exposure to

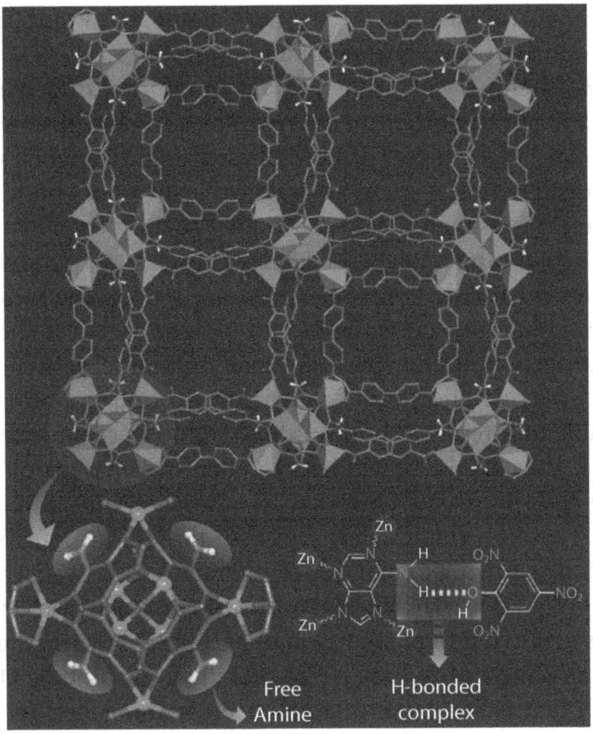

Figure 2.3 3 dimensional framework of bio-MOF-1 along the c crystallographic direction (above) and desirable H-bonding interaction between free primary amine of adenine and hydroxyl group of TNP (down) [37].

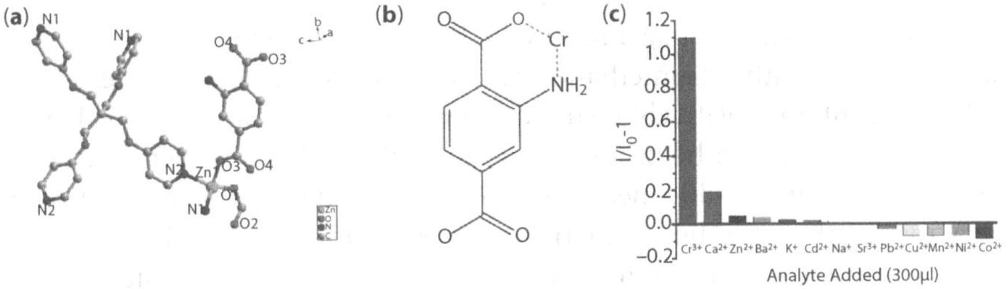

Figure 2.4 Application of $[Zn_2(TPOM)(NH_2-BDC)_2]\cdot 4H_2O$ in detection of Cr(III) ions. (a) Coordination environment of NH_2-Zn-MOF representing the free amine and carboxylate groups. (b) Proposed detection mechanism. (c) Change in emission peak of NH_2-Zn-MOF in presence of different metal ions [10].

Hg(II) metal ions reveal a new N(1 s) peak at 406.38 eV indicative of chelation interaction between N atom of amino group.

Host–guest chemistry of amine FMOFs in removal of detrimental analytes from aqueous and other media resembles to detection of metal ions or small organic molecules. So, Lewis basicity and capability to participate in hydrogen bonding are most practical features of amine function in removal of hazardous materials.

Nitrogen containing molecules like pyridine and quinoline (with N sites), pyrrole and indole (with NH sites) are among the well-known oil pollutants. So, they are potentially able to interact with the adsorbent through hydrogen bonding; pyridine and quinoline as hydrogen bond acceptor and pyrrole and indole as hydrogen bond donor. Since amine function is able to interact as both H-donor and H-acceptor site in hydrogen bonding, they could be applied for denitrogenation of oil. In this regard, Sung Hwa Jhung and coworkers synthesized some amine decorated MOFs, MIL-125-NH_2 (Figure 2.5) and UiO-66-NH_2, to purify the liquid model oil [15, 16]. Owing to mentioned interactions, host–guest chemistry of NH_2-MIL-125(Ti) is enriched and adsorption capacity toward quinoline and indole is improved in comparison of MIL-125(Ti), 103 $mg{\cdot}g^{-1}$ vs. 460 $mg{\cdot}g^{-1}$ for quinoline and 264 $mg{\cdot}g^{-1}$ vs. 502 $mg{\cdot}g^{-1}$ for indole [15]. Clearly, this improvement is attributed to presence of amine site. In next move, they protonated amine sites from $-NH_2$ to $-NH_3^+$ for additional improved adsorption capacity of quinoline (546 $mg{\cdot}g^{-1}$) and indole (583 $mg{\cdot}g^{-1}$).

Amine decorated FMOFs applied for Lewis basic catalyzed reactions like Henry and Knoevenagel reactions [38] and some of other reactions like transesterification of triglycerides [39]. Also, Amine decorated FMOFs with open metal sites applied in those kinds of tandem reactions needing both Lewis basic and Lewis acidic catalytic sites like cycloaddition reaction of CO_2 with various epoxides [20].

In Henry or Knoevenagel reactions, benzaldehyde should be activated and then react with nitromethane or malonitrile, respectively. Since, benzaldehyde contains positively-charged C-atom which is Lewis acid site, Lewis basic sites can active benzaldehyde and catalyze Henry or Knoevenagel reactions. In one possible mechanism, it is reported that benzaldehyde can be activated through the interaction with an amine group and the formation of imine through a new (amine)N=C(benzaldehyde) covalent bond that can be followed by the rearrangement and addition of malonitrile (Figure 2.6) [18]. In other mechanism it is mentioned that benzaldehyde activation in Henry reaction is done through a noncovalent Lewis base–acid interactions between Lewis basic catalytic site with the carbonyl C atom of benzaldehyde [40].

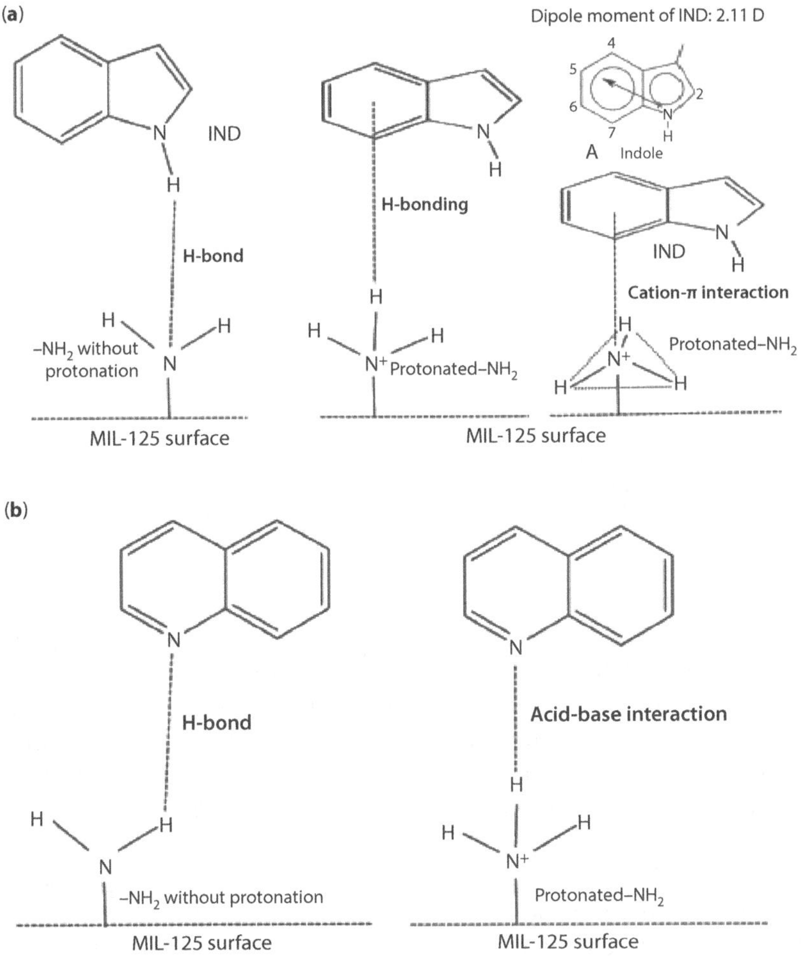

Figure 2.5 Application of NH_2-MIL-125 (Ti) and NH_3^+ -MIL-125 (Ti) in oil denitrogenation. (a) Adsorption mechanism for removal of indole. (b) Adsorption mechanism for removal of quinoline [15].

Lewis basicity and hydrogen-bond donation/accepting are common and well-known chemical properties of amine functions which applied extensively in development of functional MOFs. Anyway, there are some of other chemical properties which are interesting for fabrication of FMOFs. For example, amine function is able to interact with donor acceptor interactions. Through this mechanism, amine decorated MOFs applied for improved Li-storage capacity [41] through host-guest interactions between Li and amine groups (N atoms) and accelerated I_2 removal [42] through amine·iodine charge transfers.

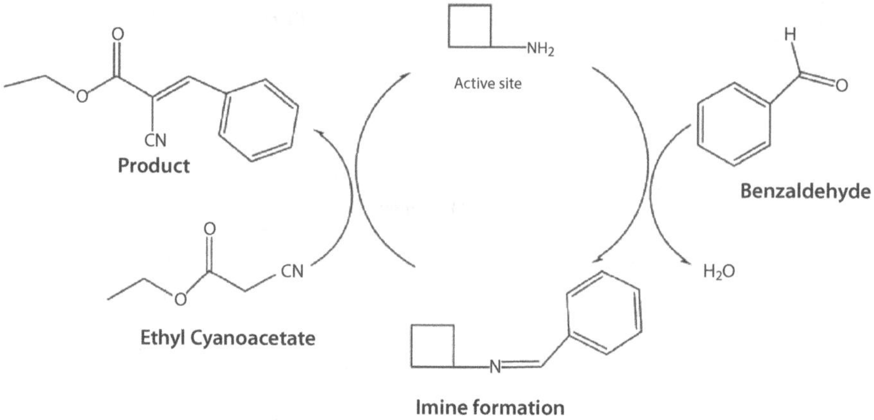

Figure 2.6 Possible catalytic mechanism by amine function in Knoevenagel reactions. In this mechanism, amine site could active benzaldehyde through formation of imine. Reaction is progressed after addition of ethyl cyanoacetate to the imine complex [18].

In synthetic organic chemistry it is known that aromatic rings with electron donor groups like amine could participate in electrophilic substitution reactions. Also, aromatic amine groups could be converted to diazonium or other products like reduction to hydrogen. These principal roles in chemistry of arylamines applied in construction of highly efficient removal and sensing of Cl_2 [43], NO [44] and NO_2 [45] gases with amine decorated MOFs.

Gregory W. Peterson and coworkers synthesized UiO-6-NH_2 and applied in for removal of chlorine gas [43]. The material could remove 1.24g·g^{-1} Cl_2(g). Using different characterization methods they proposed that there are two predominately mechanisms engaged in removal process including the loss of one carboxylate group of ligand and reaction between Cl_2(g) and Zr_6O_6 nodes as well as reaction between organic linker (2-aminoterephthlic acid) and chlorine gas through electrophilic substitution reaction in ortho and para positions. Also, produced HCl molecules are neutralized by amine functions (Figure 2.7a).

In another work by the same group, UiO-66-NH_2 applied for removal of 1.4g·g^{-1} NO_2(g) [45]. Experimental analyses show that NO_2(g) is adsorbed through different types removal mechanisms (Figure 2.7b). At low loading, NO_2(g) first adsorbs within the pores of the MOF and loading increases with decreasing the temperature indicating that physical adsorption has a major impact on removal. At higher loading, the organic ligand react with oxidant NO_2(g) molecules in multiple locations.

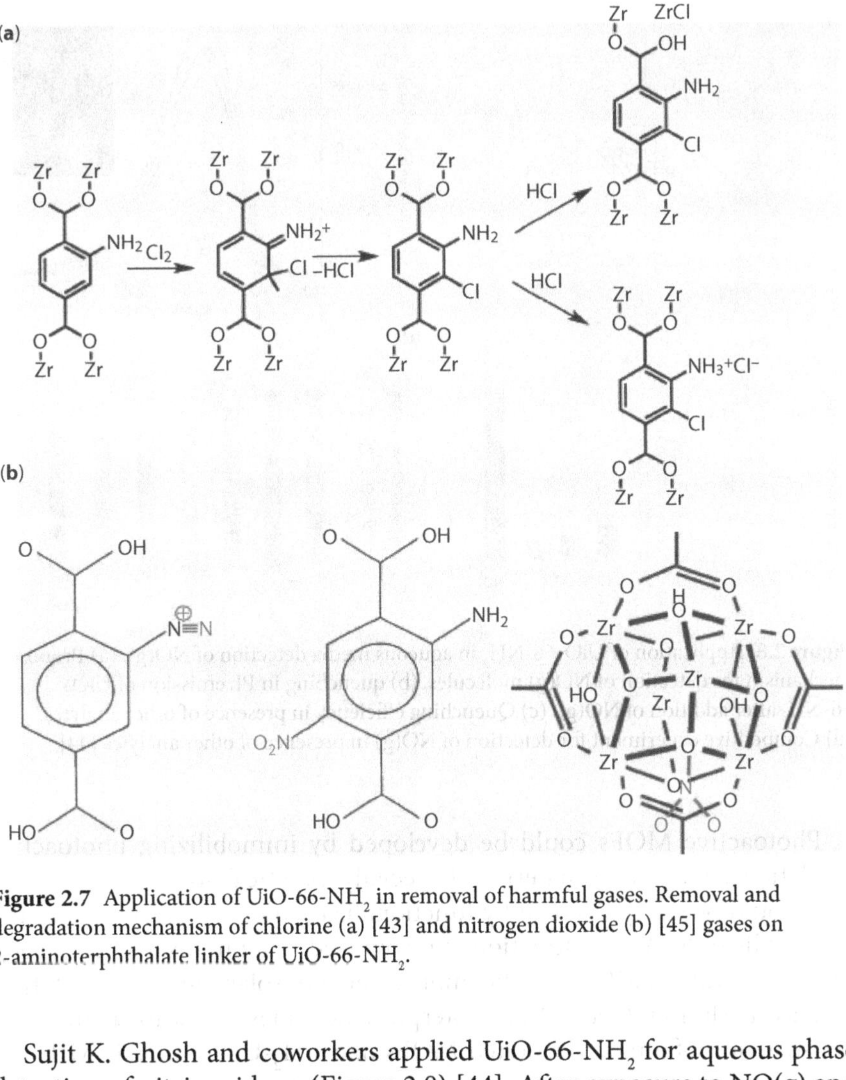

Figure 2.7 Application of UiO-66-NH_2 in removal of harmful gases. Removal and degradation mechanism of chlorine (a) [43] and nitrogen dioxide (b) [45] gases on 2-aminoterphthalate linker of UiO-66-NH_2.

Sujit K. Ghosh and coworkers applied UiO-66-NH_2 for aqueous phase detection of nitric acid gas (Figure 2.8) [44]. After exposure to NO(g) and deamination process, UiO-66-NH_2 is transformed to UiO-66. Considering this mechanism, fluorescent UiO-66-NH_2 is converted to non-fluorescent UiO-66. PL measurements reveal that UiO-66-NH_2 could detect NO(g) gas with detection limit equal to 0.575 μM and a quenching constant of $4.15 \times 10^{+5}$ M^{-1}. The UiO-66 framework do not show any change is PL emission peak which clarifies the role of the primary amine group in the NO(g) detection. Competitive experiments also show that there is no substantial change in presence of similar species while there is considerable quenching in presence of NO(g).

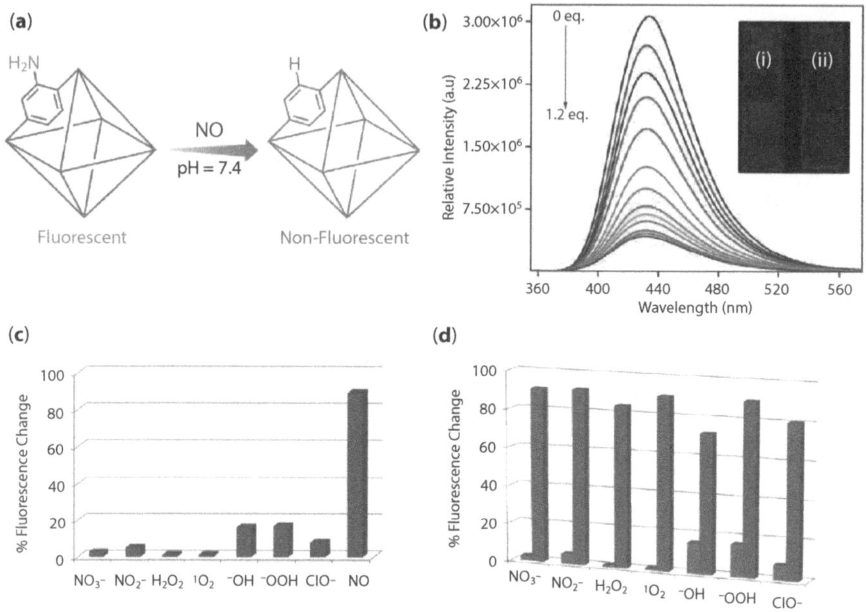

Figure 2.8 Application of UiO-66-NH_2 in aqueous media detection of NO(g). (a) Proposed mechanism for detection of NO(g) molecules. (b) quenching in PL emission of UiO-66-NH_2 after addition of NO(g). (c) Quenching efficiency in presence of other analytes. (d) Competitive experiment for detection of NO(g) in presence of other analytes [44].

Photoactive MOFs could be developed by immobilizing photoactive catalytic sites in MOF materials. Especially, practical adsorption of solar light could be easily attained by functionalization of the metal ions or the organic ligands. Amine function is recognized as a photosensitizer group in the structure of MOFs for the improvement of solar-light photocatalytic activity in MOFs [46–49]. 2-aminoterphthalic acid is well-known linker for construction of amine decorated MOFs like NH_2-UiO-66, NH_2-MIL-125 and other MOFs.

The amine function has a substantial role in the modification of optical band gap of MOFs constructed based on 2-aminoterphthalic acid ligand. In this case, HOMO of amine decorated MOFs based on 2-aminoterphthalic acid ligand composed of O, C and N 2p orbitals [50]. The insertion of N character in HOMO, or valance band, of MOFs induces the band-gap narrowing to shift the photo-absorption and lower band gap will shift the band gap to the visible light region [49]. So, the material shows an extended absorption band in the visible light region with enhanced visible-light absorption owing to introduction of photosensitizer amine function.

Moreover, upon light irradiation a ligand-to-metal charge transfer with long-lived excited charge separation could be observed [47]. This charge transfer is effectual for oxidizing of reactive substrates adsorbed on the amine site by photogenerated holes on organic ligand and reducing other reactive substrates adsorbed on the inorganic building blocks by transferring of photogenerated electrons. So amine function could intensify the photocatalytic activity of MOFs through extending in absorption band and generation of long-lived excited electron-holes.

Jinhua Ye and coworkers synthesized MIL-88(Fe) and MIL-88(Fe)-NH_2 and applied for photo-reduction of dichromate anion (Figure 2.9) [47]. The mechanism of photo-reduction is based on generation of electron-hole pairs upon light irradiation. It is essential to tune the band gap of MOFs to

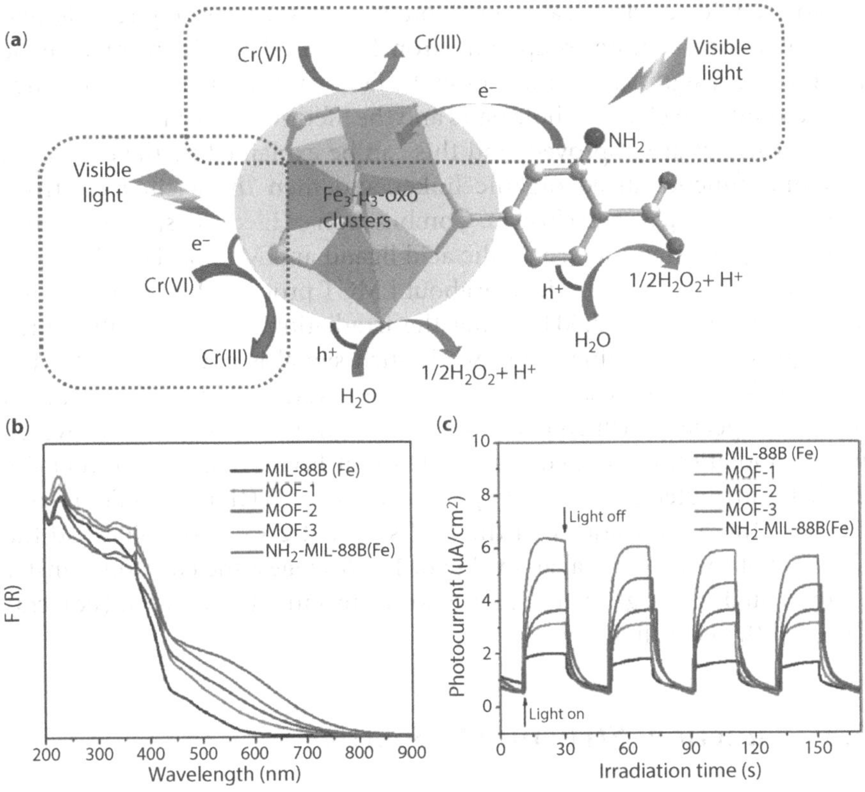

Figure 2.9 Application of MIL-88(Fe)-NH_2 in photocatalytic degradation of Cr(VI). (a) dual pass mechanism in presence of amine functionalized ligand. (b) Diffuse-reflectance UV/vis spectrum of NH_2-MIL-88B(Fe) and MIL-88B(Fe). (c) Transient photocurrent spectroscopy of NH_2-MIL-88B(Fe) and MIL-88B(Fe) [47].

optimize the photocatalytic activity of MOFs. In parent framework, MIL-88(Fe), Fe_3-μ_3-oxo clusters are directly excited and reduction of Cr(VI) and oxidation of water take place at the Fe-based oxo clusters. But in case of MIL-88(Fe)-NH_2, photo-sensitizer amine group enables organic ligand to adsorb the visible light photons, excited with generation of long-lived electron-hole pairs and then transferring photoexcited electron to Fe_3-μ_3-oxo clusters. This secondary excitation mechanism is the reason of improved photocatalytic activity of MIL-88(Fe)-NH_2 than its parent framework. Diffuse-reflectance UV/vis spectrum of these materials show that the introduction of amine group in the organic linker of iron(III)-based MOF can enhance its light absorption in the visible region. This observation indicates that the more amine group incorporated into the iron(III)-based MOF, the more electron–hole pairs can be generated via excitation of amine functionality under visible-light irradiation, which might lead to enhance photocatalytic activity. Transient photocurrent spectroscopy reveals that incorporation of amine group in the iron(III)-based MOF can enhance the photocurrent significantly indicating the fact that the separation efficiency of photoinduced electron-hole pairs and the lifetime of the photogenerated charge carriers are improved, and this can be explained by the excitation of amine-functionalized organic linker and then the excited electrons transfer to Fe 3 -μ 3 -oxo clusters. Combination of Electron spin resonance (ESR) for pure 2-aminoterohthalic acid ligand and MIL-88(Fe)-NH_2 is an effective method to gain evidence about LMCT process. Pure ligand shows an ESR signal of $g = 2.004$ during the irradiation of visible light which is originated from amine group while such signal is not detected for the MIL-88(Fe)-NH_2. ESR signal for MIL-88(Fe)-NH_2 at $g = 1.994$ is attributed to Fe(III) species in which decreases upon light irradiation and recovered when irradiation is stopped. The decrease of Fe(III) ESR signal intensity could be attributed to the trapping of electrons by Fe(III) site in Fe_3-μ_3-oxo clusters. The disappearance of the ESR signal at g value of 2.004 and the decrease of the ESR signal at g value of 1.996 suggest the electron transfer from excited amine group to Fe_3-μ_3-oxo clusters in NH_2-MIL-88B(Fe) irradiated with visible light.

2.3 Function–Structure Properties

The majority of chemistry of an amine functionalized MOF is around their capability as guest-interactive sites in host-guest chemistry. Structural features of amine FMOFs depend on the amine type as primary, secondary or tertiary.

Primary amines ($-NH_2$) mainly applied as guest interactive site in the side chain of MOFs in tethered form [21, 22]. Anyway, sometimes primary amines are able to coordinate to metal ions during self-assembly synthesis.

Secondary amines could be applied in both main chain and side chain (tethered form) of the MOFs. K. Mark Thomas and coworkers synthesized an amine FMOFs with secondary amine motifs (–NH–) in the main chain [51]. The freedom and rotation around C–N–C might give rise to flexibility of the MOF. Also, organic ligands based on secondary amines applied as V-shaped ligand for development of desired MOF [52]. In most of the time secondary amines in side chain applied for study about the Lewis basicity of the MOF for different applications [53–55].

Similar to secondary amines, tertiary amines could apply in side chain for study about the Lewis basicity of the MOF for different applications. Specific type of tritopic ligands based on tertiary ligands applied for construction of MOFs.

As we saw, amine function extensively applied with the aim of improvement in efficiency of MOFs in different type of applications. This attention is due to the fact that amine decorated MOFs benefits from several chemical features as well as easy synthesis method. Particularly, amine decorated MOFs show very good results in some application like carbon dioxide capture-release in post-combustion process. Also, they successfully applied for oil denitrogenation and detection of picric acid in aqueous solution. Though, amine decorated MOFs applied for energy and environmental purposes, but these achievements have to develop and tailor for real-life applications.

References

1. Vaidhyanathan, R., Iremonger, S.S., Shimizu, G.K.H., Boyd, P.G., Alavi, S., Woo, T.K., Competition and Cooperativity in Carbon Dioxide Sorption by Amine-Functionalized Metal–Organic Frameworks. *Angew. Chem.*, 124, 1862–1865, 2012.
2. Planas, N., Dzubak, A.L., Poloni, R., Lin, L.-C., McManus, A., McDonald, T.M., Neaton, J.B., Long, J.R., Smit, B., Gagliardi, L., The Mechanism of Carbon Dioxide Adsorption in an Alkylamine-Functionalized Metal–Organic Framework. *J. Am. Chem. Soc.*, 135, 7402–7405, 2013.
3. Vaidhyanathan, R., Iremonger, S.S., Shimizu, G.K.H., Boyd, P.G., Alavi, S., Woo, T.K., Direct Observation and Quantification of CO_2 Binding Within an Amine-Functionalized Nanoporous Solid. *Science*, 330, 650–653, 2010.
4. McDonald, T.M., Lee, W.R., Mason, J.A., Wiers, B.M., Hong, C.S., Long, J.R., Capture of Carbon Dioxide from Air and Flue Gas in the

Alkylamine-Appended Metal–Organic Framework mmen-Mg2(dobpdc). *J. Am. Chem. Soc.*, 134, 7056–7065, 2012.

5. McDonald, T.M., D'Alessandro, D.M., Krishna, R., Long, J.R., Enhanced carbon dioxide capture upon incorporation of N, N′-dimethylethylenediamine in the metal–organic framework CuBTTri. *Chem. Sci.*, 2, 2022–2028, 2011.
6. Demessence, A., D'Alessandro, D.M., Foo, M.L., Long, J.R., Strong CO2 Binding in a Water-Stable, Triazolate-Bridged Metal–Organic Framework Functionalized with Ethylenediamine. *J. Am. Chem. Soc.*, 131, 8784–8786, 2009.
7. Yeon, J.S., Lee, W.R., Kim, N.W., Jo, H., Lee, H., Song, J.H., Lim, K.S., Kang, D.W., Seo, J.G., Moon, D., Wiers, B., Hong, C.S., Homodiamine-functionalized metal–organic frameworks with a MOF-74-type extended structure for superior selectivity of CO2 over N2. *J. Mater. Chem. A*, 3, 19177–19185, 2015.
8. Lee, W.R., Jo, H., Yang, L.-M., Lee, H., Ryu, D.W., Lim, K.S., Song, J.H., Min, D.Y., Han, S.S., Seo, J.G., Park, Y.K., Moon, D., Hong, C.S., Exceptional CO2 working capacity in a heterodiamine-grafted metal–organic framework. *Chem. Sci.*, 6, 3697–3705, 2015.
9. Vaesen, S., Guillerm, V., Yang, Q., Wiersum, A.D., Marszalek, B., Gil, B., Vimont, A., Daturi, M., Devic, T., Llewellyn, P.L., Serre, C., Maurin, G., De Weireld, G., A robust amino-functionalized titanium(iv) based MOF for improved separation of acid gases. *Chem. Commun.*, 49, 10082–10084, 2013.
10. Lv, R., Wang, J., Zhang, Y., Li, H., Yang, L., Liao, S., Gu, W., Liu, X., An amino-decorated dual-functional metal–organic framework for highly selective sensing of Cr(iii) and Cr(vi) ions and detection of nitroaromatic explosives. *J. Mater. Chem. A*, 4, 15494–15500, 2016.
11. Wen, L., Zheng, X., Lv, K., Wang, C., Xu, X., Two Amino-Decorated Metal–Organic Frameworks for Highly Selective and Quantitatively Sensing of HgII and CrVI in Aqueous Solution. *Inorg. Chem.*, 54, 7133–7135, 2015.
12. Nagarkar, S.S., Desai, A.V., Samanta, P., Ghosh, S.K., Aqueous phase selective detection of 2,4,6-trinitrophenol using a fluorescent metal–organic framework with a pendant recognition site. *Dalton Trans.*, 44, 15175–15180, 2015.
13. Joarder, B., Desai, A.V., Samanta, P., Mukherjee, S., Ghosh, S.K., Selective and Sensitive Aqueous-Phase Detection of 2,4,6-Trinitrophenol (TNP) by an Amine-Functionalized Metal–Organic Framework. *Chem. Eur. J.*, 21, 965–969, 2015.
14. Das, P. and Mandal, S.K., Strategic Design and Functionalization of an Amine-Decorated Luminescent Metal Organic Framework for Selective Gas/Vapor Sorption and Nanomolar Sensing of 2,4,6-Trinitrophenol in Water. *ACS Appl. Mater. Interfaces*, 10, 25360–25371, 2018.
15. Ahmed, I., Khan, N.A., Yoon, J.W., Chang, J.-S., Jhung, S.H., Protonated MIL-125-NH2: Remarkable Adsorbent for the Removal of Quinoline and Indole from Liquid Fuel. *ACS Appl. Mater. Interfaces*, 9, 20938–20946, 2017.

16. Ahmed, I. and Jhung, S.H., Effective adsorptive removal of indole from model fuel using a metal-organic framework functionalized with amino groups. *J. Hazard. Mater.*, 283, 544–550, 2015.
17. Seo, P.W., Ahmed, I., Jhung, S.H., Adsorption of indole and quinoline from a model fuel on functionalized MIL-101: Effects of H-bonding and coordination. *Phys. Chem. Chem. Phys.*, 18, 14787–14794, 2016.
18. Gascon, J., Aktay, U., Hernandez-Alonso, M.D., van Klink, G.P.M., Kapteijn, F., Amino-based metal-organic frameworks as stable, highly active basic catalysts. *J. Catal.*, 261, 75–87, 2009.
19. Chen, J., Liu, R., Gao, H., Chen, L., Ye, D., Amine-functionalized metal-organic frameworks for the transesterification of triglycerides. *J. Mater. Chem. A*, 2, 7205–7213, 2014.
20. Verma, A., De, D., Tomar, K., Bharadwaj, P.K., An Amine Functionalized Metal–Organic Framework as an Effective Catalyst for Conversion of CO2 and Biginelli Reactions. *Inorg. Chem.*, 56, 9765–9771, 2017.
21. Ali Akbar Razavi, S. and Morsali, A., Linker functionalized metal-organic frameworks. *Coord. Chem. Rev.*, 399, 213023, 2019.
22. Lin, Y., Kong, C., Chen, L., Amine-functionalized metal–organic frameworks: Structure, synthesis and applications. *RSC Adv.*, 6, 32598–32614, 2016.
23. Zhang, Z., Yao, Z.-Z., Xiang, S., Chen, B., Perspective of microporous metal–organic frameworks for CO 2 capture and separation. *Energy Environ. Sci.*, 7, 2868–2899, 2014.
24. Si, X., Jiao, C., Li, F., Zhang, J., Wang, S., Liu, S., Li, Z., Sun, L., Xu, F., Gabelica, Z., Schick, C., High and selective CO2 uptake, H2storage and methanol sensing on the amine-decorated 12-connected MOF CAU-1. *Energy Environ. Sci.*, 4, 4522–4527, 2011.
25. Cmarik, G.E., Kim, M., Cohen, S.M., Walton, K.S., Tuning the Adsorption Properties of UiO-66 *via* Ligand Functionalization. *Langmuir*, 28, 15606–15613, 2012.
26. Jiao, J., Dou, L., Liu, H., Chen, F., Bai, D., Feng, Y., Xiong, S., Chen, D.-L., He, Y., An aminopyrimidine-functionalized cage-based metal–organic framework exhibiting highly selective adsorption of C2H2 and CO2 over CH4. *Dalton Trans.*, 45, 13373–13382, 2016.
27. Cabello, C.P., Berlier, G., Magnacca, G., Rumori, P., Palomino, G.T., Enhanced CO2 adsorption capacity of amine-functionalized MIL-100(Cr) metal–organic frameworks. *CrystEngComm*, 17, 430–437, 2015.
28. Jo, H., Lee, W.R., Kim, N.W., Jung, H., Lim, K.S., Kim, J.E., Kang, D.W., Lee, H., Hiremath, V., Seo, J.G., Jin, H., Moon, D., Han, S.S., Hong, C.S., Fine-Tuning of the Carbon Dioxide Capture Capability of Diamine-Grafted Metal–Organic Framework Adsorbents Through Amine Functionalization. *ChemSusChem*, 10, 541–550, 2016.
29. Montoro, C., García, E., Calero, S., Pérez-Fernández, M.A., López, A.L., Barea, E., Navarro, J.A.R., Functionalisation of MOF open metal sites with pendant amines for CO2 capture. *J. Mater. Chem.*, 22, 10155–10158, 2012.

30. Wang, X., Li, H., Hou, X.-J., Amine-Functionalized Metal Organic Framework as a Highly Selective Adsorbent for CO2 over CO. *J. Phys. Chem. C*, 116, 19814–19821, 2012.
31. Lee, W.R., Hwang, S.Y., Ryu, D.W., Lim, K.S., Han, S.S., Moon, D., Choi, J., Hong, C.S., Diamine-functionalized metal–organic framework: Exceptionally high CO2 capacities from ambient air and flue gas, ultrafast CO2 uptake rate, and adsorption mechanism. *Energy Environ. Sci.*, 7, 744–751, 2014.
32. Choi, S., Watanabe, T., Bae, T.-H., Sholl, D.S., Jones, C.W., Modification of the Mg/DOBDC MOF with Amines to Enhance CO2 Adsorption from Ultradilute Gases. *J. Phys. Chem. Lett.*, 3, 1136–1141, 2012.
33. Drisdell, W.S., Poloni, R., McDonald, T.M., Pascal, T.A., Wan, L.F., Pemmaraju, C.D., Vlaisavljevich, B., Odoh, S.O., Neaton, J.B., Long, J.R., Prendergast, D., Kortright, J.B., Probing the mechanism of CO2 capture in diamine-appended metal–organic frameworks using measured and simulated X-ray spectroscopy. *Phys. Chem. Chem. Phys.*, 17, 21448–21457, 2015.
34. McDonald, T.M., D'Alessandro, D.M., Krishna, R., Long, J.R., Enhanced carbon dioxide capture upon incorporation of N,N′-dimethylethylenediamine in the metal–organic framework CuBTTri. *Chem. Sci.*, 2, 2022–2028, 2011.
35. Kizzie, A.C., Wong-Foy, A.G., Matzger, A.J., Effect of humidity on the performance of microporous coordination polymers as adsorbents for CO2 capture. *Langmuir*, 27, 6368–6373, 2011.
36. Hu, M.-L., Razavi, S.A.A., Piroozzadeh, M., Morsali, A., Sensing organic analytes by metal–organic frameworks: A new way of considering the topic. *Inorg. Chem. Front.*, 7, 1598–1632, 2020.
37. Joarder, B., Desai, A.V., Samanta, P., Mukherjee, S., Ghosh, S.K., Selective and Sensitive Aqueous-Phase Detection of 2,4,6-Trinitrophenol (TNP) by an Amine-Functionalized Metal–Organic Framework. *Chem. Eur. J.*, 21, 965–969, 2014.
38. Zhu, L., Liu, X.-Q., Jiang, H.-L., Sun, L.-B., Metal–organic frameworks for heterogeneous basic catalysis. *Chem. Rev.*, 117, 8129–8176, 2017.
39. Zhang, Z., Xiang, S., Rao, X., Zheng, Q., Fronczek, F.R., Qian, G., Chen, B., A rod packing microporous metal–organic framework with open metal sites for selective guest sorption and sensing of nitrobenzene. *Chem. Commun.*, 46, 7205–7207, 2010.
40. Razavi, S.A.A. and Morsali, A., Function–Structure Relationship in Metal–Organic Frameworks for Mild, Green, and Fast Catalytic C–C Bond Formation. *Inorg. Chem.*, 58, 14429–14439, 2019.
41. Lin, Y., Zhang, Q., Zhao, C., Li, H., Kong, C., Shen, C., Chen, L., An exceptionally stable functionalized metal–organic framework for lithium storage. *Chem. Commun.*, 51, 697–699, 2015.
42. Safarifard, V. and Morsali, A., Influence of an amine group on the highly efficient reversible adsorption of iodine in two novel isoreticular interpenetrated pillared-layer microporous metal–organic frameworks. *CrystEngComm*, 16, 8660–8663, 2014.

43. DeCoste, J.B., Browe, M.A., Wagner, G.W., Rossin, J.A., Peterson, G.W., Removal of chlorine gas by an amine functionalized metal-organic framework *via* electrophilic aromatic substitution. *Chem. Commun.*, 51, 12474–12477, 2015.
44. Desai, A.V., Samanta, P., Manna, B., Ghosh, S.K., Aqueous phase nitric oxide detection by an amine-decorated metal–organic framework. *Chem. Commun.*, 51, 6111–6114, 2015.
45. Peterson, G.W., Mahle, J.J., DeCoste, J.B., Gordon, W.O., Rossin, J.A., Extraordinary NO2 Removal by the Metal-Organic Framework UiO-66-NH2. *Angew. Chem. Int. Ed.*, 55, 6235–6238, 2016.
46. Fu, Y., Sun, D., Chen, Y., Huang, R., Ding, Z., Fu, X., Li, Z., An Amine-Functionalized Titanium Metal–Organic Framework Photocatalyst with Visible-Light-Induced Activity for CO2 Reduction. *Angew. Chem. Int. Ed.*, 51, 3364–3367, 2012.
47. Shi, L., Wang, T., Zhang, H., Chang, K., Meng, X., Liu, H., Ye, J., An Amine-Functionalized Iron(III) Metal–Organic Framework as Efficient Visible-Light Photocatalyst for Cr(VI) Reduction. *Adv. Sci.*, 2, 1500006, 2015.
48. Wu, P., He, C., Wang, J., Peng, X., Li, X., An, Y., Duan, C., Photoactive Chiral Metal–Organic Frameworks for Light-Driven Asymmetric α-Alkylation of Aldehydes. *J. Am. Chem. Soc.*, 134, 14991–14999, 2012.
49. Wang, H., Yuan, X., Wu, Y., Zeng, G., Chen, X., Leng, L., Wu, Z., Jiang, L., Li, H., Facile synthesis of amino-functionalized titanium metal-organic frameworks and their superior visible-light photocatalytic activity for Cr(VI) reduction. *J. Hazard. Mater.*, 286, 187–194, 2015.
50. Long, J., Wang, S., Ding, Z., Wang, S., Zhou, Y., Huang, L., Wang, X., Amine-functionalized zirconium metal–organic framework as efficient visible-light photocatalyst for aerobic organic transformations. *Chem. Commun.*, 48, 11656–11658, 2012.
51. Zhou, J., Li, H., Zhang, H., Li, H., Shi, W., Cheng, P., A Bimetallic Lanthanide Metal–Organic Material as a Self-Calibrating Color-Gradient Luminescent Sensor. *Adv. Mater.*, 27, 7072–7077, 2015.
52. Wang, Z.-J., Han, L.-J., Gao, X.-J., Zheng, H.-G., Three Cd(II) MOFs with Different Functional Groups: Selective CO2 Capture and Metal Ions Detection. *Inorg. Chem.*, 57, 5232–5239, 2018.
53. Hahm, H., Ha, H., Kim, S., Jung, B., Park, M.H., Kim, Y., Heo, J., Kim, M., Synthesis of secondary and tertiary amine-containing MOFs: C–N bond cleavage during MOF synthesis. *CrystEngComm*, 17, 5644–5650, 2015.
54. Keenan, L.L., Hamzah, H.A., Mahon, M.F., Warren, M.R., Burrows, A.D., Secondary amine-functionalised metal–organic frameworks: Direct syntheses versus tandem post-synthetic modifications. *CrystEngComm*, 18, 5710–5717, 2016.
55. Burrows, A.D. and Keenan, L.L., Conversion of primary amines into secondary amines on a metal–organic framework using a tandem post-synthetic modification. *CrystEngComm*, 14, 4112–4114, 2012.

3

Azo and Azine Decorated Metal–Organic Frameworks

Abstract
The focus of this chapter is on chemical properties of azine and azo functions and how these chemical properties, like Lewis basicity and hydrogen bond accepting character, work in designing functional MOFs based on azine and azo functions. Also, light sensitivity of azo function discussed for designing stimuli-responsive metal-organic frameworks. Finally, structural features of azine and azo decorated metal-organic frameworks are discussed.

Keywords: Azine, azo, azobenzene, functional metal-organic frameworks, light responsive frameworks, picric acid detection, oil denitrogenation, Lewis basic catalysis

3.1 General Chemical Properties of Azine and Azo Functions

Imine (–N=), azine (=N–N=) and azo (–N=N–) are other non-cyclic functional groups based on nitrogen atom which repetitively applied in the structure of MOFs. Cleary, these functions are Lewis basic and H-bond acceptor sites. So, FMOFs with these functions could apply as Lewis basic heterogeneous catalyst, Lewis basic/H-bond acceptor sensor or adsorbent for small organic molecules which are enable to interact trough their positive H-atom(s) and Lewis acid metal ions.

3.2 Function–Application Properties

Azine decorated MOFs can act as a heterogeneous catalyst in Lewis basic interactions [1–5]. Recently, we study on Lewis basicity of azine decorated MOFs in Knoevenagel reaction by applying azine decorated TMU-4

Ali Morsali and Sayed Ali Akbar Razavi. Functional Metal-Organic Frameworks: Structure, Properties and Applications, (37–54) © 2021 Scrivener Publishing LLC

($[Zn_2(OBA)_2(BPDP)]_n$), methyl-azine decorated TMU-5 ($[Zn(OBA)(BPDH)_{0.5}]_n$) and imine decorated TMU-6 ($[Zn(OBA)(BPMB)_{0.5}]_n$) where H_2OBA, BPDB, BPDH, BPMB are 4,4'-oxybisbenzoic acid, 1,4-bis(4-pyridyl)-2,3-diaza-1,3-butadiene, 2,5-bis(4-pyridyl)-3,4-diaza-2,4-hexadiene and N^1,N^4-bis-((pyridin-4-yl)methylene)benzene-1,4-diamine (Figure 3.1a) [3]. The results reveal that the functionality of each TMU-framework and solvent are of vital roles in conversion percentage. The results show that the role of solvents in reaction conversion is H_2O > MeOH > EtOH > CH_3CN ≈ CH_2Cl_2 ≈ n-Hexane ≈ Toluene. It is clear that this trend is according

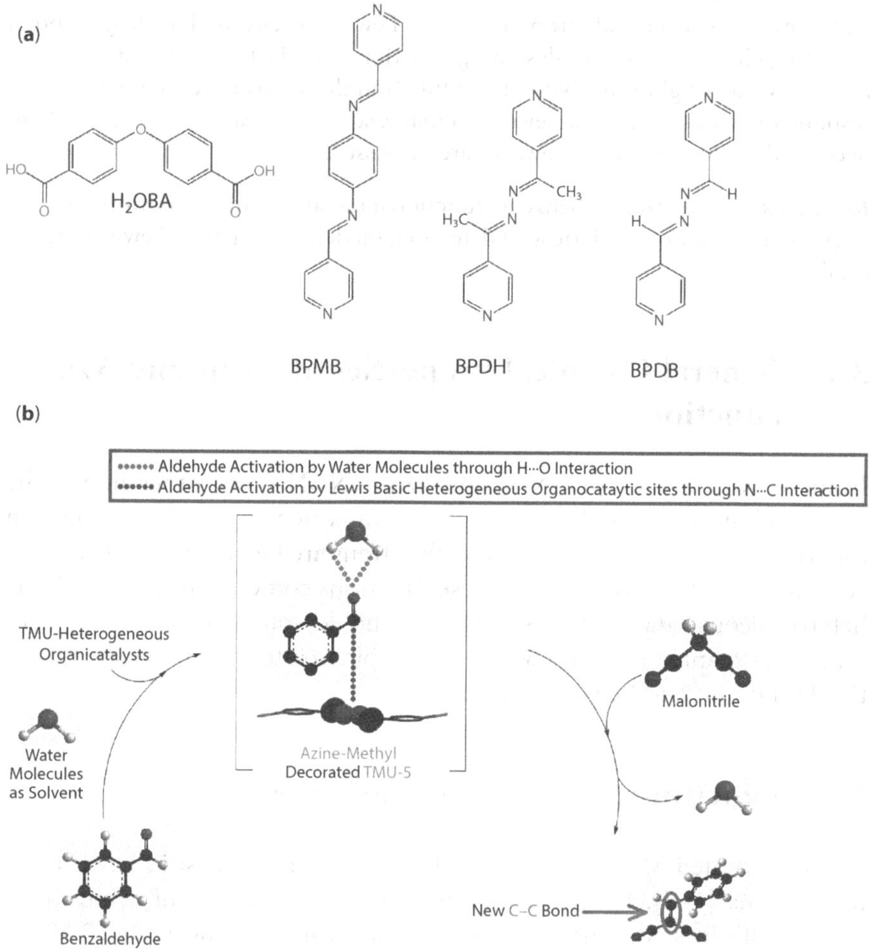

Figure 3.1 Application of TMU-4, TMU-5 and TMU-6 in knoevenagel reaction. (a) Depiction of applied ligand in the structure of applied TMU-MOFs. (b) Proposed mechanism for activation of benzaldehyde by TMU-5.

to solvent polarity ability and participating in Hydrogen bonding. The results in water reveal that after 30 min TMU-4, TMU-5 and TMU-6 reach to 100, 45 and 38% conversion percentage. TMU-4 and TMU-5 are of higher catalytic activity than TMU-6. These MOFs have adjacent N atoms in azine function giving rise to lone pair–lone pair electron repulsion and the alpha effect. This phenomenon leads to higher basicity and higher reaction conversion for TMU-4 and TMU-5 rather TMU-6. Moreover, presence of electron methyl group in the structure of BPDH improves the basicity of azine group in pore walls of TMU-5. So, TMU-5 is stronger catalyst rather TMU-4.

In next study, we tried to examine the possible catalytic mechanism by TMU-frameworks. Two types of base applied to catalyze Knoevenagel interactions; strong bases which are able to abstract an acidic H atom from methylene compound and moderately or weakly bases which are able to accelerate the reaction through activation of aldehyde or ketone and then addition of methylene compound [6]. Since, the Lewis basicity of these materials is not much enough to abstract an H from methylene compound; we propose that TMU-5 could catalyze the reaction through benzaldehyde activation. As mentioned in Chapter 2, it was proposed that Knoevenagel reaction between benzaldehyde and malonitrile (methylene compound) is catalyzed through reaction between amine site and benzaldehyde molecule to form an imine compound following by addition of malonitrile [7–9]. However, this proposed mechanism is not makes sense for activation of benzaldehyde by TMU-5 because TMU-5 contain azine site, which lack $-NH_2$ groups and so formation of imine compound. So, TMU-MOFs cannot activate benzaldehyde through the formation of imine. Instead of it, we proposed that TMU-5 could activate benzaldehyde through a noncovalent Lewis base–acid interaction between the azine N atom of TMU-5 and the carbonyl C atom of benzaldehyde (Figure 3.1b). Also based on the remarkable effects of solvent in reaction conversion, we proposed that the H atom of water molecules can activate the aldehyde through aldehyde (C=O)···(H) water hydrogen bonding (Figure 3.1b) [1]. This behavior in activation of benzaldehyde by water molecules is similar in activation of carbonyl group of benzaldehyde through a Lewis acid (3d metal)-base (benzaldehyde) interaction in presence of catalytic metal centers [8]. The results show that methyl-azine decorated TMU-5 could catalyze reaction between benzaldehyde and malonitrile in 30 min at room temperature and atmospheric pressure in water as the greenest solvent.

Azine decorated MOFs applied for removal and detection of small organic molecules by hydrogen bonding like picric acid detection [10], removal and degradation of phenol [11], and model oil denitrogenation through removal of neutral nitrogen containing contaminants [12]. Also, they applied for removal and detection of metal ions [13].

Since Azine can potentially accept a H-bond, we applied RhB@TMU-5 composite (RhB = Rhodamine B) as a 2D ratiometric host-guest fluorescent sensor for highly sensitive and selective detection of picric acid among the other NACs and VOCs [10] We expect that picric acid potentially could interact with TMU-5 through hydrogen bonding, while we did not expect for any interaction between trapped RhB into the pores of TMU-5 and picric acid (Figure 3.2). Rhb@TMU-5 composite has a dual emissive PL emission peak at 485 and 583 nm related to the PL emission band of TMU-5 and Rhb, respectively. Application of this composite in detection

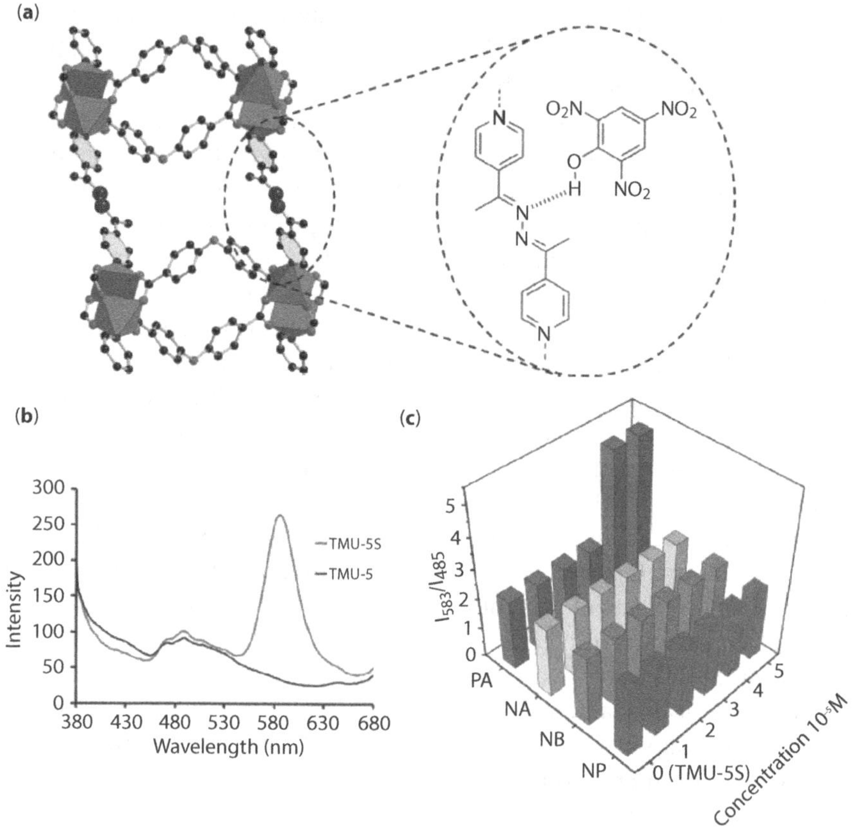

Figure 3.2 Application of RhB@TMU-5 composite in 2D detection of picric acid. (a) Azine decorated pores of TMU-5 and hydrogen bonding interaction between picric acid and TMU-5. (b) Emission spectra of TMU-5 and Rhb@TMU-5 dispersed in acetonitrile upon excitation at 355 nm. (c) Comparison of responses of RhB@TMU-5 in the presence of different nitro aromatic analytes at different concentrations [10].

of NACs show that as a result of hydrogen bonding between picric acid and azine group of TMU-5, I_{485} decreases (related to TMU-5) while I_{583} (related to RhB) did not change significantly. So, the response of sensor which is defined as I_{583}/I_{485} increases in presence of picric acid (4×10^{-5} M) while there is no specific change in presence of other NACs and VOCs.

In other work, we applied methyl-azine decorated TMU-5 for removal of neutral nitrogen containing contaminants (NNCCs) from model oil (Figure 3.3) [12]. NNCCs include hydrogen bond donor (–NH–) site in their molecular skeleton and TMU-5 is decorated with azine function which can act as hydrogen bond acceptor site. Experimental data reveal that TMU-5 has high adsorption capacity to pyrrole (518 mg·g^{-1}) and indole (578 mg·g^{-1}). Pursuant to FT-IR analysis we found that there is a strong hydrogen bonding between N atoms of azine function and NH group of pyrrole as candidate NNCC. The intense NH peak of pure pyrrole is centered at 3402 cm^{-1} and there is no observable peak in this region of wavenumbers for TMU-5 before adsorption process. After adsorption of pyrrole by TMU-5, a broadened peak which is centered at 3,305 cm^{-1} appeared in the PRR@TMU-5 FT-IR peak, which can be related to the (NH) group of adsorbed PRR molecules after denitrogenation. If we consider that the peak centered at 3,305 cm^{-1} is related to the adsorbed pyrrole molecules by TMU-5, then we see that compare to pyrrole molecules, this peak was broadened noticeably and shifted from 3,402 to 3,305 cm^{-1}. Such broadening and shift in FT-IR peak of pyrrole before and after adsorption by TMU-5 reveals a hydrogen bonding between TMU-5 and pyrrole molecules.

Host-guest chemistry of azine (and azo) functions is centered at their electron-rich N atoms. Since N atoms in azine and azo functions are adjacent together, they could destabilize each other through lone pair-lone pair interactions. So, they could interact with any guest molecules containing

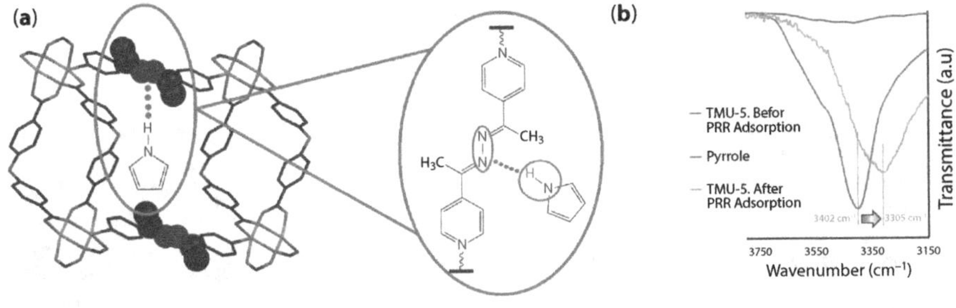

Figure 3.3 Application of TMU-5 in oil adsorptive denitrogenation. (a) Possible hydrogen bonding between azine function inside the pores of TMU-5 and NH site of pyrrole molecule. (b) Changes in FT-IR of pyrrole and TMU-5 before and after adsorption process [12].

positively charged atoms like C in CO_2, positively charged H atoms like nitroaniline, phenol [14], picric acid [11], pyrrole [10] and indole [12]. However, it is necessary to notice that the selectivity of azine and azo decorated MOFs toward these guests is attributed to some factors like pore size and shape and energy levels of the functions in the structure of MOFs.

Owing to their Lewis basicity, azine and azo decorated MOFs applied for separation of carbon dioxide from nitrogen gases [15–21]. To this aim, we applied our TMU-frameworks, including methyl-azine decorated TMU-5 and tetrazine decorated TMU-34(-2H) ($[Zn(OBA)(DPT)_{0.5}]_n$, DPT = 3,6-di(pyridin-4-yl)-1,2,4,5-tetrazine) for carbon dioxide and methane adsorption (Figure 3.4) [15, 19].

TMU-5 and TMU-34(-2H) are isostructure with three dimensional (3D) interconnected pores. Owing to presence of methyl group in the pore walls of TMU-5, this framework has smaller pore size (3.8 × 5.6 Å) than TMU-34(-2H) (4.4 × 8.1 Å) which leads to lower surface area for TMU-5 rather TMU-34(-2H) (580 vs. 670 $m^2 \cdot g^{-1}$). These frameworks applied in CO_2 adsorption and the results show that TMU-5 has higher zero coverage enthalpy (43 $kJ \cdot mol^{-1}$) in comparison with TMU-34(-2H) (30 $kJ \cdot mol^{-1}$) which reveal that there is a stronger interaction between methyl-azine groups of TMU-5 rather tetrazine groups of TMU-34(-2H). As a result of this stronger interaction, TMU-5 has relatively higher selectivity rather TMU-34(-2H) (32 vs. 29) toward carbon dioxide.

The results of methane adsorption reveal that TMU-5 can adsorb higher amount of methane rather TMU-34(-2H) is same conditions (20.3 $cm^3 \cdot g^{-1}$

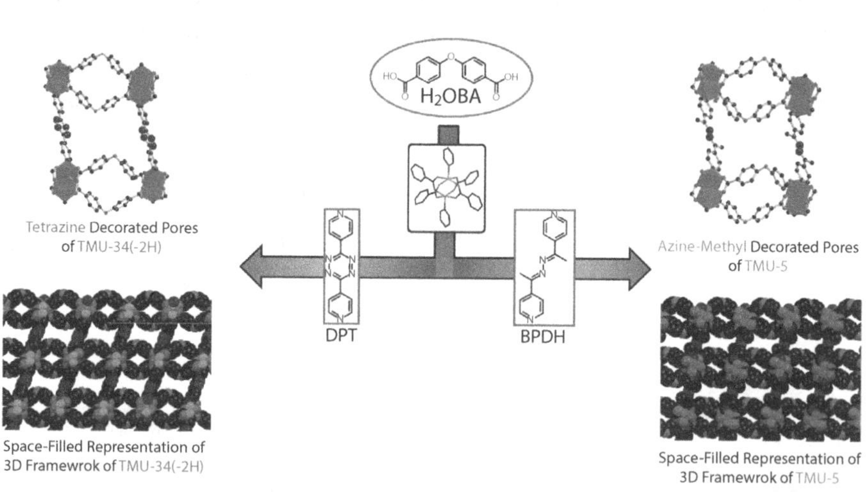

Figure 3.4 Representation of applied ligand, functional pores and 3D frameworks of TMU-5 and TMU-34(-2H).

(2.54 wt %) for TMU-5 and 15.5 $cm^3 \cdot g^{-1}$ (1.96 wt %) for TMU-34(-2H)). This observation is due to the different pore size and functionality of these frameworks. It is necessary to mention that optimal pore size for methane uptake are in very good agreement with the kinetic diameter of one (4 Å) and two (8 Å) methane molecules which is well-matched with pore size of TMU-5 and TMU-34(-2H). Also, it is proven that Lewis basic sites are effective in improvement of methane adsorption capacity. As a result of such Lewis basic functionalized pores with optimal size, TMU-5 and TMU-34(-2H) represent strong affinity to methane molecules (27 $kJ \cdot mol^{-1}$ for TMU-5 and 22 $kJ \cdot mol^{-1}$ for TMU-34(-2H)) and so noticeable methane uptake at ambient temperature and pressure, even higher than those MOFs with higher surface area.

The Lewis basicity of TMU-5, as an azine decorated MOF show itself in both Knoevenagel condensation and CO_2 adsorption. Also, TMU-5 shows its potential Lewis basicity in other applications like detection of picric acid and removal of indole and pyrrole from model oil. These results are consistent together and reveal that a rationally functionalized MOF with azine functions could represent its potentials even in different applications.

As mentioned earlier, owing to similar properties with azine function, azo functional group could apply in different type of application needing Lewis basic assistant like removal and detection of metal ions through azo(N)·metal ion coordination interaction [22–24], detection of picric acid through azo(N)·(H–O)picric acid hydrogen bonding [25], CO_2 adsorption [26, 27] and Lewis basic assistant catalysis [28, 29]. Moreover than these applications, azo decorated MOFs show some interesting chemical properties including redox activity, photo responsivity and explosive character.

Explosivity of azo functionalized materials depends on their molecular structure. If azo materials are based on aliphatic chain, they apply with some of other functions like tetrazole and nitro to design explosive materials while aromatic azo compounds are stable in this view. In this regard, almost all azo decorated MOFs are based on aromatic azo linkers.

Photo/heat responsivity maybe is the most important chemical property of azo function. Structurally, azo function resembles to alkene and similarly could transform between cis and trans conformations. This possibility in conformational change is the reason for photo or heat responsivity of azo group. In heat conduction with special temperature and light irradiation with certain wavelength, the conformation of azo group could be changed that made azo function at center of attention in development of new photoresponsive materials like molecules, chelators, and molecular machines for artificial switchable catalysts and other kind of remote-controllable materials [30]. In this regard, synthesis of azo decorated MOFs gain lots of interest among scientists [31–37].

Alexander Knebel and coworkers synthesized a fluorinated-azobenzene functionalized MOF ($Cu_2(F_2AzoBDC)_2(dabco)$, dabco = 1,4-diazabicyclo[2.2.2]octane) and studied photoisomerization properties of this MOF using UV/Vis and infrared-reflection absorption spectroscopy (Figure 3.5) [38]. Cis-trans isomerization introduces mark differences in absorption spectra of the MOF. UV/Vis absorbance spectra of the MOF on quartz reveal that upon violent-light irradiation at 400 nm, π–π^* band increases while upon green-light irradiation (530 nm) this peak is decreased. The conformational change conducted for five times with retainability in absorption properties. UV/Vis absorbance spectra in ethanol show that the n–π^* band (or shoulder) at 450 nm was shifted to larger wavelengths when irradiated with violet light and to smaller wavelengths when irradiated with green light, respectively. These changes show cis-to-trans and trans-to-cis isomerization could happen upon violent and green light irradiation, respectively. Moreover, FT-IR absorption spectra of the MOF show distinctive changes upon photoisomerization. Upon violet-light irradiation, the intensities of the vibration bands at 1,030 and 960 cm^{-1} increase, whereas the band at 1,015 cm^{-1} decreases while upon green-light irradiation, the bands at 1,030 and 960 cm^{-1} decrease, whereas the band at 1,015 cm^{-1} increases.

One delicate application of azo functionalized MOFs is for remote-control cargo release [39–48]. In this application, the adsorption performance of loaded guest molecules like organic or gas molecules could be tuned. In this regard Omar Yaghi and coworkers applied of an azobenzene functionalized

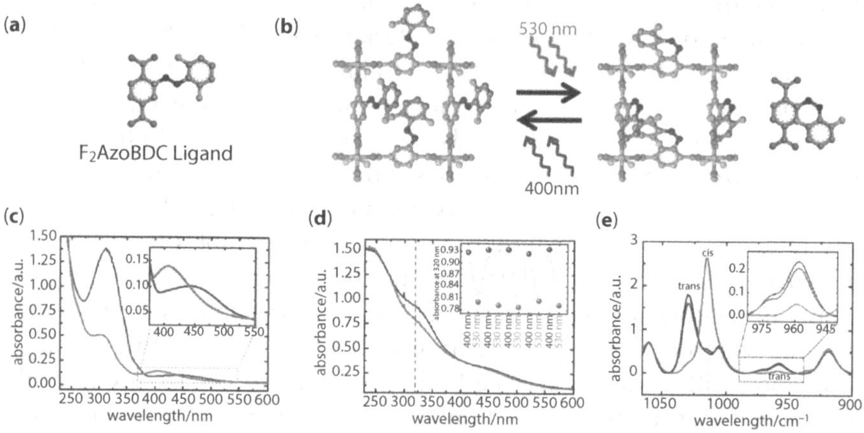

Figure 3.5 Photoisomeriation characterization of $Cu_2(F_2AzoBDC)_2(dabco)$. (a) Structure of $F_2AzoBDC$ ligand. (b) Reversible cis-trans conformational change upon light irradiation. (c) UV–Vis absorption of the MOF in ethanol. (d) UV-Vis absorption of the MOF on quartz. (e) FT-IR absorption of the MOF [38].

isoreticular metal–organic framework, IRMOF-74-III, to investigate the under-control capture-release of cargo in to pores of IRMOF-74-III in presence of light as external stimuli (Figure 3.6) [49]. Since 1D pores of IRMOF-74-III are decorated with azobenzene groups and they are headed to the middle of the pore, the pore aperture of the material is dominated by cis or trans conformation of the azobenzene group. With all azobenzene functionalities in the trans configuration the idealized pore aperture is 8.3 °A, but if all functionalities are cis, the aperture is larger 10.3 °A. Propidium iodide luminescent dye applied to evaluate the remote-control capture and release ability of stimuli-responsive IRMOF-74-III up on light irradiation. This dye was specifically chosen because its size, 8 × 11 × 16 °A without counter-ion, matches well with the pore aperture of azo-IRMOF- 74-III, between 8.3 and 10.3 °A depending on conformation of azobenzene linker. Comparison between pore diameter of the IRMOF-74-III and the dye kinetic diameter show that dye cannot easily diffuse into the pores of IRMOF-74-III when azobenzene groups are in trans conformation. Certain type of photoluminescence experiment and laser activation applied for on-command cargo release. Upon excitation at 408 nm and laser activation, cis-trans isomerization could occur which results in releasing of the dye while there is no increase in

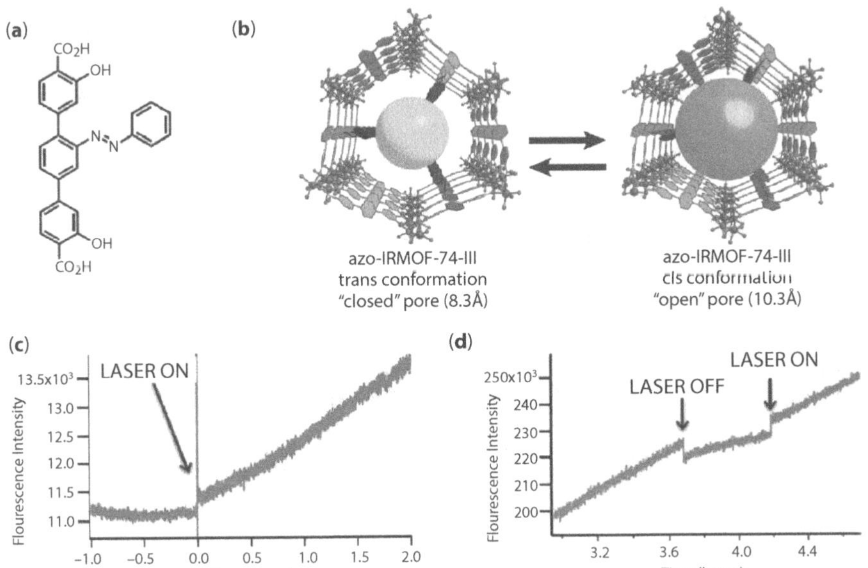

Figure 3.6 Application of IRMOF-74 in on-command cargo release of dye molecules. (a) Azobenzene functionalized linker applied for synthesis of IRMOF-74. (b) 1D pores of IRMOF-74 in cis and trans conformation of azobenzene groups. (c) Fluorescence intensity of the released dye before and after laser activation. (d) On-off trial in laser activation [49].

propidium iodide intensity prior to laser activation. This observation shows that light irradiation give rise to conformational change of azobenzene and so on-command dye release. Off-on laser activation experiments show that up on temporary stop in laser activation, the kinetic of release is reduced while after another laser activation cycle, the release kinetic is increased.

There is a debate about mechanism of cargo-release by stimuli-responsive azo decorated MOFs. This mechanism can be based on strict or polar differences in the structure of cis or trans conformation of azo group. Hiroshi Sato and coworkers show that this could be polar mechanism which is effective in on-demand capture-release of CO_2 molecules (Figure 3.7) [50]. This MOF is isostructural with UiO-68 but functionalized with azobenzene groups, named as azo-MOF. They applied azo-MOF with different ratio of cis and trans conformation for adsorption of different guests like Ar and CO_2 with close kinetic diameter and different polarity. Conformational trans-to-cis isomerization happen upon UV light irradiation (365 nm). Azo-MOF-21% (21% refers to cis-isomer content) can be synthesized through 30 min light irradiation and

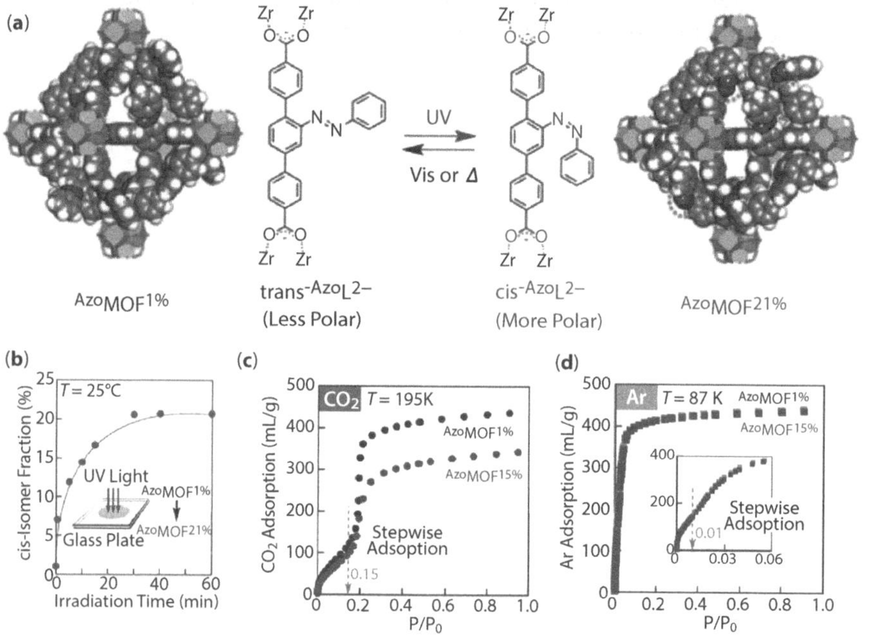

Figure 3.7 Application of azo-MOF in photo-responsive capture-release of Ar and CO_2. (a) Structural representation of azo-MOF and its azobenzene containing ligand (L^{2-} = 2'-phenyldiazenyl-1,1':4',1'-terphenyl-4,4"-dicarboxylate). (b) Cis-trans fraction of the azobenzene group in the irradiated azo-MOF after 30 min which is determined by ^{1}H-NMR. (c) Adsorption isotherms of CO_2 at 195 K. (d) Adsorption isotherms of Ar at 195 K [50].

could undergo reverse isomerization with heating or visible light irradiation (420–480 nm) to synthesize azo-MOF-1%. Adsorption profile of Ar at 87 K show rare change for azo-MOF-1% and azo-MOF-15%. Absorption profile of CO_2 shows that azo-MOF-1% and azo-MOF-15% display stepwise adsorption isotherm 195 K with a clear inflection point at $P/P_0 = 0.15$ in which azo-MOF-1% with lower content of cis-conformation could adsorb higher amount of CO_2. Calculation of isosteric heat of CO_2 adsorption shows that azo-MOF-15% has higher affinity than azo-MOF-1% to CO_2 molecules (38 vs. 29 $kJ \cdot mol^{-1}$). This observation show that the less polar azo-MOF with higher amount of trans azobenzene group can interact more weakly via CO_2 through a quadrupole/quadrupole interactions. If the release mechanism was based on strict effects, both Ar and CO_2 must show resemble behavior before and after conformation transformation. These results clearly reveal the dominant roles of a polar effect rather than a steric effect in CO_2 capture-release by azo-MOF.

Although azine and azo represent similar chemical features like Lewis basicity and H-bond accepting character, azine decorated MOFs successfully in mentioned types of application while azo group represent special characters in photo-responsivity which absorb scientists in designing stimuli-responsive platforms.

3.3 Function-Structure Properties

As we discussed azo function can locate on the main-chain and side-chain of the MOF. Azobenzene and fluorinated azobenzene are mostly applied in the side chain of the MOF and their presence does not introduce any specific change into the structure of MOF. However, their ability in cis-trans isomerization greatly impact on pore size and accessible pore volume so that this technique applied for on-demand cargo release upon light irradiation. It is important to notice that such cis-trans isomerization may not be possible due to the strict effects. In this regard, Lars Heinke and coworkers synthesized two azobenzene decorated MOFs, [Cu_2(AzoBPDC)$_2$(BiPy)] and [Cu_2(NDC)$_2$(AzoBiPy)] (where AzoBPDC = 3-azobenzene 4,4'-biphenyldicarboxylic acid, BiPy = 4,4'-bipyridine, NDC = 2,6-naphthalenedicarboxylic acid, AzoBiPy = 3-azobenzene-4,4'-bipyridine), and show that in the Cu_2(AzoBPDC)$_2$(BiPy) MOF structure, the azobenzene side groups undergo photoisomerization when irradiated with UV or visible light while Cu_2(NDC)$_2$(AzoBiPy) this could not happen (Figure 3.8) [51]. Theoretical studies show that there is possible collision between bulky NDC ligands and trans-to-cis isomerized azobenzene group and for a successful isomerization in Cu_2(NDC)$_2$(AzoBiPy) framework, structural rearrangement

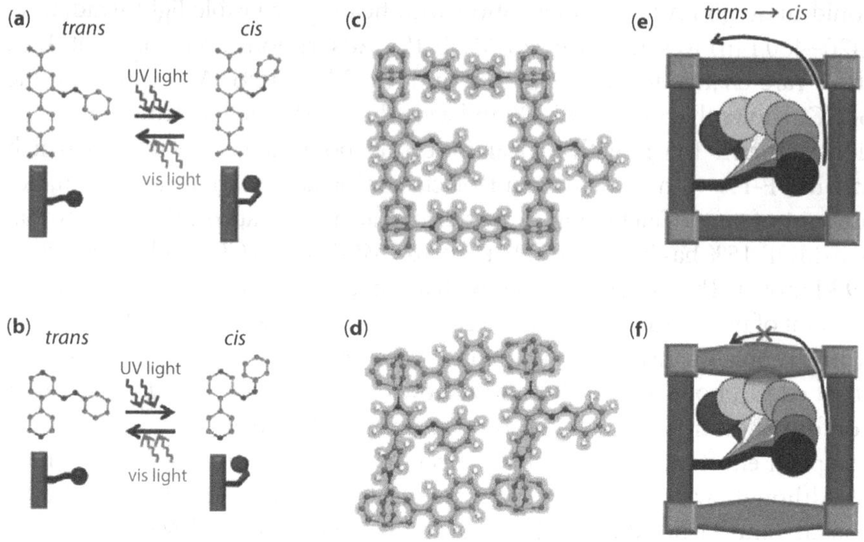

Figure 3.8 Structural strict hindrance in [$Cu_2(AzoBPDC)_2(BiPy)$] and [$Cu_2(NDC)_2(AzoBiPy)$] during azobenzene trans-to-cis photoisomerization [51].

as quick as isomerization process are essential to provide more free space around the azobenzene moiety. Since such structural rearrangement is not possible, naphthalene linker in $Cu_2(NDC)_2(AzoBiPy)$ blocks the photoisomerization due to structural hindrance.

On the other hand, it seems that azo functions which locate in the main-chain can not change the overall structure of the MOF. This statement is true when the MOF do not expose to light irradiation and if so, structural change can happen. An azo functionazlied MOF isostructure with UiO-67 is synthesized using azobenzene-4,4′-dicarboxylic acid ($H_2AzoBDC$) ligand [29, 52]. It seems that presence of azo group in the main chain of the MOF has not remarkable effects on the structure of MOF. This is maybe due to the fact that presence of azo function in the main chain of the organic ligand does not change the relative angle between coordinating sites (Figure 3.9a). In other work, Myunghyun Paik Suh and coworkers synthesized an azo decorated MOF with 1,1'-azobenzene-3,3',5,5'-tetracarboxylic acid ligand (H_4abtc) which is isostructure with NOTT-100 (Figure 3.9b) [53]. Again, presences of azo group in the main chain of the organic ligand do not change significantly the structure of the MOF. In other work, Hong-Cai Zhou and coworkers applied a mixed ligand approach to synthesis azo decorated MOFs which results in similar result [54].

Upon light irradiations, azo functions in the main-chain show different stimuli-responsivity. Azobenzene groups in the side-chain isomerize upon

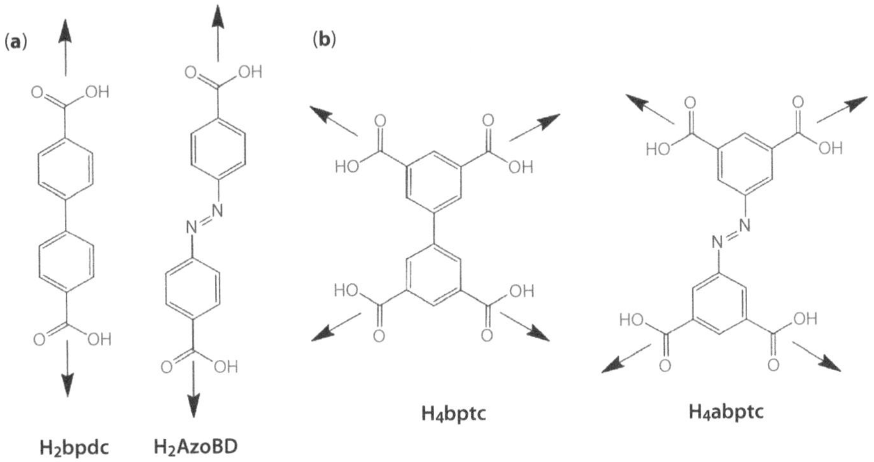

Figure 3.9 Representation of the non-functional and azo functionalized ligands applied for synthesis of isostructure MOFs. H_2bpdc = [1,1'-biphenyl]-4,4'-dicarboxylic acid and H_4bptc = Biphenyl-3,3',5,5'-tetracarboxylic acid.

light irradiation which could rise to remote release of the loaded guests. But, such cis-trans isomerization does not occur for azo groups in the main chain. Matthew R. Hill and coworkers prove that owing to strong bonding between metal clusters and azo ligands in the main chain, azo moieties are forced to undergo a suppressed localized bending around azo group could occur upon light irradiation [45, 46]. Such bending results in accumulated stress. When the light irradiation is stopped, the accumulated stress is released, accompanying with the recovery of the ground state of the ligand. As a result, the structure of MOF is retained (Figure 3.10).

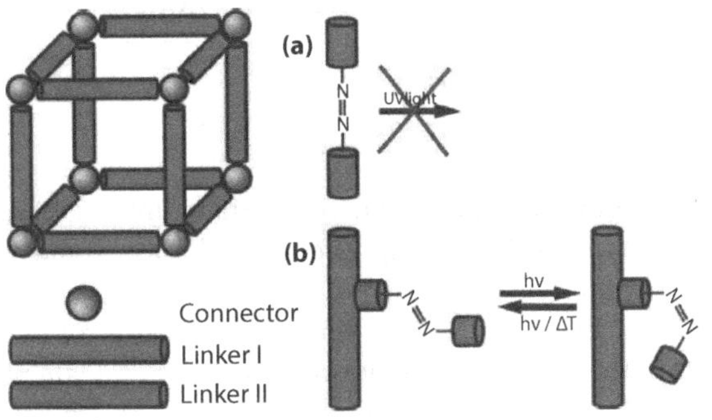

Figure 3.10 Two ways to introduce azo-functionalized molecules into MOFs. (a) Azo-linker molecules as part of the backbone of the MOF, (b) azo-group covalently attached to the inner pore wall and protruding into the pore [35].

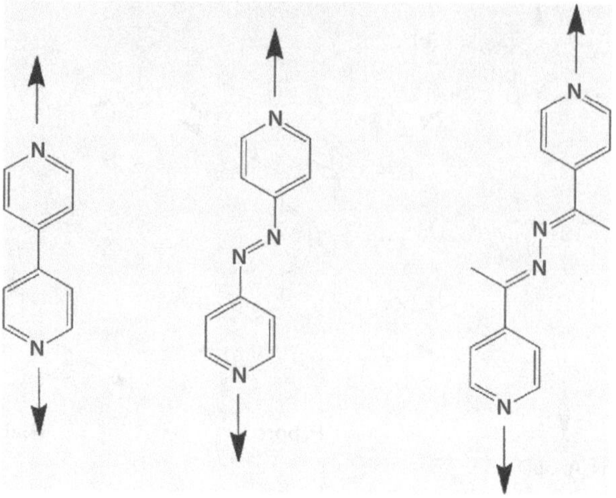

Figure 3.11 Structural combination of azine and azo pillar spacers.

Structural features of azine group are similar to azo group. The presence of azine group in the main-chain of the MOF do not change the relative angle of coordinating sites to each other (Figure 3.11). Rahul Banerjee and coworkers synthesized two pillar layer MOFs with azine and azo functionalized pillars ([Cd(L1)(suc)] and [Cd(L2)(suc)] where L1 = 2,5-bis-(4-pyridyl)-3,4-diaza-2,4-hexadiene, L2 = trans 4,4′-azobispyridine and suc = succinate dianion) [20]. These MOFs with straight azine L1 ligand and azo L2 ligand form honeycomb-like three-dimensional (3D) architecture. In other two different works, $[Zn(OBA)(L3)_{0.5}]$ and $[Zn(OBA)(L4)_{0.5}]$ with azine L3 and azo L4 ligands are synthesized (H_2OBA=4, 4'-oxybisbenzoic acid, L3 =2,5-bis(3-pyridyl)-3,4-diaza-2,4-hexadiene and L4 = 3,3'-azobis(pyridine)) [16, 26]. These MOFs form a 3D network with mab topology and 1D rhombus-shaped channels along crystallographic b axis with paddle-wheel nodes.

References

1. Razavi, S.A.A. and Morsali, A., Function–Structure Relationship in Metal–Organic Frameworks for Mild, Green, and Fast Catalytic C–C Bond Formation. *Inorg. Chem.*, 58, 14429–14439, 2019.
2. Masoomi, M.Y., Beheshti, S., Morsali, A., Shape control of Zn (II) metal–organic frameworks by modulation synthesis and their morphology-dependent catalytic performance. *Cryst. Growth Des.*, 15, 2533–2538, 2015.
3. Masoomi, M.Y., Beheshti, S., Morsali, A., Mechanosynthesis of new azine-functionalized Zn (II) metal–organic frameworks for improved catalytic performance. *J. Mater. Chem. A*, 2, 16863–16866, 2014.

4. Joharian, M., Morsali, A., Tehrani, A.A., Carlucci, L., Proserpio, D.M., Water-stable fluorinated metal–organic frameworks (F-MOFs) with hydrophobic properties as efficient and highly active heterogeneous catalysts in aqueous solution. *Green Chem.*, 20, 5336–5345, 2018.
5. Joharian, M., Abedi, S., Morsali, A., Sonochemical synthesis and structural characterization of a new nanostructured Co (II) supramolecular coordination polymer with Lewis base sites as a new catalyst for Knoevenagel condensation. *Ultrason. Sonochem.*, 39, 897–907, 2017.
6. Zhu, L., Liu, X.-Q., Jiang, H.-L., Sun, L.-B., Metal–organic frameworks for heterogeneous basic catalysis. *Chem. Rev.*, 117, 8129–8176, 2017.
7. Gascon, J., Aktay, U., Hernandez-Alonso, M.D., van Klink, G.P.M., Kapteijn, F., Amino-based metal–organic frameworks as stable, highly active basic catalysts. *J. Catal.*, 261, 75–87, 2009.
8. Yang, Y., Yao, H.-F., Xi, F.-G., Gao, E.-Q., Amino-functionalized Zr(IV) metal–organic framework as bifunctional acid–base catalyst for Knoevenagel condensation. *J. Mol. Catal. A: Chem.*, 390, 198–205, 2014.
9. Hartmann, M. and Fischer, M., Amino-functionalized basic catalysts with MIL-101 structure. *Micropor. Mesopor. Mat.*, 164, 38–43, 2012.
10. Bagheri, M. and Masoomi, M.Y., Sensitive Ratiometric Fluorescent Metal-Organic Framework Sensor for Calcium Signaling in Human Blood Ionic Concentration Media. *ACS Appl. Mater. Interfaces*, 12, 4625–4631, 2020.
11. Masoomi, M.Y., Bagheri, M., Morsali, A., Junk, P.C., High photodegradation efficiency of phenol by mixed-metal–organic frameworks. *Inorg. Chem. Front.*, 3.7, 944–951, 2016.
12. Razavi, S.A.A. and Morsali, A., High Capacity Oil Denitrogenation over Azine- and Tetrazine-Decorated Metal–Organic Frameworks: Critical Roles of Hydrogen Bonding. *ACS Appl. Mater. Interfaces*, 11, 21711–21719, 2019.
13. Tahmasebi, E., Masoomi, M.Y., Yamini, Y., Morsali, A., Application of Mechanosynthesized Azine-Decorated Zinc (II) Metal–Organic Frameworks for Highly Efficient Removal and Extraction of Some Heavy-Metal Ions from Aqueous Samples: A Comparative Study. *Inorg. Chem.*, 54, 425–433, 2014.
14. Bagheri, M., Masoomi, M.Y., Morsali, A., Highly Sensitive and Selective Ratiometric Fluorescent Metal–Organic Framework Sensor to Nitroaniline in Presence of Nitroaromatic Compounds and VOCs. *Sens. Actuators, B, Chemical*, 243, 353–360, 2016.
15. Razavi, S.A.A., Masoomi, M.Y., Islamoglu, T., Morsali, A., Xu, Y., Hupp, J.T., Farha, O.K., Wang, J., Junk, P.C., Improvement of Methane–Framework Interaction by Controlling Pore Size and Functionality of Pillared MOFs. *Inorg. Chem.*, 56, 2581–2588, 2017.
16. Pal, A., Lin, J.B., Chand, S., Das, M.C., A 3D Microporous MOF with mab Topology for Selective CO2 Adsorption and Separation. *ChemistrySelect*, 3, 917–921, 2018.
17. Alduhaish, O., Wang, H., Li, B., Arman, H.D., Nesterov, V., Alfooty, K., Chen, B., A Threefold Interpenetrated Pillared-Layer Metal–Organic Framework

for Selective Separation of C2H2/CH4 and CO2/CH4. *ChemPlusChem*, 81.8, 764, 2016.

18. Bhattacharya, B., Maity, D.K., Mondal, R., Colacio, E., Ghoshal, D., Two series of isostructural coordination polymers with isomeric benzenedicarboxylates and different azine based N, N′-donor ligands: syntheses, characterization and magnetic properties. *Cryst. Growth Des.*, 15, 4427–4437, 2015.
19. Masoomi, M.Y., Stylianou, K.C., Morsali, A., Retailleau, P., Maspoch, D., Selective CO2 capture in metal–organic frameworks with azine-functionalized pores generated by mechanosynthesis. *Cryst. Growth Des.*, 14, 2092–2096, 2014.
20. Bhattacharya, B., Dey, R., Pachfule, P., Banerjee, R., Ghoshal, D., Four 3D Cd (II)-based metal organic hybrids with different N, N′-donor spacers: syntheses, characterizations, and selective gas adsorption properties. *Cryst. Growth Des.*, 13, 731–739, 2012.
21. Parshamoni, S. and Konar, S., Selective CO2 adsorption in four zinc (ii)-based metal organic frameworks constructed using a rigid N, N′-donor linker and various dicarboxylate ligands. *CrystEngComm*, 18, 4395–4404, 2016.
22. Liu, S., Luo, M., Li, J., Luo, F., Ke, L., Ma, J., Adsorption equilibrium and kinetics of uranium onto porous azo-metal–organic frameworks. *J. Radioanal. Nucl. Chem.*, 310, 353–362, 2016.
23. Pankajakshan, A., Kuznetsov, D., Mandal, S., Ultrasensitive Detection of Hg(II) Ions in Aqueous Medium Using Zinc-Based Metal–Organic Framework. *Inorg. Chem.*, 58, 1377–1381, 2019.
24. Chand, S., Mondal, M., Pal, S.C., Pal, A., Maji, S., Mandal, D., Das, M.C., Two azo-functionalized luminescent 3D Cd(ii) MOFs for highly selective detection of Fe3+ and Al3+. *New J. Chem.*, 42, 12865–12871, 2018.
25. Goel, N. and Kumar, N., A dual-functional luminescent Tb(iii) metal–organic framework for the selective sensing of acetone and TNP in water. *RSC Adv.*, 8, 10746–10755, 2018.
26. Chand, S., Pal, A., Das, M.C., A Moisture-Stable 3D Microporous CoII-Metal–Organic Framework with Potential for Highly Selective CO2 Separation under Ambient Conditions. *Chem. Eur. J.*, 24, 5982–5986, 2018.
27. Singh, R., Ahmad, M., Bharadwaj, P.K., Coordination Polymers of Copper and Zinc Ions with a Linear Linker Having Imidazole at Each End and an Azo Moiety in the Middle: Pedal Motion, Gas Adsorption, and Emission Studies. *Cryst. Growth Des.*, 12, 5025–5034, 2012.
28. Li, Y.-P., Zhang, L.-J., Ji, W.-J., Synthesis, characterization, crystal structure of magnesium compound based 3, 3′, 5, 5′-azobenzentetracarboxylic acid and application as high-performance heterogeneous catalyst for cyanosilylation. *J. Mol. Struct.*, 1133, 607–614, 2017.
29. Hoang, L.T.M., Ngo, L.H., Nguyen, H.L., Nguyen, H.T.H., Nguyen, C.K., Nguyen, B.T., Ton, Q.T., Nguyen, H.K.D., Cordova, K.E., Truong, T., An azobenzene-containing metal–organic framework as an efficient heterogeneous

catalyst for direct amidation of benzoic acids: synthesis of bioactive compounds. *Chem. Commun.*, 51, 17132–17135, 2015.
30. Choudhury, J., Recent developments on artificial switchable catalysis. *Tetrahedron Lett.*, 59, 487–495, 2018.
31. Kanj, A.B., Müller, K., Heinke, L., Stimuli-responsive metal–organic frameworks with photoswitchable azobenzene side groups. *Macromol. Rapid Commun.*, 39, 1700239, 2018.
32. Knebel, A., Zhou, C., Huang, A., Zhang, J., Kustov, L., Caro, J., Smart Metal–Organic Frameworks (MOFs): Switching Gas Permeation through MOF Membranes by External Stimuli. *Chem. Eng. Technol.*, 41, 224–234, 2018.
33. Mukhopadhyay, R.D., Praveen, V.K., Ajayaghosh, A., Photoresponsive metal–organic materials: exploiting the azobenzene switch. *Mater. Horiz.*, 1, 572–576, 2014.
34. Castellanos, S., Goulet-Hanssens, A., Zhao, F., Dikhtiarenko, A., Pustovarenko, A., Hecht, S., Gascon, J., Kapteijn, F., Bléger, D., Structural Effects in Visible-Light-Responsive Metal–Organic Frameworks Incorporating ortho-Fluoroazobenzenes. *Chem. Eur. J.*, 22, 746–752, 2015.
35. Modrow, A., Zargarani, D., Herges, R., Stock, N., The first porous MOF with photoswitchable linker molecules. *Dalton Trans.*, 40, 4217–4222, 2011.
36. Modrow, A., Zargarani, D., Herges, R., Stock, N., Introducing a photo-switchable azo-functionality inside Cr-MIL-101-NH2 by covalent post-synthetic modification. *Dalton Trans.*, 41, 8690–8696, 2012.
37. Hermann, D., Emerich, H., Lepski, R., Schaniel, D., Ruschewitz, U., Metal–Organic Frameworks as Hosts for Photochromic Guest Molecules. *Inorg. Chem.*, 52, 2744–2749, 2013.
38. Müller, K., Knebel, A., Zhao, F., Bléger, D., Caro, J., Heinke, L., Switching Thin Films of Azobenzene-Containing Metal–Organic Frameworks with Visible Light. *Chem. Eur. J.*, 23, 5434–5438, 2017.
39. Fan, C.B., Liu, Z.Q., Gong, L.L., Zheng, A.M., Zhang, L., Yan, C.S., Wu, H.Q., Feng, X.F., Luo, F., Photoswitching adsorption selectivity in a diarylethene-azobenzene MOF. *Chem. Commun.*, 53, 763–766, 2017.
40. Huang, R., Hill, M.R., Babarao, R., Medhekar, N.V., CO2 Adsorption in Azobenzene Functionalized Stimuli Responsive Metal–Organic Frameworks. *J. Phys. Chem. C*, 120, 16658–16667, 2016.
41. Park, J., Yuan, D., Pham, K.T., Li, J.-R., Yakovenko, A., Zhou, H.-C., Reversible Alteration of CO2 Adsorption upon Photochemical or Thermal Treatment in a Metal–Organic Framework. *J. Am. Chem. Soc.*, 134, 99–102, 2012.
42. Prasetya, N., Donose, B.C., Ladewig, B.P., A new and highly robust light-responsive Azo-UiO-66 for highly selective and low energy post-combustion CO2 capture and its application in a mixed matrix membrane for CO2/N2 separation. *J. Mater. Chem. A*, 6, 16390–16402, 2018.
43. Wang, Z., Knebel, A., Grosjean, S., Wagner, D., Bräse, S., Wöll, C., Caro, J., Heinke, L., Tunable molecular separation by nanoporous membranes. *Nat. Commun.*, 7, 13872, 2016.

44. Meng, X., Gui, B., Yuan, D., Zeller, M., Wang, C., Mechanized azobenzene-functionalized zirconium metal-organic framework for on-command cargo release. *Sci. Adv.*, 2, e1600480, 2016.
45. Li, H., Martinez, M.R., Perry, Z., Zhou, H.C., Falcaro, P., Doblin, C., Lim, S., Hill, A.J., Halstead, B., Hill, M.R., A Robust Metal–Organic Framework for Dynamic Light-Induced Swing Adsorption of Carbon Dioxide. *Chem.-Eur. J.*, 22, 11176–11179, 2016.
46. Lyndon, R., Konstas, K., Ladewig, B.P., Southon, P.D., Kepert, P.C.J., Hill, M.R., Dynamic Photo-Switching in Metal–Organic Frameworks as a Route to Low-Energy Carbon Dioxide Capture and Release. *Angew. Chem.*, 125, 3783–3786, 2013.
47. Gong, L.L., Feng, X.F., Luo, F., Novel azo-Metal–Organic Framework Showing a 10-Connected bct Net, Breathing Behavior, and Unique Photoswitching Behavior toward CO2. *Inorg. Chem.*, 54, 11587–11589, 2015.
48. Song, W.-C., Cui, X.-Z., Liu, Z.-Y., Yang, E.-C., Zhao, X.-J., Light-triggered Supramolecular Isomerism in a Self-catenated Zn(II)-organic Framework: Dynamic Photo-switching CO2 Uptake and Detection of Nitroaromatics. *Sci. Rep.*, 6, 34870, 2016.
49. Brown, J.W., Henderson, B.L., Kiesz, M.D., Whalley, A.C., Morris, W., Grunder, S., Deng, H., Furukawa, H., Zink, J.I., Stoddart, J.F., Yaghi, O.M., Photophysical pore control in an azobenzene-containing metal–organic framework. *Chem. Sci.*, 4, 2858–2864, 2013.
50. Huang, H., Sato, H., Aida, T., Crystalline Nanochannels with Pendant Azobenzene Groups: Steric or Polar Effects on Gas Adsorption and Diffusion? *J. Am. Chem. Soc.*, 139, 8784–8787, 2017.
51. Wang, Z., Heinke, L., Jelic, J., Cakici, M., Dommaschk, M., Maurer, R.J., Oberhofer, H., Grosjean, S., Herges, R., Bräse, S., Reuter, K., Wöll, C., Photoswitching in nanoporous, crystalline solids: an experimental and theoretical study for azobenzene linkers incorporated in MOFs. *Phys. Chem. Chem. Phys.*, 17, 14582–14587, 2015.
52. Yang, Q., Guillerm, V., Ragon, F., Wiersum, A.D., Llewellyn, P.L., Zhong, C., Devic, T., Serre, C., Maurin, G., CH4 storage and CO2 capture in highly porous zirconium oxide based metal–organic frameworks. *Chem. Commun.*, 48, 9831–9833, 2012.
53. Lee, Y.G., Moon, H.R., Cheon, Y.E., Suh, M.P., A comparison of the H2 sorption capacities of isostructural metal–organic frameworks with and without accessible metal sites:[{Zn2 (abtc)(dmf) 2} 3] and [{Cu2 (abtc)(dmf) 2} 3] versus [{Cu2 (abtc)} 3]. *Angew. Chem.*, 120, 7855–7859, 2008.
54. Feng, D., Wang, K., Wei, Z., Chen, Y.-P., Simon, C.M., Arvapally, R.K., Martin, R.L., Bosch, M., Liu, T.-F., Fordham, S., Kinetically tuned dimensional augmentation as a versatile synthetic route towards robust metal–organic frameworks. *Nat. Commun.*, 5, 1–9, 2014.

4
Imidazolium and Pyridinium Decorated Metal-Organic Frameworks

Abstract
Imidazolium and Pyridinium are cationic functional groups with enriched host-guest chemistry. In this chapter we explain that combination of crystalline structure of metal-organic frameworks and chemical properties these cationic functional groups is an ideal strategy for synthesis of metal-organic frameworks with charge separated nature to apply in large variety of applications like gas separation, stimuli-responsive materials, detection and purification and heterogeneous catalysis.

Keywords: Imidazolium, pyridinium, viologen, functional metal-organic frameworks, light responsive frameworks, proton conductivity, cationic frameworks

Ionic N-based functions are those functions containing N, C and H atoms in their molecular structure such as ammonium (quaternary amine), imidazolium, and pyridinium. This section is classified in two major parts; first part belongs to imidazolium functionalized MOFs and second part is about pyridinium decorated MOFs.

4.1 Imidazolium Functionalized Metal–Organic Frameworks

4.1.1 General Chemical Properties of Imidazolium Function

Imidazolium (which also called azolium) is a functional group with rich host-guest chemistry as well as its organometallic chemistry. Imidazolium ring is electron deficient and positively charged which make it suitable core for interaction with polar or quadruple guest molecules through different interactions like charge–charge electrostatic or charge–dipole/quadrupole

Ali Morsali and Sayed Ali Akbar Razavi. Functional Metal-Organic Frameworks: Structure, Properties and Applications, (55–78)

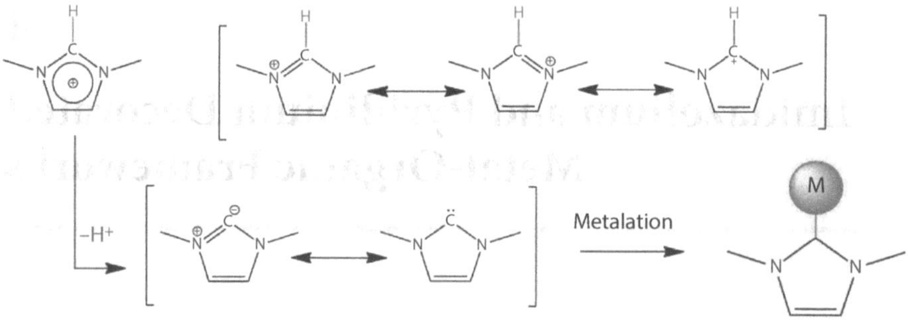

Figure 4.1 Resonance forms of imidazolium ring with carbene formation and metalation on carbenic C atom.

interactions. Moreover, there is an acidic H atom on carbenic C atom of imidazolium ring which make it suitable to interact through $(C–H)^{+}\cdots X^{n-}$ interaction with (partially) negatively charged species. Also, imidazolium ring applied in designing proton-conductive materials owing to presence of this acidic hydrogen (Figure 4.1). These advantages in host-guest chemistry of imidazolium put it under attention of scientists for designing imidazolium functional MOFs and applying them in removal and adsorption applications.

Based on imidazolium organometallic chemistry, the carbenic H atom could be eliminated to form N-heterocyclic carbenes which could coordinate to metal ions and p-block elements or apply as organocatalytic site [1]. NHC–metal complexes can be synthesized and stabilized into the structure of MOFs through deprotonation of imidazolium ring in carbene site by applying a base. This carbenic site can be stabilized by the two nitrogen atoms near the carbene center and nature of NHC-metal bond is very critical in designing related catalysts [2]. Carbenic site coordinate to metal ion through $\sigma \rightarrow d$ donating bond to form NHC–metal organometallic catalytic site (Figure 4.2). Moreover, NHC fragment involved in NHC-to-metal $\pi \rightarrow d$ donating and metal-to-NHC $d \rightarrow \pi^*$ accepting bonds when bonded to a metal giving rise to stabilization of wide range of metal (Figure 4.2). Electron deficient metals could be stabilized through a $\pi \rightarrow d$ donating bond while electron-rich metal are stabilized by $d \rightarrow \pi^*$ accepting bond. Owing to stabilization of wide range of metals, different type of reaction could be catalyzed on NHC–metal sites through right selection of metal ions.

4.1.2 Function–Application Properties

Imidazolium functionalized MOFs applied in heterogeneous catalytic applications using different strategies. In first strategy, they applied as

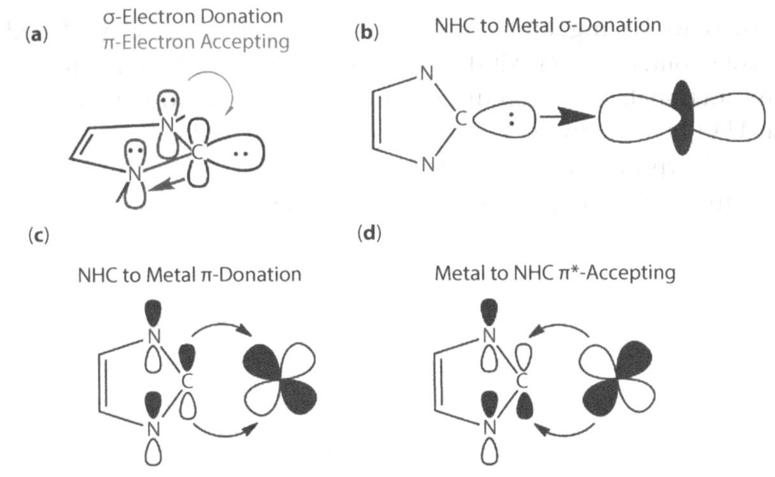

Figure 4.2 Representation of possible interactions between metal ion and carbenic C atom in NHC–metal complex. (a) Electron donor–acceptor interactions between metal ion and carbenic C atom in NHC–metal complex. (b) NHC to metal σ-donation bond. (c) NHC to metal π-donation bond. (d) Metal to NHC π*-backbond.

cationic imidazolium ring to fix the substrate into the pores of MOFs. In this strategy imidazolium FMOFs applied in reactions such as cycloaddition of CO_2 with epoxides [3–5] and catalytic phase transfer reactions [6]. In second strategy they transform into the NHC–metal catalytic centers and applied in different reactions [7–11] especially in coupling reactions [8, 9, 12, 13]. Three synthesis methods applied for carbene site metalation at the imidazolium ring and synthesis of NHC-metal catalytic center (Figure 4.3). These three methods are (I) in situ deprotonation or one-pot reaction through MOF synthesis; (II) post-synthesis modification; (III) pre-functionalization the organic linker with metal centers.

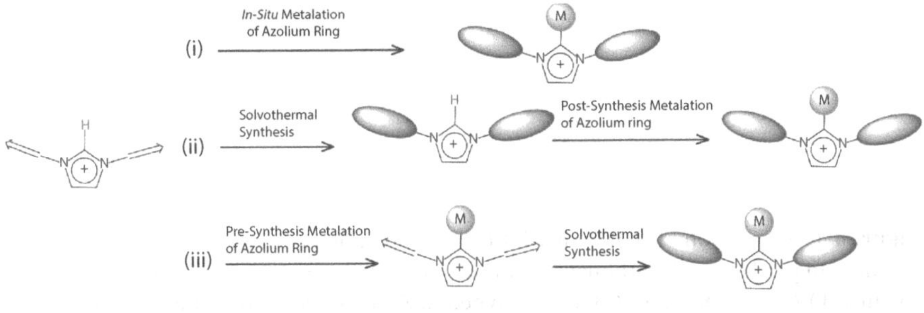

Figure 4.3 Applied synthesis approaches for construction of NHC–metal centers inside the pores of MOFs.

Yuan-Biao Huang and coworkers post-synthetically modified an imidazole-containing Zr-MOF, Im-UiO-66, into a bifunctional imidazolium functionalized zirconium metal–organic framework, named as (I^-) Meim-UiO-66 containing Brønsted acid sites and iodide ions (Figure 4.4) [14]. N_2 adsorption measurement at 77 K show that (I^-)Meim-UiO-66 BET surface area is 328 $m^2{\cdot}g^{-1}$. CO_2 adsorption measurement at 298 K indicate

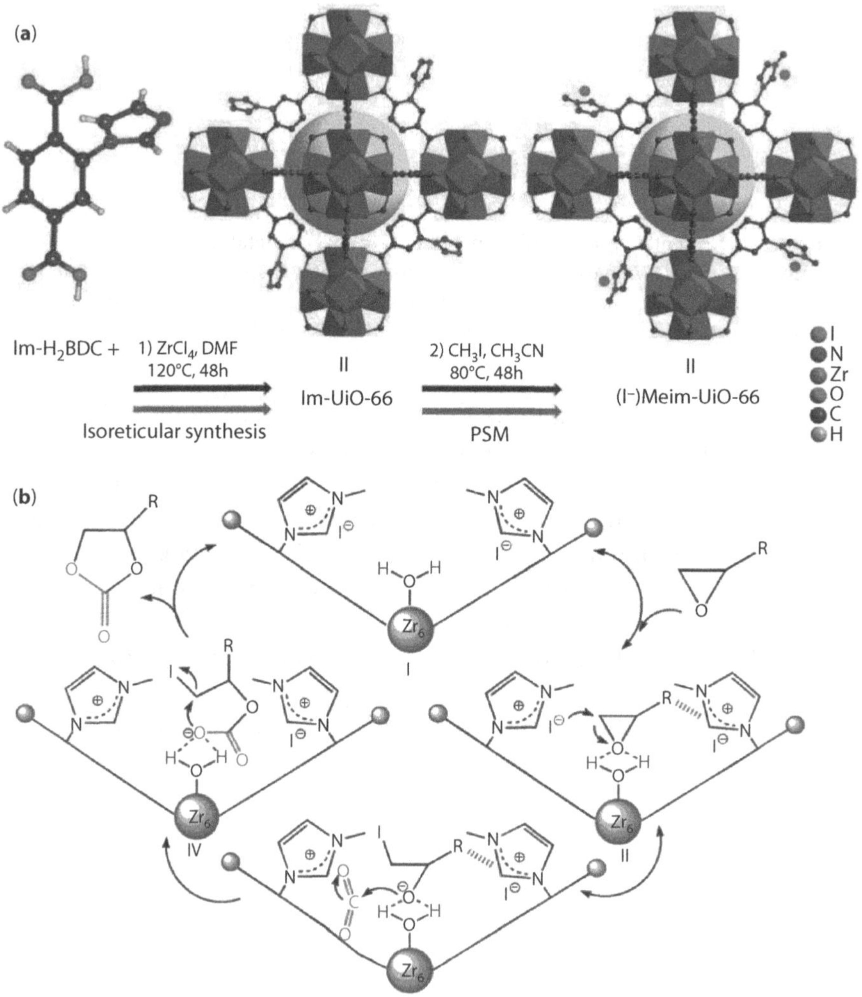

Figure 4.4 Application of (I-)Meim-UiO-66 in cycloaddition reaction of CO_2 with epoxides to produce cyclocarbonates. (a) Post-modification method for synthesis of (I^-) Meim-UiO-66 and Im-UiO-66. Green polyhedra: Zr_6 clusters. Yellow ball: micropores. (b) Proposed catalytic mechanism. The hashed bonds represent π–π, cationic-π or hydrogen bonding interactions [14].

that there is a moderate interaction between CO_2 molecules and (I^-)Meim-UiO-66 with zero-coverage enthalpy of -44.2 kJ·mol^{-1}. Since (I^-)Meim-UiO-66 benefits from presence of basic-acidic sites as well as good affinity to CO_2 molecules, it applied in the cycloaddition of CO_2 with epoxides to produce cyclocarbonates. Optimal condition for reaction is epoxide (10 mmol), catalyst (0.050 g, 0.745 mol%), P(CO_2) = 0.1 MPa, temperature (120 °C) and time (24 h). CO_2 cycloaddition with epichlorohydrin reached maximum conversion (100%) with high selectivity (93%) which is much better than that of other MOF systems where co-catalysts are generally required for this reaction. They proposed catalytic mechanism based on activation of epoxide on Lewis acid Zr clusters and then addition of anionic intermediate species formed by epoxide and iodide to CO_2 molecules.

Sonogashira cross-coupling reaction is based on coupling of vinyl or aryl halides with terminal acetylenes in category of sp^2–sp bond formation reactions in organic synthesis (Figure 4.5a). This reaction is very pivotal owing to its applications in production of natural molecules, pharmaceuticals, biologically active molecules, heterocycles, herbicides, dendrimers and conjugated polymers or nanostructures. Francis Verpoort and coworkers synthesized imidazolium functionalized $[(Zn_{0.25})_8(O)]Zn_6(L)_{12}$ framework (ZnL, where $H_2L^+Cl^-$ = 1,3-Bis(4-carboxyphenyl)imidazolium Chloride) and through a post-synthesis method imidazolium core is converted to NHC–Pd(II) catalytic center to form ZnL–Pd framework with high density and uniform distribution of the active site, NHC–Pd(II), which applied in Sonogashira cross-coupling reaction (Figure 4.5b) [9]. The optimum catalytic conditions are >0.003 mmol based on Pd when hydroxide and cesium carbonate base is used in polar protic solvents and DMF at 80 °C for at least 8 h. In this work, Pd(II) centers are catalytically

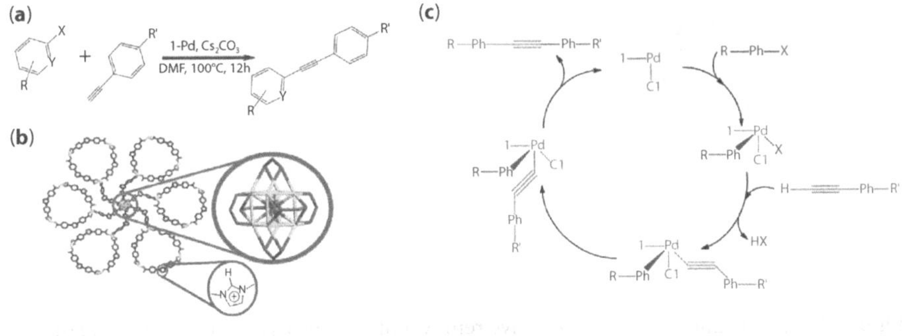

Figure 4.5 Application of ZnL–Pd in Sonogashira cross-coupling reaction. (a) Sonogashira cross-coupling reaction. (b) Structure of ZnL [9]. (c) Proposed mechanism for Sonogashira cross-coupling reaction by NHC–Pd sites [13].

active sites in C–C coupling with well-stability and durability of ZnL-Pd framework which results in Sonogashira cross-coupling reaction. Same group proposed a mechanism for Sonogashira cross-coupling reaction with NHC–Pd(II) centers (Figure 4.5c).

On simple strategy in construction of ionic MOFs is founded on application of ionic ligands in the structure of MOFs. Since functional ligands with imidazolium cores are cationic, they could apply in construction of cationic MOFs. Also, imidazolium decorated MOFs are ideal for removal and adsorption of anionic species like pollutant dyes [8, 15] owing to anionic counter ion exchange and electrostatic interactions between negatively charged species and positive imidazolium core. Xin-ong Wang and coworkers synthesized a cationic MOF (denoted as In-L with formula $[In(OHL]_5(NO_3)_5{\cdot}33H_2O{\cdot}14DMF$ where $H_2L^+Cl^-$ is 1,3-bis(4-carboxyphenyl)imidazolium chloride) based on imidazolium linker and applied it for selective removal of anionic dyes (Figure 4.6) [16]. Owing to cationic nature of the In-L framework, some dyes with a similar backbone but different charge and size had been selected to study its ion exchange properties. These dyes are methylene blue: MLB^+; sudan I: SDI^0; acid orange 7: $AO7^-$; orange G: OG^{2-}; and new coccine: NC^{3-}.

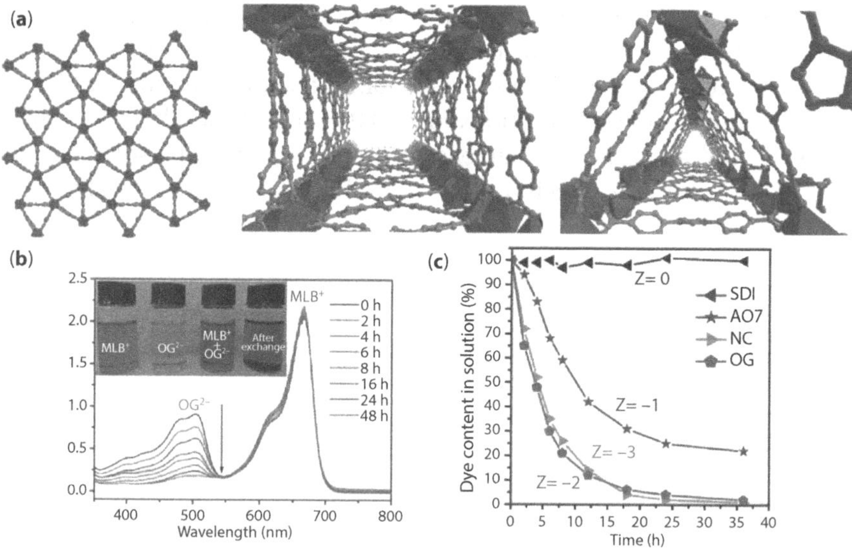

Figure 4.6 Application of In-L is selective removal of anionic dyes. (a) 3D structure of In-L along the a axis (left), the square (middle) and triangular (right) nanotubes of In-L. (b) UV–Vis absorption spectral changes of equimolar MLB^+-OG^{2-} in the presence of In-L. (c) Evaluation of the effects of dye charges on the removal process [16].

Competitive adsorption of cationic MLB^+ dye and OG^{2-} dye show that there is not any specific change in absorption band of MLB^+ while the concentration of OG^{2-} in the DMF solution drastically decreased. This observation clearly shows that the cationic In-L framework has selectivity toward anionic dyes. In next step, the effect of negative charge on the adsorption process is examined. The results show that the higher negative charge results in kinetically faster adsorption process, although the size of the dyes must be considered.

Imidazolium decorated MOFs applied for adsorption of gas molecules like CO_2 and H_2 owing to interaction between cationic imidazolium cores in the structure of MOF and quadruple moment of gas molecules [17–22].

Two imidazolium decorated FMOFs NU-301 (with formula Zn_2(2H-BTBA)$_{0.5}$(IMTA)]·nDMF where 2H-BTBA = benzenetetrabenzoicacid and H_2IMTA = N,N'-bis(2,6-dimethyl-3,5-carboxylphenyl)imidazolium) and NU-302 (with formula Zn_2(2Br-BTBA)$_{0.5}$(IMTA) where 2Br-BTBA = dibromo-benzenetetrabenzoicacid) synthesized by Hupp and coworkers applied in hydrogen adsorption [21]. The isosteric heat of hydrogen adsorption at low loading for NU-301 and NU-302 frameworks (7.0 and −6.9 kj·mol^{-1} respectively) clarify the relatively strong interaction between imidazolium decorated framework and hydrogen molecules. Theoretical studies clarified that there is zwitterionic region between two negatively charged $Zn_2(CO_2)_5$ nodes (net charge: −1.0) and positively charged imidazolium ring (net charge: +0.5) which H_2 molecules prefer to be adsorbed at this zwitterionic region since host-guest interaction between quadrupole moment of hydrogen molecules and strongly charged atoms on the zwitterionic framework is maximized in this region.

In another work, Cao and coworkers applied 2-(3-Ethyl-imidazol-1-yl)-terephthalic acid [(Etim-H_2BDC)$^+$(Br^-)] as an ionic ligand to develop a imidazolium functionalized isostructure MOF with UiO-66, named as (I^-)Meim-Uio-66, for CO_2 cycloaddition reaction with epoxides. Thanks to quadrupolar nature of carbon dioxide and positively charged nature of imidazolium ring, there is a high affinity between (I^-)Meim-Uio-66 and CO_2 molecules ($Qst_{(CO2)}$ = −44.2 kj−mol^{-1}). Such high affinity between imidazolium ring and CO_2 molecules results in adsorption and activation of CO_2 molecules for catalytic reactoin.

As mentioned, imidazolium ring contains an active proton on carbenic C atom. This H atom could participate in hydrogen bonding. Also, this active H atom could be the origination of proton conduction in imidazolium based materials. In this regard, imidazolium functionalized MOFs synthesized with different approaches [23–26].

Hong-Cai Zhou and coworkers synthesized a Fe-MOF ([$Fe_3(\mu_3$-O)$(L)_2(H_2O)_3$], where H_3L is [1,1′:3′,1″-terphenyl]-4,4″,5′-tricarboxylic acid) and modified it with imidazole (Im) molecules in different post-synthesis processes (Figure 4.7) [26]. In first strategy, Im molecules are physically adsorbed into the pores of Fe-MOF with disorder arrangement to form Im@Fe-MOF. This composite was obtained via loading imidazole into the pores of the framework. In next strategy, Im molecules are chemically coordinated to the $Fe_3(\mu_3$-O)](carboxylate)$_6$ clusters with ordered arrangement to form Im-Fe-MOF. Im-Fe-MOF was obtained via anchoring imidazole to metal nodes by in situ reaction. The structure of Im-Fe-MOF is similar to that of Fe-MOF except that coordinated water molecules in the metal nodes were fully substituted by imidazole molecules. Proton conductivity of these three frameworks was measured by alternating current impedance spectroscopy (ACIS). Generally, ACIS show that the conductivity of these three materials are boosted with increase in temperature and relative humidity. At 25 °C and 98% relative humidity, proton conductivity

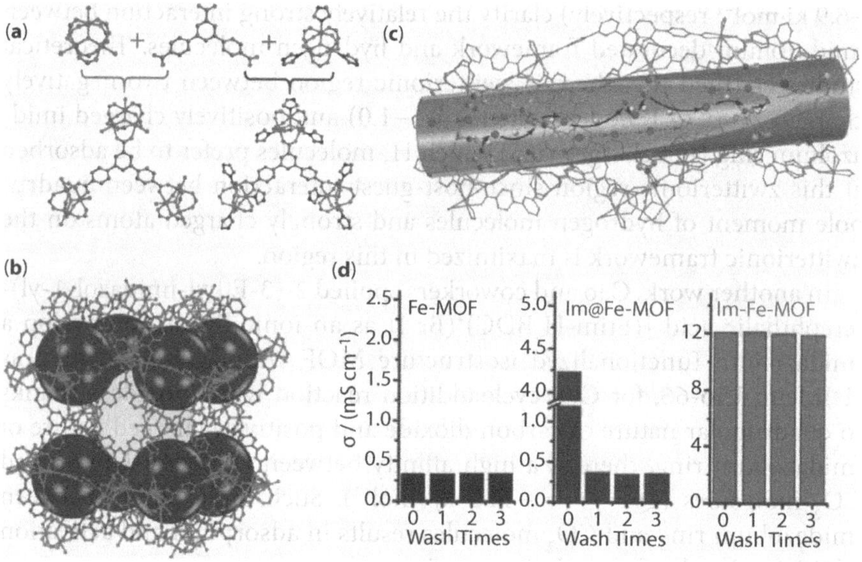

Figure 4.7 Application of Fe-MOF, Im@Fe-MOF and Im-Fe-MOF in proton conductivity. (a) Applied ligand in the synthesis of Fe-MOF and connection mode of ligands with metal clusters in Fe-MOF and Im-Fe-MOF. (b) Representation of pores of Fe-MOF. (c) Proposed proton-conductive pathways in Im-Fe-MOF. Red: Water molecules, blue arrows: protons hop along hydrogen-bonding networks formed by coordinated imidazole and absorbed water, red arrows: transport of protons through self-diffusion of protonated water. (d) Proton-conductive stability of the materials [26].

of Fe-MOF, Im@Fe-MOF and Im-Fe-MOF are 2.56×10^{-5} S·cm^{-1}, 8.41×10^{-5} S·cm^{-1} and 2.03×10^{-3} S·cm^{-1}, respectively. Upon increasing temperature from 25 to 60 °C the proton conductivity of these materials increased up to 1.25×10^{-4}, 4.23×10^{-3}, and 1.21×10^{-2} S·cm^{-1}, respectively. The remarkable proton conductivity of Im-Fe-MOF is comparable to that of Nafion as well as several of the highest performing MOFs materials under similar conditions. Activation energy is a sign of possible proton conduction mechanism in a way that reported activation energies for Grotthus and Vehicle mechanism are 0.1–0.4 eV and 0.5–0.9 eV, respectively. Calculated activation energy values for Fe-MOF, Im@Fe-MOF, and Im-Fe-MOF are 0.385, 0.573, and 0.436 eV, respectively. Low activation Grotthus mechanism for Fe-MOF is suggested because coordinated water molecules to Fe clusters and adsorbed water molecules in the pores could easily construct rich hydrogen-bonded networks in the channels of the framework for proton hopping. High activation energy and Vehicle mechanism for Im@Fe-MOF in proton conduction is suggested because the pores of Im@Fe-MOF are occupied with disordered imidazole molecules and accompanied by much lower water vapor absorption resulting in the blockage of formation of successive hydrogen bonding networks serving as proton conduction pathways. Theoretical studies in case of Im-Fe-MOF reveal that imidazole molecules can effectively replace the coordinated water in the metal nodes of Fe-MOF so that it is easier to contribute proton for the weak N–H bond by coordinated imidazole rings. Also, stable performance of these materials in proton conduction evaluated through washing each sample with water for three times and then measuring the impedance each time. The results show that the proton conductivity of Fe-MOF and Im-Fe-MOF has no apparent changes after three times of washing while the conductivity of Im@Fe–MOF decreased largely.

4.1.3 Function–Structure Properties

Introduction of imidazolium core into the structure of MOFs occurs through both main and side chains functionalization. Introduction of imidazolium group in the main-chain of the MOF has some structural effects as well as control over host-chemistry of the MOF. Introduction of azolium in the main-chain could rise to development of non-interpenetrated [19] or multi-interpenetrated MOFs [27] and flexible MOFs through single crystal-to-single crystal transformation [28]. On one hand, cationic nature of the imidazolium ring and consequent repulsion between ligands during self-assembly could prevent from interpenetration. On the other hand, angular nature of the bent imidazolium ligand and partial rotation of the

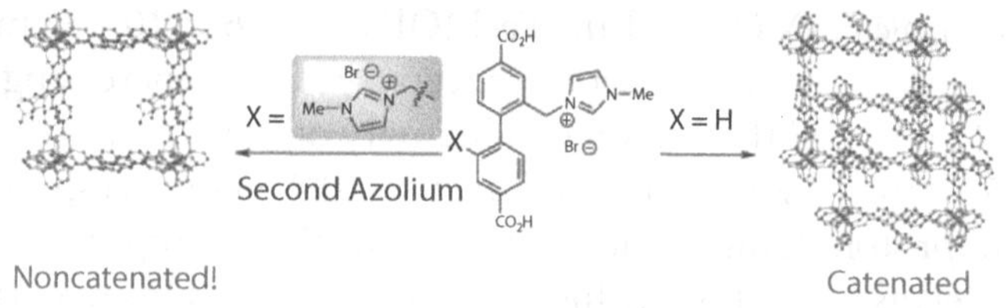

Figure 4.8 Effects of the number of imidazolium rings on the structure [29].

imidazolium moiety with respect to the benzene moiety can give rise to multi-interpenetrated MOFs.

Charge separated nature of imidazolium ring can be useful for control over catenation and morphology of FMOFs. Rational selection of imidazolium ring with optimized number of imidazolium ring in the side chain can result in dielectric repulsion between organic ligands to remove or gain control over the degree of interpenetration. Hupp and coworkers reached this conclusion through rational selection of the number of imidazolium ring in the side chain [29]. They find that control over electrostatic interaction in self-assembly process during synthesis of FMOFs as well as occupied volume by the function is the side chain can be considered as one way to gain control over catenation of the FMOFs. Hupp and coworkers reached this conclusion through rational selection of the number of imidazolium ring in the side chain (Figure 4.8) [29].

Despite the multiple advantages of imidazolium FMOFs in function-application properties, this group of functional MOFs suffers from some structural features. First, owing to their positively charged nature, there is a counter ion always near them to neutralize the positive charge of imidazolium ring. As a result, the accessible pore volume of MOF is blocked giving rise to noticeably reduced porosity. Moreover, there is strong interaction between anionic counter ions and polar solvents in the pores. So, the elimination of solvent molecules during activation process is much harder. So, this group of functional MOFs is of low-to-moderate surface area.

4.2 Pyridinium Functionalized Metal–Organic Frameworks

4.2.1 General Chemical Properties of Pyridinium Function

As pyridine conjugated acid, pyridinium function applied extensively in supramolecular and modern chemistry as building block for decades

[30, 31]. Positively charged aromatic ring of pyridinium organic functional group has several chemical properties including: Lewis acidity, π-deficiency, responsivity to photon, heat and electricity. As a result of such characters, pyridinium functionalized materials have been extensively studied to help understand the acid-base interaction and electron transfer in charge-transfer materials, and they have also been considered for various applications such as electroresponsive and photoresponsive smart chromic displays, sensing, adsorption and separation, and molecular magnets.

Derivative of 4,4'-bipyridine and pyridinium, commonly known as viologen/bipyridinium possess excellent redox ability and can act as good electron acceptor because of Lewis acidity and π-deficiency to form charge–transfer (CT) complexes with electron-rich species and in some cases with redox capability to undergo electron transfer via chemical, electrical or optical stimuli [32]. The viologen (V) exists in three main oxidation states, namely V^{2+}, $V^{+\cdot}$ and V^0. Upon one-electron reduction to produce intensely colored free radicals, obvious color changes can be observed for these compounds. First reduction step is highly reversible and can be cycled many times without significant side reaction. The further reduction to the fully reduced state is less reversible. Change on the oxidation state of viologen through viologen–guest interaction or responsivity to light or electricity induces a color change which is highly suitable for designing smart chromic devices. Besides mentioned applications, the bipyridinium unit has also been widely used in the field of supramolecular chemistry as the key building block for the assembly of mechanically interlocked molecules and molecular machines.

4.2.2 Function–Application Properties

Considering the Lewis acidity, π-deficiency, stimuli-responsivity and photoelectrochemical properties of pyridinium and viologen, the host-guest chemistry of MOFs with crystalline structure can be broadened through development of pyridinium/viologen functional MOFs. Review of the literature reveals that pyridinium functionalized MOFs are of improved physical and chemical properties that have not been observed in a single bipyridinium.

Stimuli-responsive materials are electro-active and undergo a change under external stimuli like light irradiation, electricity, heat, solvents and pressure. Under exposure to stimuli's resource, electro-active structures of stimuli-responsive materials are excited and give rise to a stimuli-induced electron transfer process which in itself could rise in color change or chromic mechanism. Owing to the nature of changes of excitation process, stimuli-responsive behavior of material could be classified in photochromism, electrochromism, thermochromism, solvatochromism especially hydrochromism and mechanochromism. If the electronic structures or chemical compositions of the material be able to transform reversibly between ground state (chemical or electronical state before exposure to stimuli's resource) and excited states (chemical or electronical state after exposure to stimuli's resource), they are switchable stimuli-responsive materials.

Electron donor–acceptor systems are special kind of switchable stimuli-responsive materials which could transfer electron from electron donor site into electron acceptor sites. This reversibility in electronic structure could be easily characterized under and without stimuli's resource. Owing to convenient instrumentation and effectivity, photon irradiation is a desirable kind of stimuli. Designing switchable photon-responsive electron donor–acceptor systems is of essential importance because they could be controlled remotely and on-demand.

Owing to their chemical and structural characteristics, MOFs are ideal platform for designing switchable stimuli-responsive electron donor–acceptor systems. MOFs are constructed based on organic ligands which could be enriched with electron donor or electron acceptor functional groups. Also, the regular crystalline structure of MOF is an excellent feature which provides scientific bases to develop and study on this kind of material with repeatable responses. To develop MOF-based electron donor-acceptor materials, scientists tried to synthesis pyridinium decorated MOFs because pyridinium is a well-known electron acceptor unit able to interact with electron donor entities like benzoate/carboxylate in a way that through their responsivity to external sources like photon [33–45], heat [44, 45], pressure [46] and solvents [36, 43, 47], they can show switching ability in photochromism, thermochromism, photoluminescence-switching, and multi-photon adsorption.

Pyridinium salts display photochemical addition and oxidization reactions under UV light irradiation as well as relatively large nonlinear two-photon absorption cross-section. Considering this point, Banglin Chen and coworkers tried to synthesis a MOF based on photoactive zwitterionic pyridinium functionalized $H_4L1{\cdot}OH$ linker ($H_4L1{\cdot}OH$ = 2,5-bis(3,5-dicarboxyphenyl)-1-methylpyridinium hydroxide) (Figure 4.9) [33]. Their efforts to direct synthesis of desirable MOF was not possible, so they applied multi-variant method (DMV) strategy to synthesis a MOF based on $H_4L1{\cdot}OH$ linker. In this regard, they applied H_4L2 ligand (H_4L2 = 5,5'-(pyridine-2,5-diyl)diisophthalic acid) which is base ligand for synthesis of ZJU-56 (with formula $Zn(L_2){\cdot}22H_2O{\cdot}$solvents) with 3D network of fof topology. Same synthetic procedure applied with 20 mol% of $H_4L1{\cdot}OH$

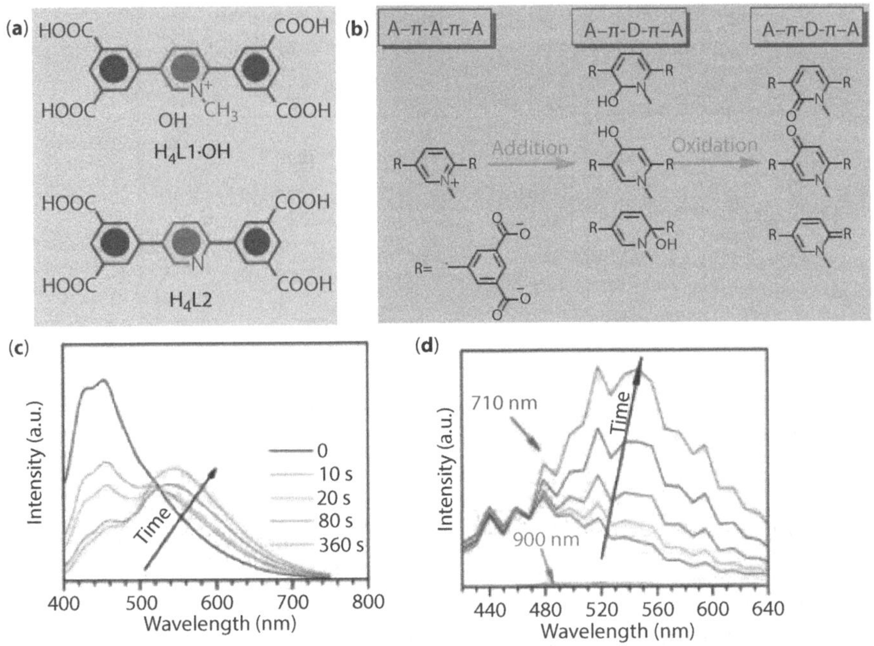

Figure 4.9 Application of ZJU-56-0.20 in two-photon adsorption. (a) Structure of applied ligands. (b) Possible photochemical reactions on pyridinium core. (c) Fluorescence shift of ZJU-56-0.020 up on 355 nm UV light irradiation. (d) Fluorescence change of ZJU-56-0.20 under femtosecond laser [33].

and 80 mol% of H_4L2 linkers. The resulting material is ZJU-56-0.20. So, despite the fact that 100 mol% of linkers in ZJU-56 are H_4L2, ZJU-56-0.20 is constructed based on both H_4L2 and $H_4L1{\cdot}OH$ linkers.

Fluorescence measurement displays ZJU-56-0.20 (before exposure to UV light) has a blue emission peak centered at 450 nm. Upon exposure to UV light (365 nm), the blue emission band (450 nm) attenuates and new yellow emission at 550 nm is appeared and increased. After 360 s, the blue emission is almost gone and the emission spectrum is based on yellow emission at 550 nm. So, after UV light irradiation the fluorescence color of ZJU-56-0.02 change from blue to yellow. Fluorescence measurement for ZJU-56 show that this MOF exhibits no fluorescence color change after UV light irradiation which indicate that the presence of $H_4L1{\cdot}OH$ in the structure of ZJU-56-0.20 is the reason for observed fluorescent changes. Characterization techniques, especially ^{1}H-NMR, reveal that the nature of ZJU-56-0.20 fluorescent changes is based on photochemical addition and oxidization occurs on pyridinium core under UV light irradiation. After recognition of linear on-photon absorption under 355 nm UV

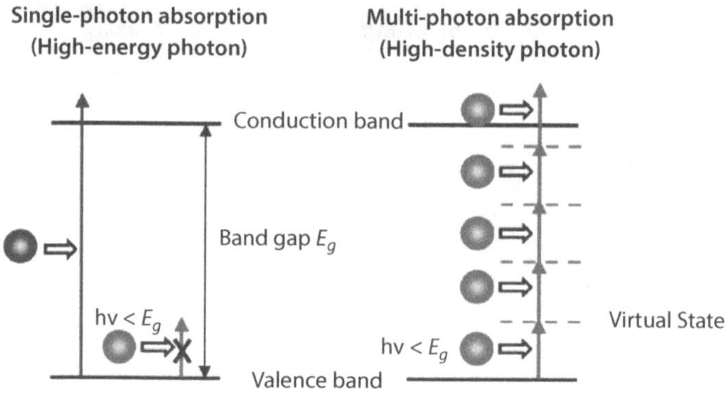

Figure 4.10 Electron excitation process in materials by single and multiphoton absorption [48].

light, authors tries to evaluate the non-linear two photon absorption of ZJU-56-0.20.

As we know, MOFs are treated to resemble like semiconductors and a semiconductor linear optical absorption can be observed only if the photon energy is at least as large as the band gap energy. Non-linear optical absorption or multiphoton absorption can occur when the energy of one photon is not enough to excite the electrons from valance band to conduction band (Figure 4.10). In other words, semiconductors multiphoton absorption could happen only when the energy of single photons is not large as the band gap. In two-photon absorption process, two photons are required to send electron from valance band to conduction band. With first photon, electron transfers from valance band to a virtual state and with the second photon, electron transfers from virtual state to the conduction band. Then the fluorescence of the material could be observed. One-photon absorption is a linear optical process which is proportional to the optical intensity while two-photon absorption is a nonlinear optical process proportional to the square of light intensity. As a result, nonlinear two-photon absorption is very small for low or moderate optical intensities, but can be major phenomena in high intensities especially with focus laser pulses.

So, one-photon absorption of ZJU-56-0.20 at 355 nm under UV light can be switched into two photon absorption at 710 nm (double wavelength and half energy) femtosecond laser. The fluorescence spectra of ZJU-56-0.20 based on two-photon absorption process show similar emission compared to one photon absorption process in a way that emission band around 550 nm is gradually enhanced. The approach of introducing nonlinear optical linkers into MOFs for 3D patterning and two photon imaging will enable MOFs to be useful materials for sensor arrays, and data storage media.

Such photo-responsivity of pyridinium functions is useful to synthesis light sensitive materials. Zhiyong Fu and coworkers synthesized a viologen-based MOF ([Zn(p-CPBPY)(HBTC)]·H_2O, (*p*-CPBPY = N-(*para*carboxyphenyl)-4,4'-bipyridinium (*p*-CPBPY), BTC = 1, 3, 5-benzenetricarboxylate) enabling convenient color change upon light irradiation for naked eye detection of instantaneous UV exposure levels. (Figure 4.11) [49]. Viologens are a pyridinium based electro-active function Applied in photochromic systems due to their reversible color change ability upon reduction and oxidation. A new absorption band (676 nm) appeared under UV light irradiation (5 mw·cm^{-2}) which intensifies with increasing in irradiation time. This new band belongs to photo-induced electron transfer from electron rich carboxyl donor sites to electron deficient viologen acceptor sites. ESR measurement for the colored crystal gives a *g* value of 2.0085. The intensity of color change (colorless to green) is attributed to different intensities of UV irradiation, which corresponds

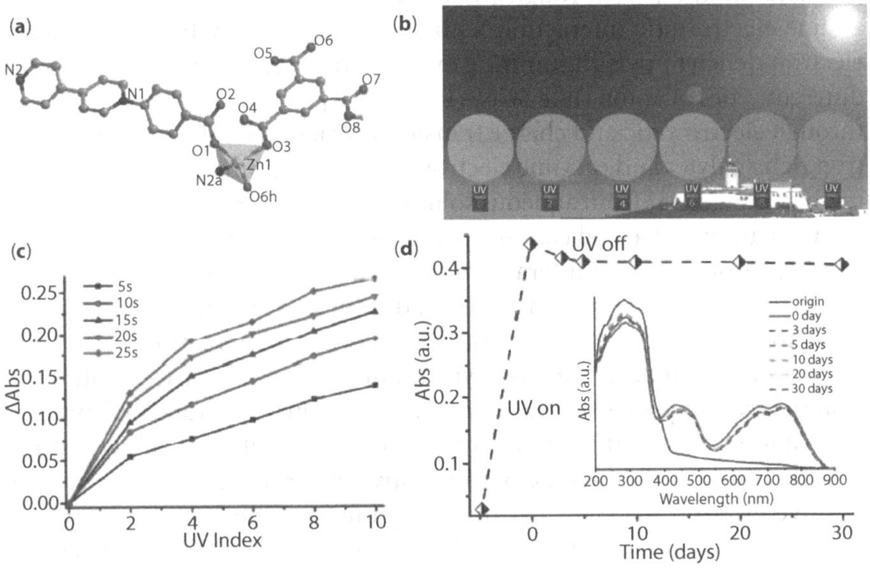

Figure 4.11 Application of [Zn(p-CPBPY)(HBTC)]·H_2O in UV light detection. (a) the asymmetric unit. (b) photographs indicate different UV exposure levels via different color change. (c) Plot of absorption change at 676 nm in line with irradiation time at the following UV index (from bottom to top): 2, 4, 6, 8 and 10; (d) Vis absorption changes at 676 nm for the colored sample stored in the dark [49].

to the UV index from 0 to 10. UV index calculated based on linear color changing under UV irradiation. Photochromic response observes just after less than 0.1 s with intensity as low as 0.001 $mw{\cdot}cm^{-2}$. Despite the high sensitivity to UV light, the materials did not show sensitivity to Vis light (>400 nm). Diffuse reflectance spectroscopy show that optical band gap of the MOF is 3.2 eV lying in UV range.

Pyridinium is positively charged core with high π-deficiency, Lewis acid organic site and electronically active in donor–acceptor and redox-active systems. These chemical properties made this function known as a desirable guest-interactive site into the structure of MOFs for sensing, removal and separation of hazardous materials. So, this function could interact with different analytes through different host–guest chemistry like electrostatic interaction with (partially) negatively charged species or molecules containing electron-rich atom in their molecular structure, π(deficient)–π(rich) stacking interactions with molecules with extended π-rich system, formation of electron donor–acceptor complex by charge transfer [50].

In this regard pyridinium decorated MOFs applied in detection of organic molecules with electron rich sites like phenol [51], picric acid [52] and aniline [51] through pyridinium·(OH)TNP/(OH)phenol/(NH_2) aniline electrostatic interaction with charge and energy transfer between electron deficient pyridinium ring and electron rich analytes, detection of different types of anions like SCN^-, I^-, Cl^-, N^{3-} [47] and $Cr_2O_7^{2-}$ [35, 53, 54] through electrostatic and charge transfer interaction between anionic electron rich analytes and cationic electron poor pyridinium ring, separation of methyl orange [55] from aqueous solution through electrostatic interaction between anionic dye and cationic pyridinium ring, separation of ammonia [56, 57] through strong interaction between polarity of Lewis basic guests and cationic pyridine ring and separation of benzene from cyclohexane [58] through π-poor(pyridinium)·(benzene)π-rich interaction.

As an electron acceptor site, pyridinium could interact well with electron rich carboxylate site. This is very ideal chemical character of pyridinium ring which could apply in construction of MOFs. Since pyridinium organic ligand could include both negative electron rich carboxylate and positive electron deficient pyridinium groups, zwitterionic carboxylate–pyridinium ligands could apply in construction of MOF with charge separated nature. Such charge separated zwitterionic nature is effectual in polarizing guest and hence enrichment of host–guest chemistry of the MOF. Figure 4.12 illustrate the charge separated surface of a MOF-framework based on 1,1'-bis(3,5-dicarboxybenzyl)-4,4'-bipyridiniumdibromide ligand [36].

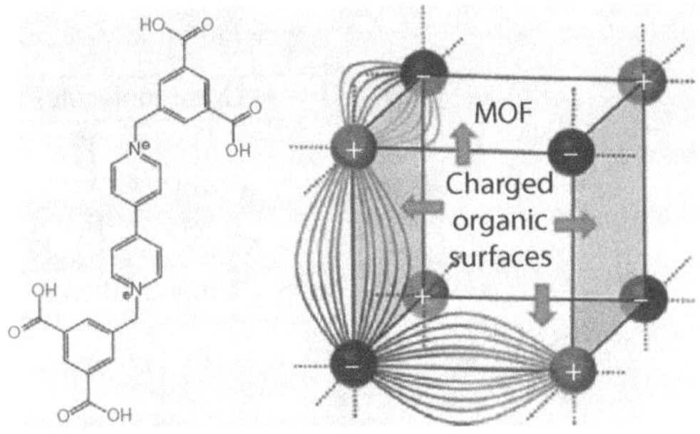

Figure 4.12 Charge separated surface of a MOF-framework based on 1,1'-bis(3,5-dicarboxybenzyl)-4,4'-bipyridiniumdibromide ligand [36].

In addition to the charge separated organic surface, zwitterionic carboxylate–pyridinium ligands could be practical for introduction of photo-responsible donor–acceptor site into the structure of MOFs. So, carboxylate-pyridinium ligands applied to synthesis MOFs with charge separated surface and photo-responsible electron donor-acceptor sites. These two characters of pyridinium decorated MOF could apply in on-demand guest adsorption release. In charge separated mode of the pyridinium decorated MOF, it is possible to adsorb guest molecules with dipole or quadrupole moment. Upon light irradiation and electron transfer from electron rich site to the electron deficient site, these charge separated centers are disappeared owing to formation of radicals on electro active sites. In this condition, the charge separated nature of MOF change to charge-less surface which could be useful for remote control and on demand removal of the guest molecules.

Mario Wriedt and coworker applied this hypothesis with synthesis of a MOF ([CdBr(**L**)]·(ClO_4)·2DMF) based on a tritopic zwitterionic pyridinium–carboxylate ligand (H_3**L**Br_3 = 1,1',1"(benzene-1,3,5-triyl) tris(methylene)tris(4-carboxypyridinium)tribromide) to examine the effect of separated charge nature on CO_2 adsorption before and after UV light irradiation (Figure 4.13) [59]. The combination of observations before and after light irradiation show that isosteric heat of CO_2 adsorption (from 40.5 to 27.3 kJ·mol^{-1}) reduced significantly after light irradiation as well as 43.2% reduction in CO_2 uptake at 273 K and 1 bar.

4.2.3 Function–Structure Properties

As mentioned pyridinium function is among the rare Lewis acidic organic functional groups. Also, it is familiar that introduction of open metal sites is most common strategy to develop MOFs with Lewis acid sites.

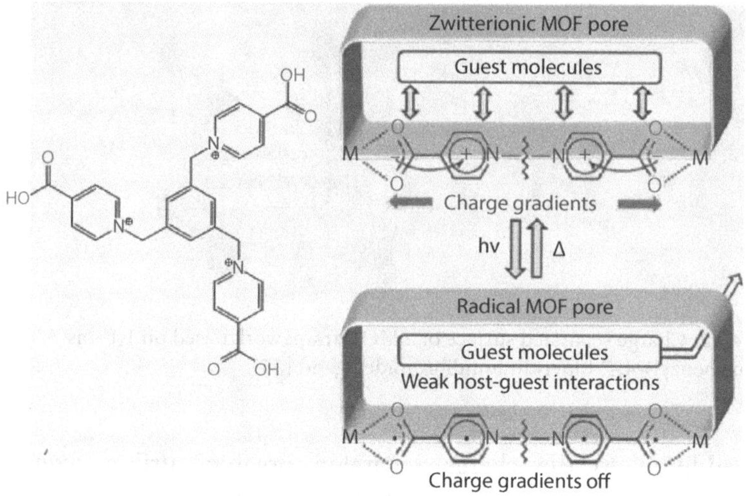

Figure 4.13 The ligand used for construction of zwitterionic MOF for CO_2 adsorption and the effects of light on host-guest interaction through reversible conversion of pyridinium ring to its radical homolog [59].

Combination of these two sentences clearly elucidate that synthesis of pyridinium functionalized MOF is novel strategy to construct FMOFs with organic Lewis acid sites.

As we discussed synthesis of FMOFs based on zwitterionic pyridinium–carboxylate ligand are of interest because electron rich carboxylate functions and electron deficient pyridinium functions are ideal donor-acceptor terminals in designing electroactive FMOFs. Despite this fascinating property, synthesis of zwitterionic MOFs based on pyridinium–carboxylate ligands is a challenge. During the synthesis process, there is a strong electrostatic interaction between carboxylate groups of one zwitterionic pyridinium–carboxylate ligand with pyridinium functions of another ligand through $\left(CO_2^-\right)\cdot(N^+)$ electrostatic interactions. This type of unfavorable interaction works against synthesis of MOFs and as a result of this interaction dense coordination polymers can be achieved. But it is necessary to mention that many efforts were done to develop zwitterionic MOFs based on zwitterionic pyridinium–carboxylate ligands.

References

1. Hopkinson, M.N., Richter, C., Schedler, M., Glorius, F., An overview of N-heterocyclic carbenes. *Nature*, 510, 485–496, 2014.

2. Ezugwu, C.I., Kabir, N.A., Yusubov, M., Verpoort, F., Metal–organic frameworks containing N-heterocyclic carbenes and their precursors. *Coord. Chem. Rev.*, 307, 188–210, 2016.
3. Liu, D., Li, G., Liu, H., Functionalized MIL-101 with imidazolium-based ionic liquids for the cycloaddition of CO2 and epoxides under mild condition. *Appl. Surf. Sci.*, 428, 218–225, 2018.
4. Liang, J., Xie, Y.-Q., Wang, X.-S., Wang, Q., Liu, T.-T., Huang, Y.-B., Cao, R., An imidazolium-functionalized mesoporous cationic metal–organic framework for cooperative CO2 fixation into cyclic carbonate. *Chem. Commun.*, 54, 342–345, 2018.
5. Adam, F., Appaturi, J.N., Ng, E.-P., Halide aided synergistic ring opening mechanism of epoxides and their cycloaddition to CO2 using MCM-41-imidazolium bromide catalyst. *J. Mol. Catal. A: Chem.*, 386, 42–48, 2014.
6. Wang, J.-C., Ma, J.-P., Liu, Q.-K., Hu, Y.-H., Dong, Y.-B., Cd(ii)-MOF-IM: Post-synthesis functionalization of a Cd(ii)-MOF as a triphase transfer catalyst. *Chem. Commun.*, 52, 6989–6992, 2016.
7. Dong, Y., Li, Y., Wei, Y.-L., Wang, J.-C., Ma, J.-P., Ji, J., Yao, B.-J., Dong, Y.-B., A N-heterocyclic tetracarbene Pd(ii) moiety containing a Pd(ii)–Pb(ii) bimetallic MOF for three-component cyclotrimerization *via* benzyne. *Chem. Commun.*, 52, 10505–10508, 2016.
8. Kong, G.-Q., Ou, S., Zou, C., Wu, C.-D., Assembly and Post-Modification of a Metal–Organic Nanotube for Highly Efficient Catalysis. *J. Am. Chem. Soc.*, 134, 19851–19857, 2012.
9. Ezugwu, C.I., Mousavi, B., Asraf, M.A., Luo, Z., Verpoort, F., Post-synthetic modified MOF for Sonogashira cross-coupling and Knoevenagel condensation reactions. *J. Catal.*, 344, 445–454, 2016.
10. Carson, F., Martínez-Castro, E., Marcos, R., Miera, G.G., Jansson, K., Zou, X., Martín-Matute, B., Effect of the functionalisation route on a Zr-MOF with an Ir–NHC complex for catalysis. *Chem. Commun.*, 51, 10864–10867, 2015.
11. Burgun, A., Crees, R.S., Cole, M.L., Doonan, C.J., Sumby, C.J., A 3-D diamondoid MOF catalyst based on *in situ* generated [Cu(L)2] N-heterocyclic carbene (NHC) linkers: Hydroboration of CO2. *Chem. Commun.*, 50, 11760–11763, 2014.
12. Kong, G.-Q., Xu, X., Zou, C., Wu, C.-D., Two metal–organic frameworks based on a double azolium derivative: post-modification and catalytic activity. *Chem. Commun.*, 47, 11005–11007, 2011.
13. Ezugwu, C.I., Mousavi, B., Asrafa, M.A., Mehta, A., Vardhan, H., Verpoort, F., An N-heterocyclic carbene based MOF catalyst for Sonogashira cross-coupling reaction. *Catal. Sci. Technol.*, 6, 2050–2054, 2016.
14. Liang, J., Chen, R.-P., Wang, X.-Y., Liu, T.-T., Wang, X.-S., Huang, Y.-B., Cao, R., Postsynthetic ionization of an imidazole-containing metal–organic framework for the cycloaddition of carbon dioxide and epoxides. *Chem. Sci.*, 8, 1570–1575, 2017.

15. He, W., Li, N., Wang, X., Hu, T., Bu, X., A cationic metal–organic framework based on {Zn4} cluster for rapid and selective adsorption of dyes. *Chin. Chem. Lett.*, 29, 857–860, 2018.
16. Song, B.-Q., Wang, X.-L., Zhang, Y.-T., Wu, X.-S., Liu, H.-S., Shao, K.-Z., Su, Z.-M., Periodic tiling of triangular and square nanotubes in a cationic metal–organic framework for selective anion exchange. *Chem. Commun.*, 51, 9515–9518, 2015.
17. Yao, B.-J., Ding, L.-G., Li, F., Li, J.-T., Fu, Q.-J., Ban, Y., Guo, A., Dong, Y.-B., Chemically cross-linked MOF membrane generated from imidazolium-based ionic liquid-decorated UiO-66 type NMOF and its application toward CO2 separation and conversion. *ACS Appl. Mater. Interfaces*, 9, 38919–38930, 2017.
18. Lee, J.Y., Roberts, J.M., Farha, O.K., Sarjeant, A.A., Scheidt, K.A., Hupp, J.T., Synthesis and Gas Sorption Properties of a Metal–Azolium Framework (MAF) Material. *Inorg. Chem.*, 48, 9971–9973, 2009.
19. Sen, S., Neogi, S., Aijaz, A., Xu, Q., Bharadwaj, P.K., Construction of Non-Interpenetrated Charged Metal–Organic Frameworks with Doubly Pillared Layers: Pore Modification and Selective Gas Adsorption. *Inorg. Chem.*, 53, 7591–7598, 2014.
20. Wang, S., Yang, Q., Zhang, J., Zhang, X., Zhao, C., Jiang, L., Su, C.-Y., Two-Dimensional Charge-Separated Metal–Organic Framework for Hysteretic and Modulated Sorption. *Inorg. Chem.*, 52, 4198–4204, 2013.
21. Lalonde, M.B., Getman, R.B., Lee, J.Y., Roberts, J.M., Sarjeant, A.A., Scheidt, K.A., Georgiev, P.A., Embs, J.P., Eckert, J., Farha, O.K., Snurr, R.Q., Hupp, J.T., A zwitterionic metal–organic framework with free carboxylic acid sites that exhibits enhanced hydrogen adsorption energies. *CrystEngComm*, 15, 9408–9414, 2013.
22. Capon, P.K., Burgun, A., Coghlan, C.J., Crees, R.S., Doonan, C.J., Sumby, C.J., Hydrogen adsorption in azolium and metalated N-heterocyclic carbene containing MOFs. *CrystEngComm*, 18, 7003–7010, 2016.
23. Sen, S., Nair, N.N., Yamada, T., Kitagawa, H., Bharadwaj, P.K., High Proton Conductivity by a Metal–Organic Framework Incorporating Zn8O Clusters with Aligned Imidazolium Groups Decorating the Channels. *J. Am. Chem. Soc.*, 134, 19432–19437, 2012.
24. Sen, S., Yamada, T., Kitagawa, H., Bharadwaj, P.K., 3D coordination polymer of Cd (II) with an imidazolium-based linker showing parallel polycatenation forming channels with aligned imidazolium groups. *Cryst. Growth Des.*, 14, 1240–1244, 2014.
25. Song, B.-Q., Wang, X.-L., Yang, G.-S., Wang, H.-N., Liang, J., Shao, K.-Z., Su, Z.-M., A polyrotaxane-like metal–organic framework exhibiting luminescent sensing of Eu3+ cations and proton conductivity. *CrystEngComm*, 16, 6882–6888, 2014.

26. Zhang, F.-M., Dong, L.-Z., Qin, J.-S., Guan, W., Liu, J., Li, S.-L., Lu, M., Lan, Y.-Q., Su, Z.-M., Zhou, H.-C., Effect of imidazole arrangements on proton-conductivity in metal-organic frameworks. *J. Am. Chem. Soc.*, 139, 6183–6189, 2017.
27. Sen, S., Yamada, T., Kitagawa, H., Bharadwaj, P.K., 3D Coordination Polymer of Cd(II) with an Imidazolium-Based Linker Showing Parallel Polycatenation Forming Channels with Aligned Imidazolium Groups. *Cryst. Growth Des.*, 14, 1240–1244, 2014.
28. Sen, S., Neogi, S., Rissanen, K., Bharadwaj, P.K., Solvent induced single-crystal to single-crystal structural transformation and concomitant transmetalation in a 3D cationic Zn(ii)-framework. *Chem. Commun.*, 51, 3173–3176, 2015.
29. Roberts, J.M., Farha, O.K., Sarjeant, A.A., Hupp, J.T., Scheidt, K.A., Two Azolium Rings Are Better Than One: A Strategy for Controlling Catenation and Morphology in Zn and Cu Metal-Organic Frameworks. *Cryst. Growth Des.*, 11, 4747–4750, 2011.
30. Sun, J.-K., Yang, X.-D., Yang, G.-Y., Zhang, J., Bipyridinium derivative-based coordination polymers: From synthesis to materials applications. *Coord. Chem. Rev.*, 378, 533–560, 2019.
31. Sun, J.-K. and Zhang, J., Functional metal-bipyridinium frameworks: self-assembly and applications. *Dalton Trans.*, 44, 19041–19055, 2015.
32. Bird, C. and Kuhn, A., Electrochemistry of the viologens. *Chem. Soc. Rev.*, 10, 49–82, 1981.
33. Yu, J., Cui, Y., Wu, C.-D., Yang, Y., Chen, B., Qian, G., Two-photon responsive metal-organic framework. *J. Am. Chem. Soc.*, 137, 4026–4029, 2015.
34. Cai, L.-X., Chen, C., Tan, B., Zhang, Y.-J., Yang, X.-D., Zhang, J., A photoactive porous metal-organic complex: Synthesis, crystal morphology and the influence of photocycloaddition on fluorescence properties and adsorption behavior. *CrystEngComm*, 17, 2353–2358, 2015.
35. Zhang, C., Sun, L., Yan, Y., Shi, H., Wang, B., Liang, Z., Li, J., A novel photo- and hydrochromic europium metal-organic framework with good anion sensing properties. *J. Mater. Chem. C*, 5, 35, 8999–9004, 2017.
36. Aulakh, D., Varghese, J.R., Wriedt, M., A New Design Strategy to Access Zwitterionic Metal-Organic Frameworks from Anionic Viologen Derivates. *Inorg. Chem.*, 54, 1756–1764, 2015.
37. Sun, J.-K., Cai, L.-X., Chen, Y.-J., Li, Z.-H., Zhang, J., Reversible luminescence switch in a photochromic metal-organic framework. *Chem. Commun.*, 47, 6870–6872, 2011.
38. Li, H.-Y., Wei, Y.-L., Dong, X.-Y., Zang, S.-Q., Mak, T.C.W., Novel Tb-MOF Embedded with Viologen Species for Multi-Photofunctionality: Photochromism, Photomodulated Fluorescence, and Luminescent pH Sensing. *Chem. Mater.*, 27, 1327–1331, 2015.

39. Li, P.-X., Wang, M.-S., Guo, G.-C., Two New Coordination Compounds with a Photoactive Pyridinium-Based Inner Salt: Influence of Coordination on Photochromism. *Cryst. Growth Des.*, 16, 3709–3715, 2016.
40. Hu, S., You, M., Chen, S., Fu, Z., A fast-response photochromic host-guest coordination polymer with a close-packed stacking structure. *CrystEngComm*, 18, 7221–7224, 2016.
41. Yang, X.-D., Chen, C., Zhang, Y.-J., Cai, L.-X., Tan, B., Zhang, J., Halogen-bridged metal–organic frameworks constructed from bipyridinium-based ligand: Structures, photochromism and non-destructive readout luminescence switching. *Dalton Trans.*, 45, 4522–4527, 2016.
42. Sun, J.-K., Wang, P., Chen, C., Zhou, X.-J., Wu, L.-M., Zhang, Y.-F., Zhang, J., Charge-distribution-related regioisomerism of photoresponsive metal–organic polymeric chains. *Dalton Trans.*, 41, 13441–13446, 2012.
43. Zhang, C., Sun, L., Yan, Y., Liu, Y., Liang, Z., Liu, Y., Li, J., Metal–organic frameworks based on bipyridinium carboxylate: Photochromism and selective vapochromism. *J. Mater. Chem. C*, 5, 2084–2089, 2017.
44. Chen, H., Zheng, G., Li, M., Wang, Y., Song, Y., Han, C., Dai, J., Fu, Z., Photo- and thermo-activated electron transfer system based on a luminescent europium organic framework with spectral response from UV to visible range. *Chem. Commun.*, 50, 13544–13546, 2014.
45. Zhang, C., Sun, L., Zhang, C., Wan, S., Liang, Z., Li, J., Novel photo- and/or thermochromic MOFs derived from bipyridinium carboxylate ligands. *Inorg. Chem. Front.*, 3, 814–820, 2016.
46. Sun, J.-K., Chen, C., Cai, L.-X., Ren, C.-X., Tan, B., Zhang, J., Mechanical grinding of a single-crystalline metal–organic framework triggered emission with tunable violet-to-orange luminescence. *Chem. Commun.*, 50, 15956–15959, 2014.
47. Sun, J.-K., Wang, P., Yao, Q.-X., Chen, Y.-J., Li, Z.-H., Zhang, Y.-F., Wu, L.-M., Zhang, J., Solvent- and anion-controlled photochromism of viologen-based metal–organic hybrid materials. *J. Mater. Chem.*, 22, 12212–12219, 2012.
48. Sugioka, K., Hanada, Y., Midorikawa, K., *3D microchips fabricated by femtosecond laser for biomedical applications, Optomechatronic Technologies 2008*, p. 726603, *Proceedings Volume 7266, Optomechatronic Technologies.* International Society for Optics and Photonics, San Diego, California, United States, 2008.
49. Hu, S., Zhang, J., Chen, S., Dai, J., Fu, Z., Efficient Ultraviolet Light Detector Based on a Crystalline Viologen-Based Metal–Organic Framework with Rapid Visible Color Change under Irradiation. *ACS Appl. Mater. Interfaces*, 9, 39926–39929, 2017.
50. Ali Akbar Razavi, S. and Morsali, A., Linker functionalized metal-organic frameworks. *Coord. Chem. Rev.*, 399, 213023, 2019.
51. Gong, T., Li, P., Sui, Q., Chen, J., Xu, J., Gao, E.-Q., A stable electron-deficient metal–organic framework for colorimetric and luminescence sensing of phenols and anilines. *J. Mater. Chem. A*, 6, 9236–9244, 2018.

52. Wang, K.-M., Du, L., Ma, Y.-L., Zhao, Q.-H., Selective sensing of 2,4,6-trinitrophenol and detection of the ultralow temperature based on a dual-functional MOF as a luminescent sensor. *Inorg. Chem. Commun.*, 68, 45–49, 2016.
53. Rapti, S., Sarma, D., Diamantis, S.A., Skliri, E., Armatas, G.S., Tsipis, A.C., Hassan, Y.S., Alkordi, M., Malliakas, C.D., Kanatzidis, M.G., Lazarides, T., Plakatouras, J.C., Manos, M.J., All in one porous material: exceptional sorption and selective sensing of hexavalent chromium by using a Zr4+ MOF. *J. Mater. Chem. A*, 5, 14707–14719, 2017.
54. Zhang, C., Liu, Y., Sun, L., Shi, H., Shi, C., Liang, Z., Li, J., A Zwitterionic Ligand-Based Cationic Metal–Organic Framework for Rapidly Selective Dye Capture and Highly Efficient Cr2O72-Removal. *Chem. Eur. J.*, 24, 2718–2724, 2017.
55. Xu, L., Luo, Y., Sun, L., Pu, S., Fang, M., Yuan, R.-X., Du, H.-B., Tuning the properties of the metal–organic framework UiO-67-bpy *via* post-synthetic N-quaternization of pyridine sites. *Dalton Trans.*, 45, 8614–8621, 2016.
56. Tan, B., Chen, C., Cai, L.-X., Zhang, Y.-J., Huang, X.-Y., Zhang, J., Introduction of Lewis Acidic and Redox-Active Sites into a Porous Framework for Ammonia Capture with Visual Color Response. *Inorg. Chem.*, 54, 3456–3461, 2015.
57. Leroux, M., Mercier, N., Allain, M., Dul, M.-C., Dittmer, J., Kassiba, A.H., Bellat, J.-P., Weber, G., Bezverkhyy, I., Porous Coordination Polymer Based on Bipyridinium Carboxylate Linkers with High and Reversible Ammonia Uptake. *Inorg. Chem.*, 55, 8587–8594, 2016.
58. Ren, C.-X., Cai, L.-X., Chen, C., Tan, B., Zhang, Y.-J., Zhang, J., π-Conjugation-directed highly selective adsorption of benzene over cyclohexane. *J. Mater. Chem. A*, 2, 9015–9019, 2014.
59. An, W., Aulakh, D., Zhang, X., Verdegaal, W., Dunbar, K.R., Wriedt, M., Switching of Adsorption Properties in a Zwitterionic Metal–Organic Framework Triggered by Photogenerated Radical Triplets. *Chem. Mater.*, 28, 7825–7832, 2016.

5

Heterocyclic Azine Decorated Metal-Organic Frameworks

Abstract
This chapter belongs to heterocyclic azine based functions which are including pyridine, diazines, s-triazine and s-tetrazine. The number of N atoms in the aromatic ring of these functions plays critical roles in host-guest chemistry and coordination strength of these functions. This chapter deeply discuses about these features of heterocyclic azine functions.

Keywords: Pyridine, diazines, pyrazine, pyrimidine, pyradazine, triazine, tetrazine, functional metal-organic frameworks

5.1 General Chemical Properties of Heterocyclic Azine Functions

The structure of these functional groups is based on six-member benzene ring. In these functions, certain number of (–CH) motifs in benzene ring is replaced with electronegative nitrogen atoms. Heterocyclic azine N-based functions are pyridine, diazines (pyrimidine, pyrazine, and pyradazine), triazine (usually 1,3,5-triazine) and tetrazine (1,2,4,5-tetrazine). Based on the number of N atoms in the six-member ring, they show similar or different chemical properties. Chemical properties of these functions revolve around their Lewis basicity and π-deficiency. It is essential to clarify the roles of the number of nitrogen atoms to discuss about Lewis basicity and π-deficiency of heterocyclic azine N-based functions.

For pyridine with one nitrogen atom in the aromatic ring, the electron rich site is located near the nitrogen atom [1]. For diazines, the electron rich sites are only near the nitrogen atoms but the nitrogen atoms are more electron deficient than pyridine [1]. For 1,3,5-triazine, electrostatic

Ali Morsali and Sayed Ali Akbar Razavi. Functional Metal-Organic Frameworks: Structure, Properties and Applications, (79–106) © 2021 Scrivener Publishing LLC

potential surfaces reveal that the nitrogen atoms are more electron deficient rather pyridine and diazines [1]. This trend clearly show that the higher number of nitrogen atoms, the higher electron deficiency on nitrogen atoms and aromatic ring. Since the binding energies of N atoms as electron-rich Lewis basic sites are proportional to their electron density in donor–acceptor interactions, Lewis acid–base of electrostatic interactions, N-heterocyclic azine functions with higher number of N atoms are of lower basicity. This observation represents that N-heterocyclic azine functions with lower number of N atoms (diazines and pyridine) are suitable functions to interact with guests with partially positive charged or electron deficient sites like carbon dioxide, metal ions and acidic hydrogens (Figure 5.1). Inversely, N-heterocyclic azine functions with higher number of N atoms are of higher electron deficiency and are able to interact with aromatic electron-rich species (Figure 5.1). It is worthy to mention that these groups of functions with lower N atoms do not represent special mark affinity in π(rich)–π(deficient) interactions as electron deficient sites while triazine and tetrazine are able to interact weakly with metal ions or CO_2 molecules.

Considering these characters N-heterocyclic azine functions applied in the structure of MOFs. Pyridine and diazine applied as coordinating sites to develop the structure of novel MOF with or without carboxylate groups on the heterocyclic ring. On the other hand, the nitrogen atoms of N-based functions located in the aromatic skeleton of the framework not in the pores of MOFs. As a result, the porosity and accessible pore volume is almost retained. Based on published papers, heterocyclic-azine

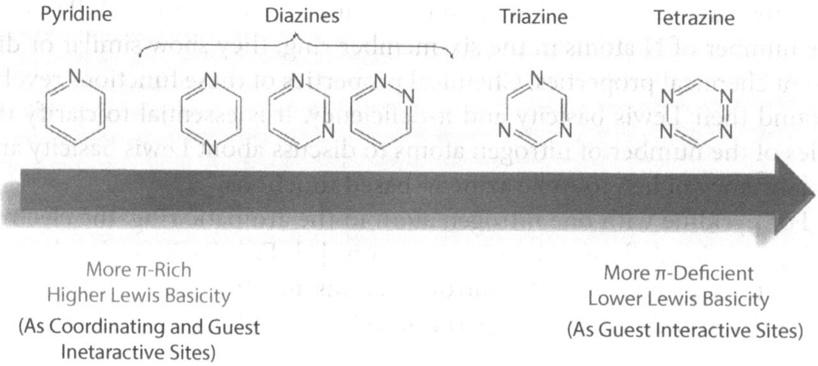

Figure 5.1 Combination of chemical properties of N-heterocyclic azine functions.

decorated MOFs are used as an adsorbent or a sensor for organic solvents [2–5], π-rich analytes [6, 7], hydrogen bond donor species [8–13] and metal ions [14–23]. Also they applied for polarization of CO_2 [24–40], CH_4 [40–47], and C2 hydrocarbons [40, 41, 48–53] and separation of benzene [53, 54]. Moreover owing to presence of N atoms, they applied as Lewis basic catalysis [55–63].

5.2 Function–Application Properties

As mentioned, the most important chemical property of heterocyclic azine rings with low number of N atoms is founded on their Lewis basicity. This chemical character makes the pyridine and diazine functions practical as both guest-interactive and coordinating sites. As guest-interactive site, pyridine and diazine functions applied in different field of applications especially gas storage and separation.

Regarding the presence of N atom(s) in the molecular structure of heterocyclic azines, especially pyridine, FMOFs bearing free pyridinic N atoms applied for selective separation of CO_2 and C_2H_2. Possible interactions between acetylene and carbon dioxide molecules with pyridine function like CO_2(C)·(N)pyridine donor–acceptor interaction and C_2H_2(H)·(N) hydrogen bonding are the reasons of affinity between CO_2 and C_2H_2 molecules and heterocyclic azines decorated MOFs. Consequently, pyridine-N atoms locating in the channels are more attractive to CO_2 and C_2H_2 than N_2 and CH_4 due to forming mentioned strong interactions.

Banglin Chen and coworkers applied ZJU-5 with NbO topology ([Cu_2(PDDI)] where H_4PDDI = 5,5'-(pyridine-2,5-diyl)diisophthalic acid) for high C_2H_2 and methane storage at room temperature (Figure 5.2) [41].

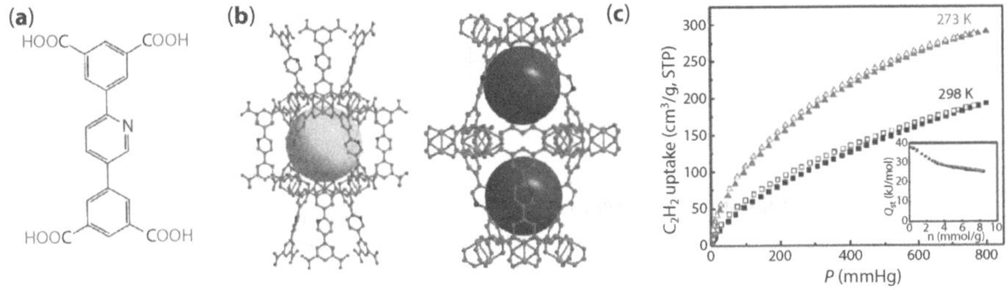

Figure 5.2 Application of ZJU-5 in acetylene storage. (a) Structure H_4PDDI ligand. (b) X-ray single crystal structure of ZJU-5 showing two types of cages, small cage of about 10.5 Å in diameter in yellow and large irregular elongated pore of about 9.5 × 22.5 Å in violet. (c) Gas sorption isotherms of ZJU-5a for C_2H_2 at 273 K and 298 K. Inset: Calculated C_2H_2 adsorption enthalpy [41].

They mentioned the uptake capacity of ZJU-5 reaches the remarkable value of 290 $cm^3.g^{-1}$ at 273 K and 1 bar. This noticeable value is attributed to presence of open metal sites, suitable pore space and Lewis basic pyridine functionalized pores. Calculation of acetylene adsorption enthalpies of activated ZJU-5 at low coverage reveals that the acetylene adsorption enthalpy is 35.8 $kJ \cdot mol^{-1}$. This acetylene–pyridine interactions is because of possible pyridine(N)·(H–CCH)acetylene interaction.

In other work, De-li Chen and coworkers reported that gas adsorption properties of MOFs decorated with heterocyclic azine nitrogen-based functions, especially pyridine and pyrazine, depend not only on the number of Lewis basic N atoms nitrogen in the functional site but more importantly on their direction in pores and their accessibility for CO_2 molecules (Figure 5.3) [27]. They synthesized a series of NbO type MOFs including ZJNU-43 (Cu_2L1 where H_4L1 is illustrated in Figure 5.2a), ZJNU-44 (Cu_2L2 where H_4L2 is illustrated in Figure 5.2a) and ZJNU-45 (Cu_2L3 where H_4L3 is illustrated in Figure 5.2a) and applied them in gas adsorption. Despite the very close surface area (2,243 $m^2 \cdot g^{-1}$ for ZJNU-43, 2,314 $m^2 \cdot g^{-1}$ for ZJNU-44 and 2,232 $m^2 \cdot g^{-1}$ for ZJNU-45) they show remarkable differences in CO_2/N_2 selectivity and CO_2 uptake capacity. They conducted quantum chemical calculations and they find that the location of N atom

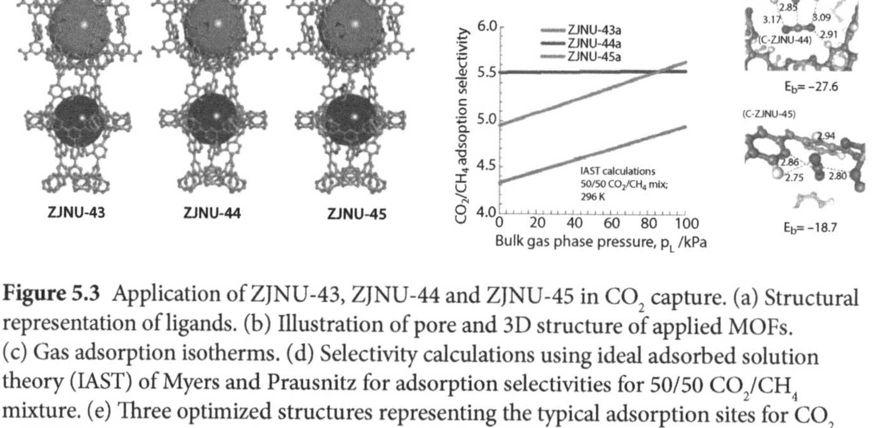

Figure 5.3 Application of ZJNU-43, ZJNU-44 and ZJNU-45 in CO_2 capture. (a) Structural representation of ligands. (b) Illustration of pore and 3D structure of applied MOFs. (c) Gas adsorption isotherms. (d) Selectivity calculations using ideal adsorbed solution theory (IAST) of Myers and Prausnitz for adsorption selectivities for 50/50 CO_2/CH_4 mixture. (e) Three optimized structures representing the typical adsorption sites for CO_2 in ZJNU-43, ZJNU-44, and ZJNU-45 at Lewis basic sites [27].

at β position is straightly directed into the pores of ZJNU-44 which make it available for stronger interaction with CO_2 molecules.

Diazines divided in three isomers: 1,4-diazine (pyrazine), 1,3-diazine (pyrimidine) and 1,2-diazine (pyradazine). Regarding two nitrogen atoms in the diazines ring, they show similar chemical characters to pyridine like Lewis basicity, polarizing ability and hydrogen bond accepting. On the other hand, due to the one more nitrogen atom in the structure of diazines rather pyridine, the electron density on nitrogens is lower and the aromatic ring has more electron deficiency. Application of diazine functions in the MOFs structure as guest interactive sites has resulted in very noticeable effects on the absorption of carbon dioxide, acetylene and especially methane.

Banglin Chen and coworkers developed some FMOFs with NbO topology and decorated with Heterocyclic azines for high capacity methane storage (Figure 5.4). These FMOFs are pyridine functionalized ZJU-5, pyradazine functionalized UTSA-75 and pyrimidine functionalized UTSA-76 which are isostructure with NOTT-101 [45]. The results of their research proved that high volumetric methane storage can be attained by functionalization of MOFs with Lewis basics heterocyclic azine sites. Reported volumetric methane storage capacities for these MOFs are 237 $cm^3(STP)\cdot cm^{-3}$ for NOTT-101, 249 $cm^3(STP)\cdot cm^{-3}$ for ZJU-5, 251 $cm^3(STP)\cdot cm^{-3}$ for UTSA-75 and 257 $cm^3(STP)\cdot cm^{-3}$ for UTSA-76. These results show that introduction of pyrimidine into the pores display the best improvement in methane storage among these isostructure MOFs. More investigations clarify that same trend is observed for methane volumetric working capacity; 181 $cm^3\cdot cm^{-3}$

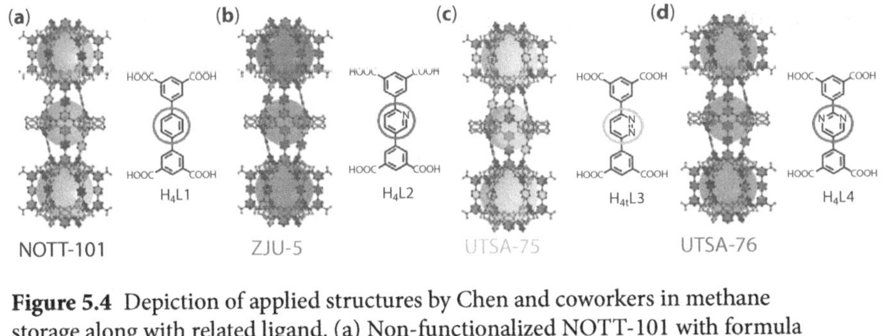

Figure 5.4 Depiction of applied structures by Chen and coworkers in methane storage along with related ligand. (a) Non-functionalized NOTT-101 with formula $[Cu_2(\mathbf{L1})(H_2O)_2]$ where $H_4\mathbf{L1}$ is Terphenyl-3,3",5,5"-tetracarboxylic acid. (b) Pyridine-functionalized ZJU-5 with formula $[Cu_2(\mathbf{L2})(H_2O)_2]$ where $H_2\mathbf{L2}$ is 5,5'-(pyridine-2,5-diyl)diisophthalic acid. (c) Pyradazine-functionalized UTSA-75 with formula $[Cu_2(\mathbf{L3})(H_2O)_2]\cdot 3DMF$ where $H_4\mathbf{L3}$ is 5, 5'-(pyridazine-2, 5-diyl)diisophthalic acid. (d) Pyrimidine-functionalized UTSA-76 with formula $[Cu_2(\mathbf{L4})(H_2O)_2]\cdot 5DMF\cdot 3H_2O$ where $H_4\mathbf{L4}$ is 5,5'-(pyrimidine-2, 5-diyl)diisophthalic acid [45].

for NOTT-101, 188 $cm^3 \cdot cm^{-3}$ for ZJU-5, 192 $cm^3 \cdot cm^{-3}$ for UTSA-75 and 197 $cm^3 \cdot cm^{-3}$ for UTSA-76. Gas adsorption analysis show that these isostructure MOFs have close surface areas and pore volumes. So, since the porosity of these MOFs is close and they are isostructure, it can be concluded that the observed difference in volumetric methane working capacity is related to the variations in functional groups. Calculation of isosteric heats of adsorption shows that there is not any significant difference in values of methane adsorption enthalpy (15.44 $kJ.mol^{-1}$ for UTSA-76 and 15.49 $kJ \cdot mol^{-1}$ for NOTT-101). The results of theoretical studies show that despite the fact that NOTT-101 and UTSA-76 display similar methane uptake, at higher pressure UTSA-76 can adsorb higher amount of methane. This is due to the fact that central pyrimidine rings within UTSA-76 can be more easily adjusted and oriented to optimize the methane packing at high pressure than the central benzene rings within NOTT-101.

In another work by Banglin Chen and coworkers, UTSA-110 has been synthesized with optimized surface area and high density of functional sites to improve the volumetric and gravimetric methane adsorption capacity. UTSA-110 is developed with the idea of expanding the UTSA-76 framework with linker-extending approach to contain higher density of functional pyrimidine sites (Figure 5.5) [44]. It is observed that UTSA-110 exhibits both very high gravimetric and volumetric working capacities of 317 $cm^3(STP) \cdot g^{-1}$ and 190 $cm^3(STP) \cdot cm^{-3}$, respectively.

The results of these researches by Chen and coworkers clearly show that rationally developed MOFs with suitable functional groups are best candidate for improving in methane working capacity even better than MOFs with open metal sites. Owing to presence of open metal sites, there is strong interaction with methane molecules and MOFs containing open

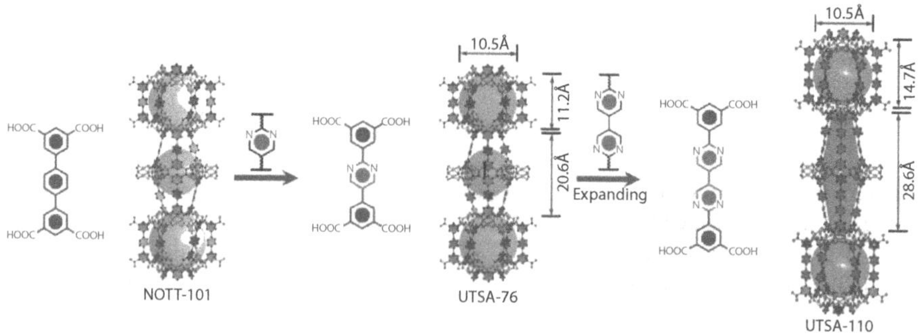

Figure 5.5 The idea of synthesis of UTSA-110 with optimized porosity and high density of pyrimidine functions for high capacity methane storage [44].

metal sites like HKUST-1 and because of such strong interaction, methane molecules cannot be completely desorbed at lower pressures. So, the working capacity is reduced. But in case of functional MOFs like UTSA-76 and UTSA-110, there is not any specific interaction between methane molecules and pyrimidine functions. So, at low pressure methane molecules could be desorbed easily. Also, owing to their reorientation at high pressures, methane molecules could be highly packed. So, methane working capacity is increased for these FMOFs.

Because of Lewis basic activity of diazines, they applied for improvement of CO_2-framework interaction to achieve highly efficient CO_2 separation (Figure 5.6) [34]. As a great example, adsorption of CO_2 in bio-MOF-11 studied by Rosi and coworkers. Bio-MOF-11 (BET surface area = 1,040 $m^2 \cdot g^{-1}$, pore

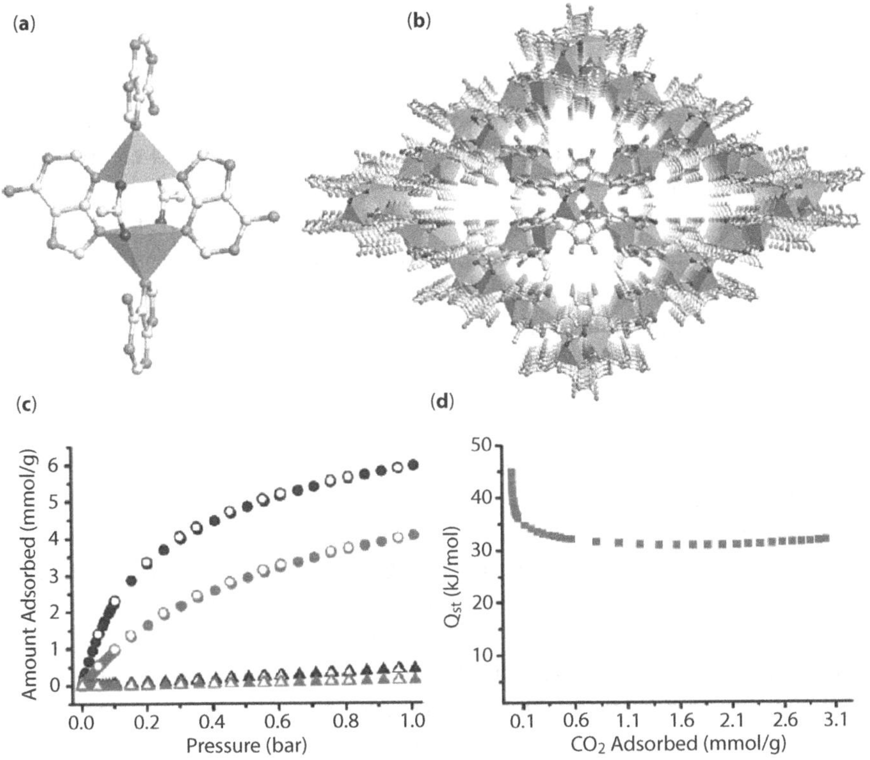

Figure 5.6 Application of bio-MF-11 in CO_2 capture. (a) Co(II)-adeninate-acetate clusters. (b) Extended 3D porous structure with channels along the *a* and *b* crystallographic directions. (c) Adsorption isotherms for CO_2 (circles) and N_2 (triangles) at 273 (black) and 298 K (red). (d) Isosteric heat of adsorption for CO_2 at different CO_2 loadings [34].

volume = 0.45 $cm^3 \cdot g^{-1}$) consists of cobalt–adeninate acetate paddle wheel clusters. Adenine has multiple Lewis basic sites including an amino group and pyrimidine nitrogens. Their achieved data reveal that bio-MOF-11 exhibits high CO_2 uptake capacity compared to many other MOFs such as amine-functionalized MOFs and imidazole based frameworks. This material adsorbs 6.0 and 4.1 $mmol \cdot g^{-1}$ of CO_2 and only 0.43 and 0.13 $mmol \cdot g^{-1}$ of N_2 at 273 and 298 K respectively. The calculated CO_2/N_2 selectivity is 81:1 at 273 K and 75:1 at 298 K. Isosteric heat of adsorption calculated for CO_2 is ~−45 $kJ \cdot mol^{-1}$, which is similar to values for some other amine-functionalized MOFs. Such a high Q_{st} is due to favorable interactions between adsorbed CO_2 molecules and the Lewis basic amine and pyrimidine functionalities decorating the pores because the pores of bio-MOF-11 are densely decorated with Lewis basic amine and pyrimidine groups. A total of four amino groups and four pyrimidine groups are directly exposed to each individual cavity. Similarly computational study reveals that interaction energy between CO_2 and adenine is higher than that between CO_2 and other nitrogen-containing MOF linker molecules. The interaction between the localized dipoles of the N-containing groups and the quadrupole moment of CO_2 molecule induces the dispersion and electrostatic forces to enhance CO_2 adsorption.

S-triazine functionalized MOFs extensively applied in the field of gas adsorption because constructed MOFs with this function have high structural symmetry, high surface area and N-rich pores. This group of functionalized applied for improvement of CO_2 capture capacity and separation (CO_2/N_2 and CO_2/CH_4) efficiency in MOFs. It can be appropriately explained with the viewpoint that the rich nitrogen atoms incorporated in triazinyl can serve as the Lewis-base centers to enhance the dipole–quadrupole interactions with the CO_2 gas. Spectroscopic techniques demonstrate the separate interactions of CO_2 with triazine groups.

Beside the CO_2 adsorption, triazine functionalized MOFs applied in adsorption and separation of light C2 hydrocarbons and benzene/cyclohexane separation. Rong Cao and coworkers synthesized an anionic MOF denoted as FJI-C1, $(Et_4N)_3[In_3(TATB)_4]$ where H_3TATB=4,4',4"-s-triazine-2,4,6-triyltribenzoic acid which not only exhibits high C_2 and C_3 hydrocarbon adsorption uptake, but also shows high separation selectivity for C_2 and C_3 hydrocarbons over CH_4; C_3H_8 towards C_2 hydrocarbons; and C_2H_2 over C_2H_4 and C_2H_6 [53]. The ligand $TATB^{3-}$ of FJI-C1 has three nitrogen atoms in the s-triazine ring, which provides a π-electron deficient and hydrophilic on the internal surface of the pores to enhance the separation of polar and π-electron rich guests from the nonpolar and non-aromatic molecules. As results FJU-C1 show high adsorption of polar vapors like methanol, water and ethanol rather nonpolar molecules and high separation selectivity for benzene over cyclohexane owing to π rich–π deficient interactions between benzene molecules and the s-triazine rings. In other similar work, π-electron deficient diaminotriazine functionalized DAT-MOF-1 for selective sorption of benzene over cyclohexane which show high capacity toward benzene (1.5 $mmol{\cdot}g^{-1}$ for benzene vs about 0.2 $mmol{\cdot}g^{-1}$ for cyclohexane) rather cyclohexane and high C_6H_6/C_6H_{12} selectivity with value in excess of about 200 [54].

As mentioned earlier, for heterocyclic azines with high number of N atoms in the aromatic ring, the electron density on N atoms is diminished rather heterocyclic azines with lower number of N atoms. On the other hand, the strength of interaction with CO_2 molecules by donor-acceptor interactions is proportional to the electron density on N atoms. So, we can decelerate that despite the higher number of N atoms in the aromatic ring of tetrazine, there is not any expected strong donor-acceptor interactions between N atoms of tetrazine ring and carbon atom of carbon dioxide. As a result, tetrazine decorated FMOFs did not show high CO_2 isosteric heat of adsorption up-to-date. But, recently, we applied two tetrazine decorated MOFs, named as TMU-34 and TMU-34(-2H), for methane adsorption at room temperature and pressure and we find that owing to high affinity between tetrazine groups and methane molecules, these FMOFs have noticeable methane adsorption capacity, even higher than MOFs with higher surface area [46].

The role of number of N atoms in the chemical properties of heterocyclic azine functions determine they behavior as guest-interactive site.

Because of basic features of pyridine and diazines, they could bind to transition metal ions as a relatively weak ligand. Moreover, Because of the accumulation of a greater negative charge on the nitrogen atoms in the pyridine ring and diazines, these functions can interact with guests with electron deficient sites.

Banglin Chen and coworkers synthesized a luminescent MOF $[Eu(pdc)_{1.5}(dmf)]{\cdot}(DMF)_{0.5}(H_2O)_{0.5}$ (Eu-pdc, pdc = pyridine-3,5-dicarboxylate) functionalized with pyridine rings. The most significant structural feature is the presence of free Lewis basic pyridyl sites within the pores, highlighting the potential for their recognition of metal ions (Figure 5.7) [14]. This MOF excited at around 321 nm ascribed to the absorption of pdc ligand and emit 590, 616, and 698 nm, which are ascribed to the $^5D_0$–$^7F_1$, $^5D_0$–$^7F_2$ and $^5D_0$–$^7F_4$ transitions of Eu(III) centers, respectively. Such photoluminescence spectrum is attributed to the energy transfer from pdc ligands to Eu(III) ions. Application of Eu-pdc in metal ion detection shows that the luminescence intensity of the Cu(II)-incorporated Eu(III)-pdc from

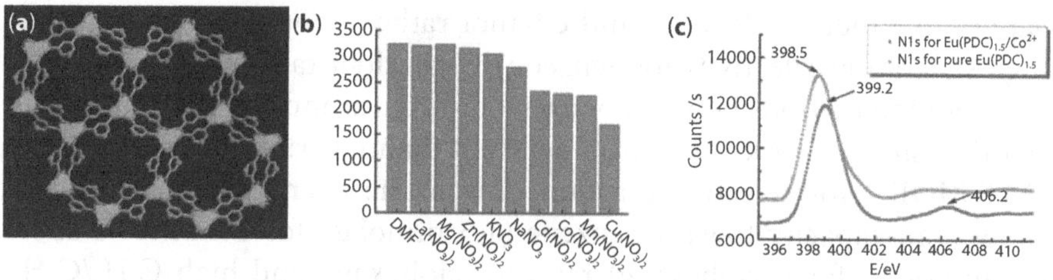

Figure 5.7 Application of Eu(III)-pdc in photoluminescence detection of Cu(II) ions. (a) Crystal structure and pyridine decorated pores of Eu(III)-pdc. (b) Luminescent change in presence of different metal ions. (c) XPS spectra of Eu(III)-pdc and Co(II)@Eu(III)-pdc [14].

a 10 mm DMF solution of $Cu(NO_3)_2$ is about half that of metal-ion free Eu(III)-pdc, indicating the potential of Eu(III)-pdc for the sensing of metal ions. N1s X-ray photoelectron spectroscopy (XPS) with Cu(II)@Eu(III)-pdc and Co(II)@Eu(III)-pdc applied to evaluate the possible detection mechanism. The N1s peak from free pyridyl nitrogen atoms at 398.5 eV in Eu(III)-pdc is shifted to 399.2 eV on the addition of Cu(II) or Co(II), indicating the weak binding of pyridyl nitrogen atoms to metal ions. A peak at 406.1 eV in the N1s XPS spectra for metal ion incorporated Eu(III)-pdc is attributed to nitrate counter ions. Such host-guest interaction between pyridine function and Cu(II) ions reduces the energy transfer efficiency in resonance energy transfer from pdc ligand to Eu(III) ions giving rise to quenching of the luminescence.

1,3,5-triazine or s-triazine function is one the most applied polyazine synthons in the supramolecular chemistry for architecting supramolecular networks through coordination or hydrogen bond and their application in host–guest chemistry, catalysis, anion recognition, sensoring, electronics and magnetism. Such an extensive application of 1,3,5-triazine in material chemistry is because of their proven characters like high electron-deficiency, π-conjugated aromatic system, tendency to form non-covalent hydrogen bonds and coordination interactions.

In comparison with diazines and pyridine, the most important feature of s-triazine ring is its high electron deficiency which makes it very interactive toward various nucleophiles and electron-rich guest molecules. This π-electron deficient nature is due to the presence of abundance number of electronegative nitrogen atoms. Moreover than π-electron deficiency of triazine ring, large number of nitrogen atoms results in reduction in electron density on nitrogen atoms. As results, ability of triazine nitrogen atoms to coordinate to Lewis acids in comparison with diazines and pyridine is noticeably reduced.

Owing to the different chemical properties of the triazine group, triazine decorated MOFs have been used to sense compounds such as, metal ions through Lewis basic donor acceptor interactions, nitrophenols through hydrogen bonding, π-conjugated aromatic hydrocarbons through π-deficient–π-rich interactions.

Lei Han *et al.* synthesized a MOF, $[Zn_3(TATB)_2(H2O)_2]_n$ (Zn-TATB, **2**) where TATB = 4,4′,4″-s-triazine-2,4,6-triyltribenzoate, and applied it as a probe for designing charge transfer (CT) materials (Figure 5.8) [64]. It can be predicted that a CT interaction would occur between a TATB-based MOF and an electron-rich DMA (N,N'-Dimethylaniline) guest molecule. DMA can be successfully included into the triazine incorporated pores of MOF Zn-TATB with color change from white for Zn-TATB to green for DMA@Zn-TATB. As results, absorption bands of Zn-TATB and DMA@ Zn-TATB are different. The changes indicate that the triazine containing TATB units interact with the DMA guests in the ground state to form a CT complex in DMA@Zn-TATB. Interestingly, the DMA@Zn-TATB emission spectrum is quite different from that of Zn-TATB and shows an intensive, broad, and red-shifted emission maximum at 480 nm. The emission with a large Stokes shift represents an efficient photoinduced CT complex, an

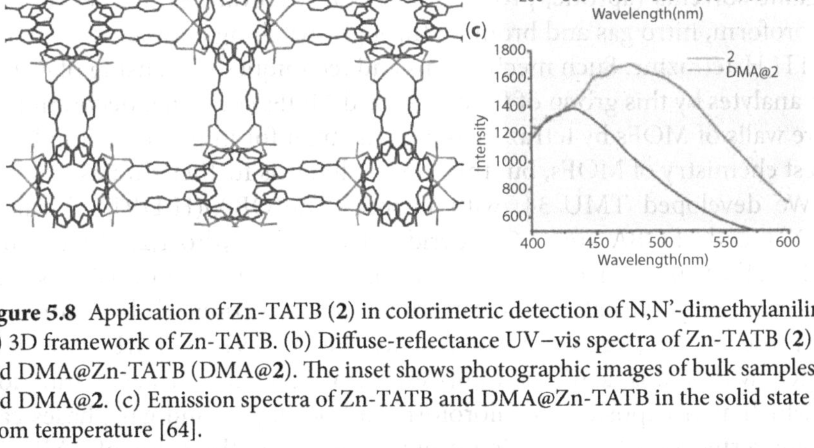

Figure 5.8 Application of Zn-TATB (**2**) in colorimetric detection of N,N'-dimethylaniline. (a) 3D framework of Zn-TATB. (b) Diffuse-reflectance UV–vis spectra of Zn-TATB (**2**) and DMA@Zn-TATB (DMA@**2**). The inset shows photographic images of bulk samples **2** and DMA@**2**. (c) Emission spectra of Zn-TATB and DMA@Zn-TATB in the solid state at room temperature [64].

exciplex between the host triazine units (electron acceptor) and the guest DMA molecules (electron donor). Upon inspection of the above results, the authors mentioned the planarity of the TATB ligand and its tendency to encourage π·π stacking in MOF proved an extraordinary ability to increase the framework stability and a strong advantage in absorption and transformation of light energy, resulting in a preferential adjustment of the luminescence through CT.

s-Tetrazines (1,2,4,5-tetrazines) are long ago discovered molecules, whose first report dates back to the end of the 19th century. In azine heterocycles, one or more CH group of benzene is replaced by π-accepting and σ-donating sp^2 N atoms. Increasing such replacement leads via diazines (pyridazine, pyrimidine, pyrazine) and the triazines to the tetrazines of which the 1,2,4,5 isomer is the most stable form. Because of the larger number of nitrogen atoms in the tetrazine ring rather pyridine, diazines and triazine, and since the higher homologue pentazine and of course hexazine have been recognized to be unstable, at least at room temperature, tetrazine ring can be considered as the nitrogen richest and electron-poorest aromatic system of the classical C, H, O, N, S chemistry.

Altogether, the tetrazine functional group is enriched with σ-donor and π-acceptor nitrogen atoms, highly π-deficient and electro active low lying π* tetrazine molecular orbitals and weak n-π* electronic transitions in the visible range, redox active and able to form stable radical-anion in non-protic media. Considering these chemical features, tetrazine functionalized MOFs repetitively applied as sensor for detection of different types of chemicals with different mechanisms and methods including cations and electron rich organics like phenol by fluorescence spectroscopy, acetone and phenyl hydrazine by change in absorbance peak, different kind of organic solvents (aprotic, protic, nonpolar) by solvatochromic effect and chloroform, nitro gas and bromine through switching between s-tetrazine and H,H-tetrazine. Such mechanisms and techniques for sensing of different analytes by this group of functionalized MOFs show that decoration of pore walls of MOFs by tetrazine is a good target for improvement in host-guest chemistry of MOFs, but this topic still needs further studies.

We developed TMU-34, with formula $[Zn(OBA)(H_2DPT)_{0.5}]$ where H_2DPT and H_2OBA are (3,6-di(pyridin-4-yl)-1,4-dihydro-1,2,4,5-tetrazine) and (4,4'-oxybis(benzoic acid), as a one of the most promising solid-state naked-eye visual chemosensors for the detection of chloroform in the presence of a large variety of analytes (Figure 5.9) [2]. TMU-34 is decorated with redox active stimuli-responsive dihydro-tetrazine function in which after exposure to chloroform in both liquid and gas phases can be converted into the tetrazine groups. As results the color of TMU-34

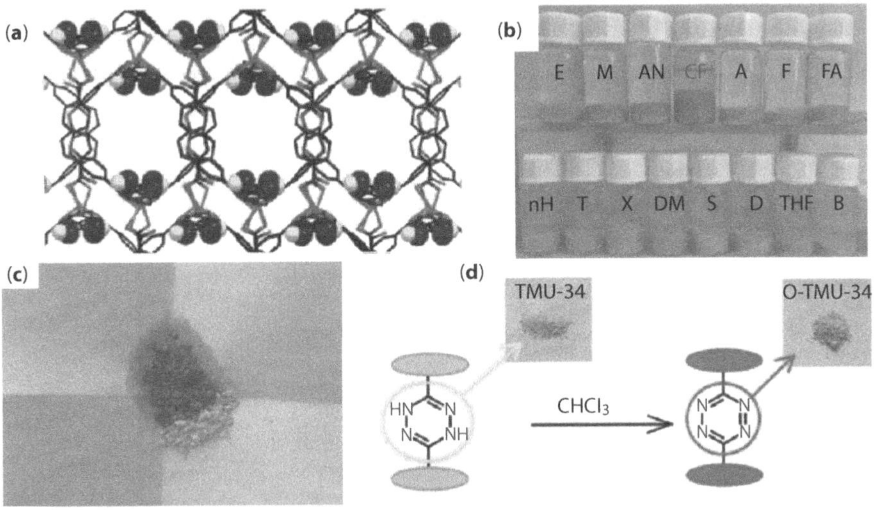

Figure 5.9 Application of TMU-34 in colorimetric detection of chloroform. (a) 3D structure of TMU-34 and dihydro-tetrazine decorated pores. (b) Color change of TMU-34 in presence of VOCs. (c) Solid state color change of chloroform after exposing to chloroform. (d) Mechanism of color change [2].

changed from yellow to pink. TMU-34 can detect chloroform in liquid and gas phases up to 2.5×10^{-5} M. TMU-34 is reversible through exposing to DMF molecules. Selectivity is one of the most important characters in designing all kind of sensors. TMU-34 shows unique selectivity toward chloroform. In comparison with other MOF-based sensors that represent color change in presence of different analytes, the color of TMU-34 just changes in presence of chloroform.

In other work by Stefan Kaskel and coworkers, dihydro-tetrazine decorated UiO-66 (UiO-66(dhtz) with formula $[Zr_6O_4(OH)_4(bdc)_{5.1}(dhtz)_{0.9}]$) applied as optical sensor for oxidizing gases (Figure 5.10) [65]. Exposure of UiO-66(dhtz) suspended in chloroform to nitrous gases results in oxidation of the dihydro-1,2,4,5-tetrazine to 1,2,4,5-tetrazine to give UiO-66(tz) by a color change from yellow to pink. In contrast, aqueous solutions of potassium dichromate or iron(III) nitrate, respectively, were not able to oxidize UiO-66(dhtz). If the material was stored under a nitrogen or argon atmosphere, no color change took place even after half a year. Moreover, gas phase detection of nitrous gases shows that a similar color change is visible. This sensor is regenerable by sodium dithionite.

We applied tetrazine functionalized TMU-34(-2H) (with formula $[Zn(OBA)(DPT)_{0.5}]\cdot DMF$ where H_2OBA and DPT are 4,4'-oxybis(benzoic acid) and 3,6-di(pyridin-4-yl)-1,2,4,5-tetrazine) applied for highly

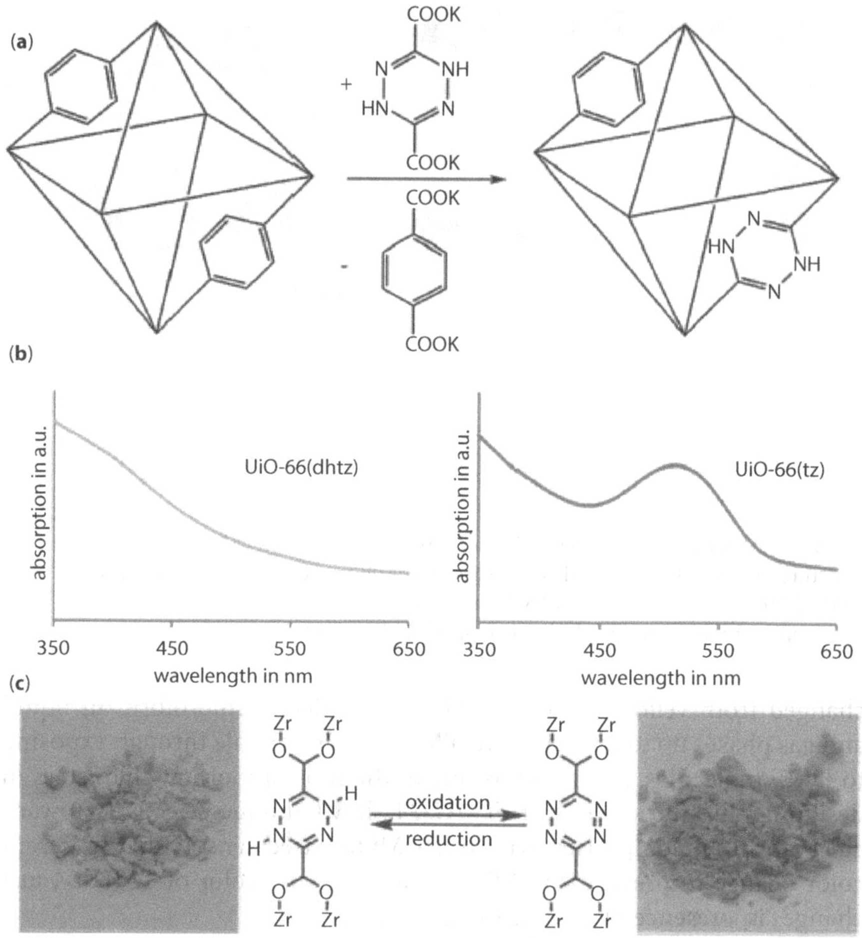

Figure 5.10 Application of UiO-66(dhtz) in detection of oxidizing gases. (a) Introduction of dihydro-1,2,4,5-tetrazine-3,6-dicarboxylate (dhtz) into the UiO-66 framework by post-synthetic exchange giving UiO-66(dhtz). (b) UV–Vis of the sensor before and after exposure to oxidizing gases. (c) Mechanism of color change [65].

sensitive detection of Hg(II) ions (Figure 5.11) [22]. Up on different photoluminescence behavior of TMU-34(-2H) in water and acetonitrile, metal ion detection using TMU-34(-2H) conducted in both solvents. For acetonitrile, the excitation and emission wavelengths are centered at 458 and 618 nm, respectively while related data for water are 504 and 648 nm. These differences in photoluminescence behavior of TMU-34(-2H) in water and acetonitrile is because of hydrogen bonding between tetrazine group and water molecules. Results of cation detection show that the emission spectra

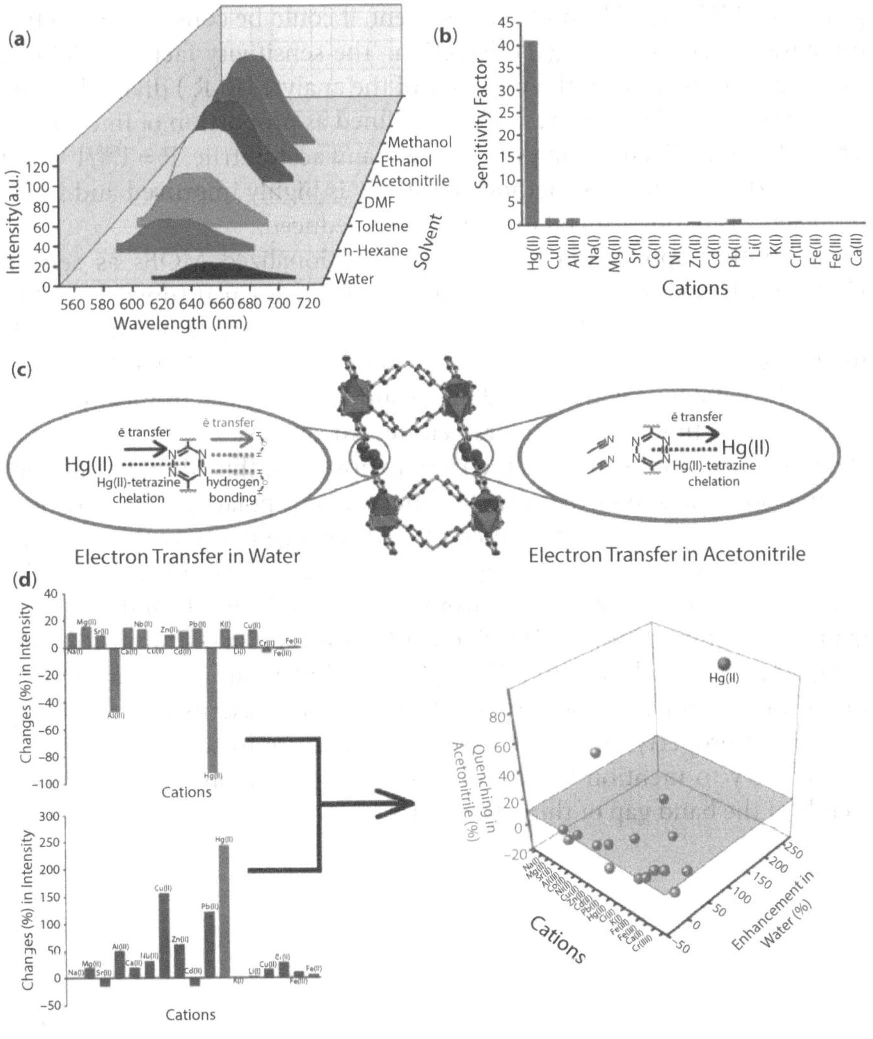

Figure 5.11 Application of TMU-34(-2H) in ion detection. (a) Effect of the solvent on the TMU-34(-2H) fluorescence emission peak. (b) Sensitivity factors for all of the cations tested. (c) Representation of Electron Transfer between Hg(II) and TMU-34(-2H) in Water and Acetonitrile. (d) 2D sensing curve for Hg(II) detection in water and acetonitrile [22].

of TMU-34(-2H) enhanced 243% in water and attenuate 90% in acetonitrile. In next step, the results of Hg(II) detection in water and acetonitrile converged in one table to construct double solvent sensing method. Double solvent sensing curve is constructed based on combination of changes in emission spectra of TMU-34(-2H) in water and acetonitrile in presence of Hg(II). Since double solvent sensing curve is based on two emission peaks of TMU-34(-2H) indifferent solvent, it could be considered as a two dimensional curve for Hg(II) detection. The sensitivity factor is defined as change in response in the presence of the analyte ($R-R_0$) divided by the initial response (R). Here, response is defined as proportion of intensity of TMU-34(-2H) PL emission peak in water and acetonitrile; $R = I^{648}/I^{618}$ and $R_0 = I_0^{648}/I_0^{618}$. Using this method sensitivity is highly improved and interfering effects of other metal ions effectively reduced.

As another application of tetrazine functionalized MOFs as sensor, He-Gen Zheng and coworkers fabricated $[(WS_4Cu_4)I_2(dptz)_3]$ 3DMF (denoted as **1**, dptz = 3,6-di-(pyridin-4-yl)-1,2,4,5-tetrazine) for colorimetric detection of organic solvents (Figure 5.12) [3]. This MOF is based on Cu(I) metal ions with filled-shell 3d sub-layer and electronically active tetrazine function. Cu(I) centers selected to observe the guest-induced electronic transitions in tetrazine ring. Exposure of this MOF to different organic solvents shows that due to differences in polarity of the solvent, solvent@**1** samples represent different color change in a way that solvents with higher polarity signify larger blue-shift in absorption band of the material. Although the solvatochromism of **1** and dptz ligand originate from different transitions (MLCT transition for **1** and $\pi \rightarrow \pi^*$ transition for dptz), the result indicates that the dptz ligand plays an important role in the solvatochromic response of **1**, which should be ascribed to its strong π-acceptor property and labile electronic structure to solvent polarity. It is necessary to mention that protic and aprotic solvents follow different trends in the band gap of the solvent@**1**.

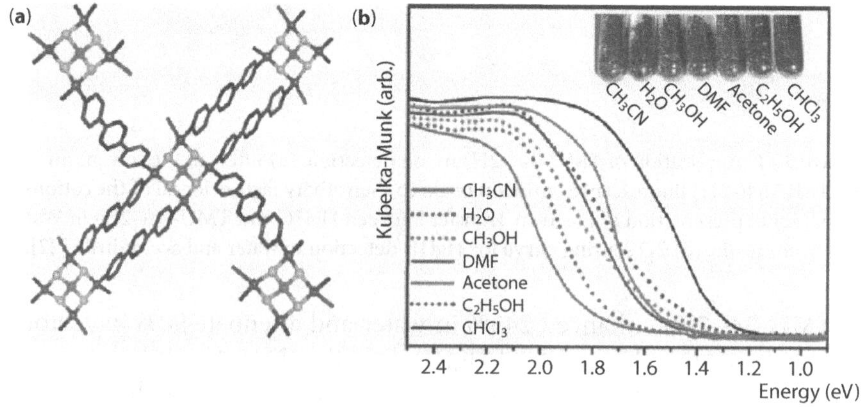

Figure 5.12 Application of $[(WS_4Cu_4)I_2(dptz)_3]$ 3DMF in solvatochromism. (a) The coordination environment of the$WS_4Cu_4^{2+}$ unit. (b) The UV–vis spectra and photograph of the inclusion Compounds solvents@$[(WS_4Cu_4)I_2(dptz)_3]$ [3].

In summary about heterocyclic azine N-based functions: (I) in this category of functions, the nitrogen atoms are located in the aromatic skeleton of MOFs but not in the pore space, which results is simultaneous retained porosity and functionalization, (II) Regarding to higher number of N atoms, the σ-donating and π-accepting character of heterocyclic azine ring is decreased and increased respectively and owing to this point pyridine and diazines usually applied and coordination sites but tetrazine and triazines applied as guest interactive sites respectively, (III) because of Lewis basicity of heterocyclic azine ring they interact with metal ions, hydrogen bond donor guests and polar/quadrupole guests, (IV) N-rich functions of this category (triazine, tetrazine) can participate in π-π interactions as π-deficient sites with π-rich guests.

5.3 Function–Structure Properties

Because of their chemical properties, s-triazine and tetrazine functions can be applied in designing functional MOFs as well as supramolecular polymers. With the same idea they repeatedly applied as guest-interactive site into the structure of MOFs. But, unlike the pyridine and diazines, because of poor coordination ability of bare 1,3,5-triazine and tetrazine rings, there is not any synthesized MOF based on 1,3,5-triazine and tetrazine as coordinating site.

Owing to their relatively strong Lewis basicity, diazines and especially pyridine applied as coordinating site in the development of novel MOFs. Especially as coordinating site, pyridine function applied in the structure of pillared MOFs. The advantages of synthesis of pillared MOFs are more convenient pre-synthesis functionalization of these ligands as well as their ability in development of metal-carboxylate 2D sheet into 3D frameworks [66]. However, this kind of MOFs suffers from low chemical stability especially in presence of Lewis bases and metal-attacker guests. Diazines especially pyrazine, applied in the synthesis of MOFs as binding sites (in bare or carboxylated forms) for construction of the frameworks.

N-heterocyclic azine functions with lower number of N atoms, applied in some way in development of MOFs. Bare pyridine is not applied for construction of MOFs, but it is used as modulator reagent to control the size and morphology of MOF particles. In case of bare diazines, pyrazine is mostly applied in the fabrication of MOFs with desirable pore size. In molecular structure of pyrazine, two nitrogens of pyrazine occupy para positions which are completely suitable for construction of 3D MOFs with N-donor pillars. In contrast to pyrazine, bare pyrimidine and especially pyradazine rings are rarely applied in the structure of MOFs because their N atoms are located in places which are not ideal for construction of porous frameworks.

In case of synthesis of MOFs based on N-heterocyclic functions, we classified these ligands in three groups including (I) heteromultitopic carboxylate-heterocyclic azine ligands which carboxy function(s) are directly connected to pyridine and diazine rings, (II) heteromultitopic carboxylate-heterocyclic azine ligands which heterocyclic-azine N atoms and carboxylate functions are located on different benzene rings and (III) heterocyclic azine ligands without coordinating carboxy groups.

In group one, it is almost impossible to construct frameworks with free N-sites by pre-synthesis design of the structure. This group of ligands usually include ligands such as pyridine-4-carboxylate, pyridine-3,5-dicarboxylate, pyridine-2,5-dicarboxylate, pyridine-2,4,6-tricarboxylate, pyrazine-2, 5-dicarboxylate, pyrazine-2,3-dicarboxylate, pyrazine-2-carboxylate, pyrazine-2,3,5,6-tetracarboxylate, pyrimidine-4, 6-dicarboxylate, pyrimidine-3-carboxylate. For example, Xiang-Jun Zheng and coworkers synthesized new three-dimensional lanthanide-manganese MOF $[Gd_2Mn_3(PZTC)_4(H_2O)_{12}] \cdot 5H_2O$ based on PZTC = 2,3,5-Pyrazinetricarboxylate ligand (Figure 5.13) [67]. The $PZTC^{3-}$ ligands adopt two coordination modes, one of which connects with two Gd(III) ions and three Mn(1) ions in pentadentate mode, while the other connects with two Gd(III) ions and one Mn(2) ion in heptadentate mode. One Gd(III) ion in an asymmetric unit is linked with Mn(1) ions in another unit by the $PZTC^{3-}$ ligand through mode (b), resulting in a 2D layer. Meanwhile Gd(III) ion is linked with Mn(2) ion by the $PZTC^{3-}$ ligand through mode (a) to produce 1D chain along the a-axis. Although in this work just one heteromultitopic ligand is applied, for some pyzdc ligands the heterocyclic-azine N atoms are coordinated to metal ions while in others they do not coordinate. A similar situation is observed in other coordination polymers like $[Co_2(pmc)_4(OH)_2]$, $[Cd(pmc)_2]$ and $[Cu(pmc)_2]$ where pmc is pyrimidine-5-carboxylate, $[Gd_2Ag_6(pzdc)_6(H_2O)_9]\cdot 8H_2O$ where pzdc is pyrazine-2,3-dicarboxylate. Therefore, it is hard to determine freedom or engagement of heterocyclic-azine N atoms before self-assembly process and structural analysis for this group of ligands.

In case of heteromultitopic carboxylate–heterocyclic azine ligands, which heterocyclic-azine N atoms and carboxylate functions are located on different benzene rings, rational selection of the ligand and metal ion is essential to synthesis MOFs with desirable coordinating or guest interactive site.

In this regard there are three major ligand structural factors that affect on the freedom or coordination of N atoms such as: (I) position of

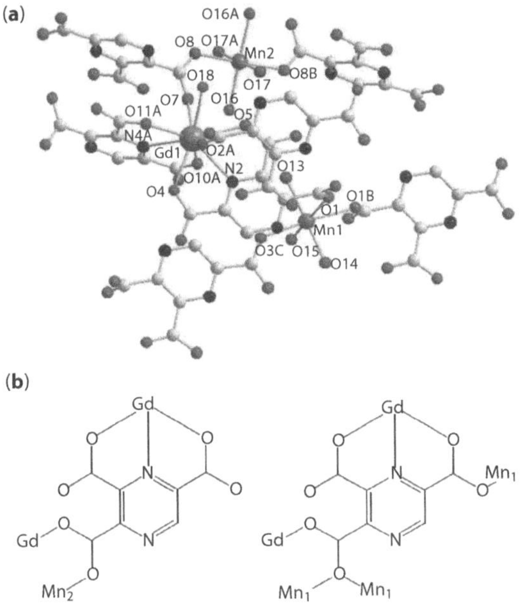

Figure 5.13 Illustration of crystal structure of $[Gd_2Mn_3(PZTC)_4(H_2O)_{12}]{\cdot}5H_2O$ [67]. (a) The coordination environments of Mn(II) and Gd(III) ions. (b) The coordination mode of the $PZTC^{3-}$ ligand.

coordinating site on terminal or central ring, (II) strength and type of coordinating sites (carboxy or N coordinating sites) and (III) ligand geometry (linear, bent, or multi-chain). In self-assembly process for the construction of a highly crystalline and regular MOF structure, the geometry of ligand plays critical roles through ligand based template effects. This template effect of ligand can be assisted by strength of coordinating site and their position on terminal ring. The strength of a coordinating site is defined based on its coordination strength. For example, carboxy coordinating sites are strong coordination agent while N-donor ones are relatively poor. Also, for N-donor functions, the number of N-atoms in the ring of coordinating function determines their strength.

Figure 5.14 illustrates the role of terminal phenyl ring in the synthesis of MOF based on heteromultitopic carboxylate heterocyclic azine linkers. Although coordination strength of N-coordinating site is lower that carboxylate group, the replacement of carboxy function with N-coordinating site reveals that it is similar to carboxyl function and the N-coordinating groups are connected to the metal ion. This observation is because of the fact that N-donor pyridine groups with relatively poor coordination

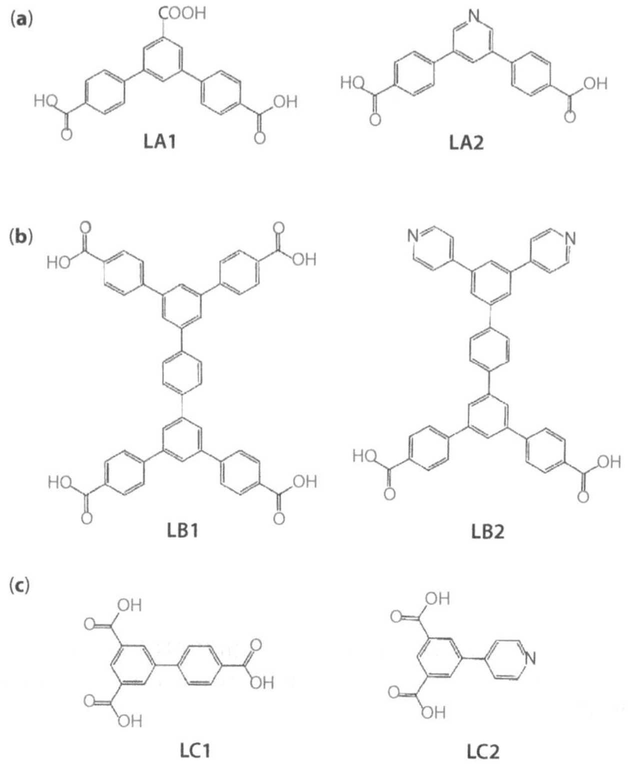

Figure 5.14 Illustration of some groups of ligands with similar coordinating connection of carboxy and pyridine sites. (a) LA bent ligands. (b) LB multi-chain ligands. (c) LC straight ligands with two terminal phenyl ring.

strength are located in terminal ring with para position with very high ligand based template effect.

Figure 5.15 is about the number and position of N-donor coordinating sites in the terminal ring of the ligand with high template effect. In LC1–LC4 ligands, owing to their bent structure and accessibility of central phenyl ring, this ring is of high template effect. For LC2 ligand the N atom is located at metha position of the central ring. Owing to the high template effect of central phenyl ring in these bent ligands especially in metha position, N atom of LC2 ligand could be coordinated to metal ions. In case of LC4 which N atom is located at internal ortho position with low template effect, this pyridinic atom could not coordinate to metal ions and remain free as guest interactive site. Overall, it worth to mention that locating the pyridinic N atom with relatively poor coordination strength at positions with low ligand based template effect makes them remain free.

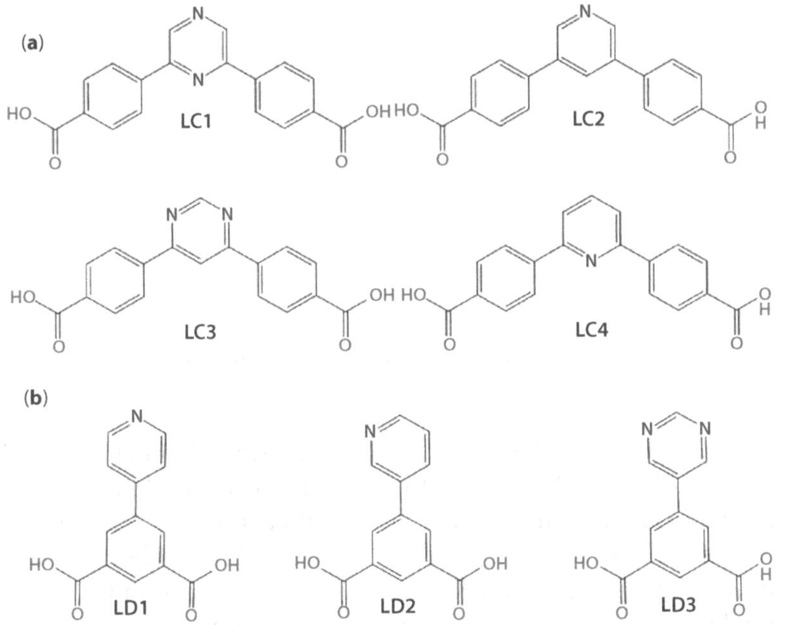

Figure 5.15 Illustration of two groups of ligands for discussion about the roles of number and strength of N-coordinating sites in the terminal ring.

This statement is true about ligand LC1. In LC1 ligand, one nitrogen is similar to LC2 and one is similar to LC4. The nitrogen atom similar to LC2 ligand could be coordinated to metal ions while the nitrogen atom similar to LC4 ligand remains free. For LC3, both N atoms are located at positions with weak ligand based template effect. So, owing to location of N atoms with relatively low coordination strength on position with weak ligand based template effect, both of them remain free. Also, it is necessary to notice that the Lewis basicity and so coordination strength of pyrimidinic N atoms are lower than pyridinic N atoms.

Figure 5.16 is about combination of some ligands with straight chain with carboxy or N-donor coordinating sites on the central and terminal ring. In combination of straight LE ligands and bent LC ligand we have to mention that owing to bent nature of LC ligands and their similarity to tritopic ligand, central ring is of high ligand template effect while for LE ligand with straight chain, the central phenyl ring has low ligand template effect. In the case of LE4 and LE5 ligands, coordinating group with strong coordinating affinity to metal ions is located on terminal phenyl ring with strong ligand template effect. So, they are coordinated to metal ions despite their different location. This is observable too for LE1, LE2 and LE3 ligands

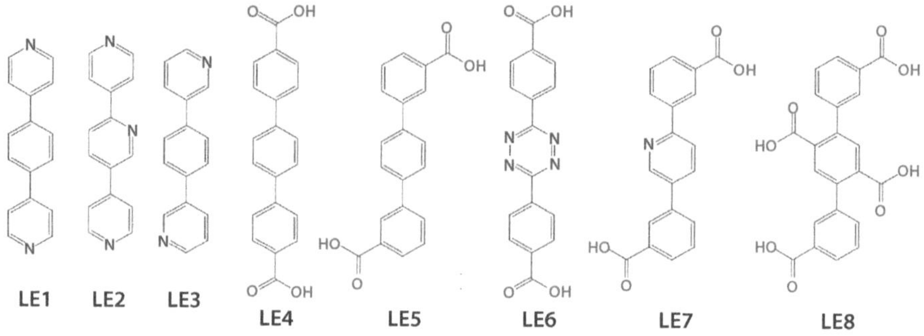

Figure 5.16 Combination of some ligands with straight chain with carboxy or N-donor coordinating sites on the central and terminal ring.

in which coordinating sites are located on the terminal ring. In case of LE2, LE6 and LE7 ligands, N atoms with relatively weak coordination strength (in comparison with carboxy) are located at central ring with low ligand template effect. So, they remain free. In case of LE8, carboxy groups are classified in two groups; those located on central ring and those located on terminal rings. As mentioned, carboxy groups at terminal phenyl rings are able to coordinate to metal ions while in case of carboxy groups at central phenyl ring, the synthesis conditions like solvent, temperature, the ratio of the ligand to metal ion and synthesis time are dominating parameters because carboxy groups with strong coordination strength are located at central phenyl ring with weak ligand template effect.

Another strategy to synthesis MOFs with desirable free N-Lewis basic sites is selection of highly positive charged metal ions (hard metal ions) like Zr(IV) and Al(III) for synthesis of MOFs based on heteromultitopic carboxylate-heterocyclic azine ligands. This strategy is specially effective when using ligands with chelating N-based Lewis basic sites like 2,2'-bipyridine-5,5'-dicarboxylic acid as an anchor site to stabilize catalytically or photo catalytically active metal centers into the pores of MOFs through post-synthesis and even pre-synthesis methods. Application of the mentioned ligand with Zn^{2+}, Co^{2+}, and Ni^{2+} leads to chelation of metal ions to N atoms. But combination of 2,2-bipyridine-5,5'-dicarboxylate and Zr^{4+} ions results in the construction of a framework with UiO-67 framework containing free N atoms.

References

1. Brea, O., Luna, A., Díaz, C., Corral, I., Molecular Modelling of the H2-Adsorptive Properties of Tetrazolate-Based Metal–Organic Frameworks:

From the Cluster Approach to Periodic Simulations. *ChemPhysChem*, 19, 1349–1357, 2018.

2. Razavi, S.A.A., Masoomi, M.Y., Morsali, A., Stimuli-Responsive MOF with Chemo-Switchable Properties for Colorimetric Detection of CHCl3. *Chem.—Eur. J.*, 23, 12559–12564, 2017.
3. Lu, Z.-Z., Zhang, R., Li, Y.-Z., Guo, Z.-J., Zheng, H.-G., Solvatochromic behavior of a nanotubular metal–organic framework for sensing small molecules. *J. Am. Chem. Soc.*, 133, 4172–4174, 2011.
4. Bezuidenhout, C.X., Esterhuysen, C., Barbour, L.J., Solvatochromism as a probe to observe the solvent exchange process in a 1-D porous coordination polymer with 1-D solvent accessible channels. *Chem. Commun.*, 53, 5618–5621, 2017.
5. Müller, P., Wisser, F.M., Freund, P., Bon, V., Senkovska, I., Kaskel, S., Optical Sensors Using Solvatochromic Metal–Organic Frameworks. *Inorg. Chem.*, 56, 14164–14169, 2017.
6. Han, L., Zhou, J., Li, X., Sun, C.-Y., Zhao, L., Zhang, Y.-T., Zhu, M., Wang, X.-L., Su, Z.-M., Recognition of harmful fused aromatic hydrocarbons *via* a metal–organic framework with hydrophobic pores. *Inorg. Chem. Commun.*, 86, 200–203, 2017.
7. Razavi, S.A.A., Masoomi, M.Y., Morsali, A., Ultrasonic assisted synthesis of a tetrazine functionalized MOF and its application in colorimetric detection of phenylhydrazine. *Ultrason. Sonochem.*, 37, 502–508, 2017.
8. Wang, Y.-Q., Tan, Q.-H., Liu, H.-T., Sun, W., Liu, Z.-L., A luminescent europium MOF containing Lewis basic pyridyl site for highly selective sensing of o-, m- and p-nitrophenol. *RSC Adv.*, 5, 86614–86619, 2015.
9. Wen, G.-X., Han, M.-L., Wu, X.-Q., Wu, Y.-P., Dong, W.-W., Zhao, J., Li, D.-S., Ma, L.-F., A multi-responsive luminescent sensor based on a super-stable sandwich-type terbium(iii)–organic framework. *Dalton Trans.*, 45, 15492–15499, 2016.
10. Zhu, M., Song, X.-Z., Song, S.-Y., Zhao, S.-N., Meng, X., Wu, L.-L., Wang, C., Zhang, H.-J., A Temperature-Responsive Smart Europium Metal-Organic Framework Switch for Reversible Capture and Release of Intrinsic Eu3+ Ions. *Adv. Sci.*, 2, 1500012, 2015.
11. He, H., Chen, S.-H., Zhang, D.-Y., Yang, E.-C., Zhao, X.-J., A luminescent metal–organic framework as an ideal chemosensor for nitroaromatic compounds. *RSC Adv.*, 7, 38871–38876, 2017.
12. Razavi, S.A.A., Masoomi, M.Y., Morsali, A.J.I.C., Host–Guest Interaction Optimization through Cavity Functionalization for Ultra-Fast and Efficient Water Purification by a Metal–Organic Framework. *Inorg. Chem.*, 57, 11578–11587, 2018.
13. Razavi, S.A.A. and Morsali, A., High Capacity Oil Denitrogenation over Azine- and Tetrazine-Decorated Metal–Organic Frameworks: Critical Roles of Hydrogen Bonding. *ACS Appl. Mater. Interfaces*, 11, 21711–21719, 2019.

14. Chen, B., Wang, L., Xiao, Y., Fronczek, F.R., Xue, M., Cui, Y., Qian, G., A luminescent metal–organic framework with Lewis basic pyridyl sites for the sensing of metal ions. *Angew. Chem. Int. Ed.*, 48, 500–503, 2009.
15. Liu, B., Hou, L., Wu, W.-P., Dou, A.-N., Wang, Y.-Y., Highly selective luminescence sensing for Cu2+ ions and selective CO2 capture in a doubly interpenetrated MOF with Lewis basic pyridyl sites. *Dalton Trans.*, 44, 4423–4427, 2015.
16. Gao, Y., Zhang, X., Sun, W., Liu, Z., A robust microporous metal–organic framework as a highly selective and sensitive, instantaneous and colorimetric sensor for Eu3+ ions. *Dalton Trans.*, 44, 1845–1849, 2015.
17. Yang, Y., Chen, L., Jiang, F., Wan, X., Yu, M., Cao, Z., Jing, T., Hong, M., Fabricating a super stable luminescent chemosensor with multi-stimuli-response to metal ions and small organic molecules through turn-on and turn-off effects. *J. Mater. Chem. C*, 5, 4511–4519, 2017.
18. Rouhani, F. and Morsali, A., Fast and Selective Heavy Metal Removal by a Novel Metal–Organic Framework Designed with *In-Situ* Ligand Building Block Fabrication Bearing Free Nitrogen. *Chem. Eur. J.*, 24, 5529–5537, 2018.
19. Tang, Q., Liu, S., Liu, Y., Miao, J., Li, S., Zhang, L., Shi, Z., Zheng, Z., Cation sensing by a luminescent metal–organic framework with multiple lewis basic sites. *Inorg. Chem.*, 52, 2799–2801, 2013.
20. Liu, B.-H., Liu, D.-X., Yang, K.-Q., Dong, S.-J., Li, W., Wang, Y.-J., A new cluster-based metal-organic framework with triazine backbones for selective luminescent detection of mercuryII. ion. *Inorg. Chem. Commun.*, 90, 61–64, 2018.
21. Razavi, S.A.A., Masoomi, M.Y., Morsali, A., Morphology-dependent sensing performance of dihydro-tetrazine functionalized MOF toward Al (III). *Ultrason. Sonochem.*, 41, 17–26, 2018.
22. Razavi, S.A.A., Masoomi, M.Y., Morsali, A., Double Solvent Sensing Method for Improving Sensitivity and Accuracy of Hg II. Detection Based on Different Signal Transduction of a Tetrazine-Functionalized Pillared Metal–Organic Framework. *Inorg. Chem.*, 56, 9646–9652, 2017.
23. Zhang, R., Liu, Y., An, Y., Wang, Z., Wang, P., Zheng, Z., Qin, X., Zhang, X., Dai, Y., Huang, B., A water-stable triazine-based metal-organic framework as an efficient adsorbent of Pb II. ions. *Colloids Surf. A: Physicochem. Eng. Asp.*, 560, 315–322, 2019.
24. Deria, P., Li, S., Zhang, H., Snurr, R.Q., Hupp, J.T., Farha, O.K., A MOF platform for incorporation of complementary organic motifs for CO 2 binding. *Chem. Commun.*, 51, 12478–12481, 2015.
25. Chen, D.-M., Xu, N., Qiu, X.-H., Cheng, P., Functionalization of Metal–Organic Framework *via* Mixed-Ligand Strategy for Selective CO2 Sorption at Ambient Conditions. *Cryst. Growth Des.*, 15, 961–965, 2015.
26. Dang, Q.-Q., Zhan, Y.-F., Duan, L.-N., Zhang, X.-M., A pyridyl-decorated MOF-505 analogue exhibiting hierarchical porosity, selective CO2 capture and catalytic capacity. *Dalton Trans.*, 44, 20027–20031, 2015.

27. Song, C., Hu, J., Ling, Y., Feng, Y., Krishna, R., Chen, D.-l., He, Y., The accessibility of nitrogen sites makes a difference in selective CO2 adsorption of a family of isostructural metal–organic frameworks. *J. Mater. Chem. A*, 3, 19417–19426, 2015.
28. Qian, J., Li, Q., Liang, L., Li, T.-T., Hu, Y., Huang, S., A microporous MOF with open metal sites and Lewis basic sites for selective CO2 capture. *Dalton Trans.*, 46, 14102–14106, 2017.
29. Zhu, Y., Wang, Y.-M., Zhao, S.-Y., Liu, P., Wei, C., Wu, Y.-L., Xia, C.-K., Xie, J.-M., Three N–H Functionalized Metal–Organic Frameworks with Selective CO2 Uptake, Dye Capture, and Catalysis. *Inorg. Chem.*, 53, 7692–7699, 2014.
30. Kim, H.-C., Huh, S., Lee, D.N., Kim, Y., Selective carbon dioxide sorption by a new breathing three-dimensional Zn-MOF with Lewis basic nitrogen-rich channels. *Dalton Trans.*, 47, 4820–4826, 2018.
31. Du, L., Lu, Z., Zheng, K., Wang, J., Zheng, X., Pan, Y., You, X., Bai, J., Fine-tuning pore size by shifting coordination sites of ligands and surface polarization of metal–organic frameworks to sharply enhance the selectivity for CO2. *J. Am. Chem. Soc.*, 135, 562–565, 2012.
32. Li, T., Chen, D.-L., Sullivan, J.E., Kozlowski, M.T., Johnson, J.K., Rosi, N.L., Systematic modulation and enhancement of CO2: N2 selectivity and water stability in an isoreticular series of bio-MOF-11 analogues. *Chem. Sci.*, 4, 1746–1755, 2013.
33. Jiao, J., Dou, L., Liu, H., Chen, F., Bai, D., Feng, Y., Xiong, S., Chen, D.-L., He, Y., An aminopyrimidine-functionalized cage-based metal–organic framework exhibiting highly selective adsorption of C2H2 and CO2 over CH4. *Dalton Trans.*, 45, 13373–13382, 2016.
34. An, J., Geib, S.J., Rosi, N.L., High and Selective CO2 Uptake in a Cobalt Adeninate Metal–Organic Framework Exhibiting Pyrimidine- and Amino-Decorated Pores. *J. Am. Chem. Soc.*, 132, 38–39, 2010.
35. Luebke, R., Eubank, J.F., Cairns, A.J., Belmabkhout, Y., Wojtas, L., Eddaoudi, M., The unique rht-MOF platform, ideal for pinpointing the functionalization and CO2 adsorption relationship. *Chem. Commun.*, 48, 1455–1457, 2012.
36. Pal, S., Bhunia, A., Jana, P.P., Dey, S., Möllmer, J., Janiak, C., Nayek, H.P., Microporous La–Metal–Organic Framework MOF. with Large Surface Area. *Chem. Eur. J.*, 21, 2789–2792, 2014.
37. Wang, X., Chen, M., Du, M., A Clear Insight into the Distinguishing CO2 Capture by Two Isostructural DyIII–Carboxylate Coordination Frameworks. *Inorg. Chem.*, 55, 6352–6354, 2016.
38. Chen, Y., Wang, H., Li, J., Lockard, J.V., In situ spectroscopy studies of CO2 adsorption in a dually functionalized microporous metal–organic framework. *J. Mater. Chem. A*, 3, 4945–4953, 2015.
39. Kim, J., Yang, S.-T., Choi, S.B., Sim, J., Kim, J., Ahn, W.-S., Control of catenation in CuTATB-n metal–organic frameworks by sonochemical synthesis and its effect on CO2 adsorption. *J. Mater. Chem.*, 21, 3070–3076, 2011.

40. Hijikata, Y., Horike, S., Sugimoto, M., Inukai, M., Fukushima, T., Kitagawa, S., Pore design of two-dimensional coordination polymers toward selective adsorption. *Inorg. Chem.*, 52, 3634–3642, 2013.
41. Rao, X., Cai, J., Yu, J., He, Y., Wu, C., Zhou, W., Yildirim, T., Chen, B., Qian, G., A microporous metal–organic framework with both open metal and Lewis basic pyridyl sites for high C 2 H 2 and CH 4 storage at room temperature. *Chem. Commun.*, 49, 6719–6721, 2013.
42. Li, L., Tang, S., Wang, C., Lv, X., Jiang, M., Wu, H., Zhao, X., High gas storage capacities and stepwise adsorption in a UiO type metal–organic framework incorporating Lewis basic bipyridyl sites. *Chem. Commun.*, 50, 2304–2307, 2014.
43. Li, B., Wen, H.-M., Wang, H., Wu, H., Tyagi, M., Yildirim, T., Zhou, W., Chen, B., A Porous Metal–Organic Framework with Dynamic Pyrimidine Groups Exhibiting Record High Methane Storage Working Capacity. *J. Am. Chem. Soc.*, 136, 6207–6210, 2014.
44. Wen, H.M., Li, B., Li, L., Lin, R.B., Zhou, W., Qian, G., Chen, B., A Metal–Organic Framework with Optimized Porosity and Functional Sites for High Gravimetric and Volumetric Methane Storage Working Capacities. *Adv. Mater.*, 30, 1704792, 2018.
45. Li, B., Wen, H.-M., Wang, H., Wu, H., Yildirim, T., Zhou, W., Chen, B., Porous metal–organic frameworks with Lewis basic nitrogen sites for high-capacity methane storage. *Energy Environ. Sci.*, 8, 2504–2511, 2015.
46. Razavi, S.A.A., Masoomi, M.Y., Islamoglu, T., Morsali, A., Xu, Y., Hupp, J.T., Farha, O.K., Wang, J., Junk, P.C., Improvement of Methane–Framework Interaction by Controlling Pore Size and Functionality of Pillared MOFs. *Inorg. Chem.*, 56, 2581–2588, 2017.
47. Sen, S., Neogi, S., Aijaz, A., Xu, Q., Bharadwaj, P.K., Structural variation in Zn ii. coordination polymers built with a semi-rigid tetracarboxylate and different pyridine linkers: Synthesis and selective CO 2 adsorption studies. *Dalton Trans.*, 43, 6100–6107, 2014.
48. Xu, H., He, Y., Zhang, Z., Xiang, S., Cai, J., Cui, Y., Yang, Y., Qian, G., Chen, B., A microporous metal–organic framework with both open metal and Lewis basic pyridyl sites for highly selective C2H2/CH4 and C2H2/CO2 gas separation at room temperature. *J. Mater. Chem. A*, 1, 77–81, 2013.
49. Song, C., Jiao, J., Lin, Q., Liu, H., He, Y., C2H2 adsorption in three isostructural metal–organic frameworks: Boosting C2H2 uptake by rational arrangement of nitrogen sites. *Dalton Trans.*, 45, 4563–4569, 2016.
50. Wen, H.-M., Wang, H., Li, B., Cui, Y., Wang, H., Qian, G., Chen, B., A Microporous Metal–Organic Framework with Lewis Basic Nitrogen Sites for High C2H2 Storage and Significantly Enhanced C2H2/CO2 Separation at Ambient Conditions. *Inorg. Chem.*, 55, 7214–7218, 2016.
51. Liu, K., Li, X., Ma, D., Han, Y., Li, B., Shi, Z., Li, Z., Wang, L., A microporous yttrium metal–organic framework of an unusual nia topology for high

adsorption selectivity of C2H2 and CO2 over CH4 at room temperature. *Mater. Chem. Front.*, 1, 1982–1988, 2017.

52. Liu, K., Ma, D., Li, B., Li, Y., Yao, K., Zhang, Z., Han, Y., Shi, Z., High storage capacity and separation selectivity for C2 hydrocarbons over methane in the metal–organic framework Cu–TDPAT. *J. Mater. Chem. A*, 2, 15823–15828, 2014.
53. Huang, Y., Lin, Z., Fu, H., Wang, F., Shen, M., Wang, X., Cao, R., Porous Anionic Indium–Organic Framework with Enhanced Gas and Vapor Adsorption and Separation Ability. *ChemSusChem*, 7, 2647–2653, 2014.
54. Manna, B., Mukherjee, S., Desai, A.V., Sharma, S., Krishna, R., Ghosh, S.K., A π-electron deficient diaminotriazine functionalized MOF for selective sorption of benzene over cyclohexane. *Chem. Commun.*, 51, 15386–15389, 2015.
55. Ye, J. and Johnson, J.K., Catalytic hydrogenation of CO2 to methanol in a Lewis pair functionalized MOF. *Catal. Sci. Technol.*, 6, 8392–8405, 2016.
56. Liu, H., Yin, B., Gao, Z., Li, Y., Jiang, H., Transition-metal-free highly chemo- and regioselective arylation of unactivated arenes with aryl halides over recyclable heterogeneous catalysts. *Chem. Commun.*, 48, 2033–2035, 2012.
57. Zhou, H.-F., Liu, B., Hou, L., Zhang, W.-Y., Wang, Y.-Y., Rational construction of a stable Zn4O-based MOF for highly efficient CO2 capture and conversion. *Chem. Commun.*, 54, 456–459, 2018.
58. Seo, J.S., Whang, D., Lee, H., Jun, S.I., Oh, J., Jeon, Y.J., Kim, K., A homochiral metal–organic porous material for enantioselective separation and catalysis. *Nature*, 404, 982, 2000.
59. Tan, Y., Fu, Z., Zhang, J., A layered amino-functionalized zinc-terephthalate metal organic framework: Structure, characterization and catalytic performance for Knoevenagel condensation. *Inorg. Chem. Commun.*, 14, 1966–1970, 2011.
60. Guo, F., Yuan, B., Shi, W., A novel 2D metal-organic framework with Lewis basic sites as a heterogeneous base catalysis. *Inorg. Chem. Commun.*, 86, 285–289, 2017.
61. Miao, Z., Luan, Y., Qi, C., Ramella, D., The synthesis of a bifunctional copper metal organic framework and its application in the aerobic oxidation/Knoevenagel condensation sequential reaction. *Dalton Trans.*, 45, 13917–13924, 2016.
62. Razavi, S.A.A. and Morsali, A., Ultrasonic-Assisted Linker Exchange (USALE): A Novel Post-Synthesis Method for Controlling the Functionality, Porosity, and Morphology of MOFs. *Chem. Eur. J.*, 25, 10876–10885, 2019.
63. Razavi, S.A.A. and Morsali, A., Function–Structure Relationship in Metal–Organic Frameworks for Mild, Green, and Fast Catalytic C–C Bond Formation. *Inorg. Chem.*, 58, 14429–14439, 2019.
64. Han, L., Qin, L., Xu, L.-P., Zhao, W.-N., Doubly interpenetrated chiral (10, 3)-a network with charge-transfer-type guest inclusion. *Inorg. Chem.*, 52, 1667–1669, 2013.

65. Nickerl, G., Senkovska, I., Kaskel, S., Tetrazine functionalized zirconium MOF as an optical sensor for oxidizing gases. *Chem. Commun.*, 51, 2280–2282, 2015.
66. ZareKarizi, F., Joharian, M., Morsali, A., Pillar-layered MOFs: functionality, interpenetration, flexibility and applications. *J. Mater. Chem. A*, 6, 19288–19329, 2018.
67. Ablet, A., Wang, D., Cao, W., Li, L.-C., Zheng, X.-J., Synthesis, structure and magnetic property of a Gd–Mn coordination polymer based on 2, 3, 5-pyrazinetricarboxylate. *Inorg. Chem. Commun.*, 52, 20–22, 2015.

6

Heterocyclic Azole Decorated Metal-Organic Frameworks

Abstract
In this chapter we discussed about host-guest chemistry and coordination chemistry of heterocyclic azole functions, including pyrazole, imidazole, triazole and tetrazole in the the structure of MOFs.

Keywords: Heterocyclic azoles, imidazole, pyrazole, tetrazole, triazole, zeolite imidazole frameworks, energetic materials

6.1 General Chemical Properties of Heterocyclic Azole Functions

Chemical properties of five-member ring heterocyclic azole functions depended on the number of N atoms in the ring. This group of functions consists of pyrazole, imidazole, triazole and tetrazole functional groups. Despite the heterocyclic azine functions containing of just pyridinic N atoms, heterocyclic azoles are composed from two types of N atoms including pyridinic (–N–) site(s) and pyrrolic (–NH–) site. All heterocyclic azole functions are of one pyrrolic (–NH–) site and the variation in the number of N atoms is related to the different number of pyridinic (–N–) atoms. Owing to presence of (–NH–) and (–N–) sites in their ring, heterocyclic azoles are Lewis basic. So, they applied as guest-interactive sites and especially coordinating site. Also, their difference in pK_a value, gives them different coordination strength. In the continuation of this chapter, these contents will be explained in full detail.

Ali Morsali and Sayed Ali Akbar Razavi. Functional Metal-Organic Frameworks: Structure, Properties and Applications, (107–132) © 2021 Scrivener Publishing LLC

6.2 Function–Application Properties

Owing to presence of pyridinic and pyrrolic N sites in the structure of heterocyclic azoles, all of them are Lewis basic. So, similar to other N-donor functions they applied in some applications like CO_2 adsorption [1–29], catalytically Lewis basic assisted reactions [30], detection and separation of hazardous chemicals like metal ions [31–35] small organic molecules enabling to interact through H-bond [36–39]. Presence of both (–NH–) and (–N–) sites enable heterocyclic azole to interact as hydrogen-bond donor or acceptor sites while for heterocyclic azine functions they only act as hydrogen acceptor site.

Pyrazole functionalized MOFs are used for improvement in CO_2 adsorption and separation through different types of interactions like pyrazole(NH)·(O)CO_2 hydrogen bond, pyrazole(π-ring)·(C)CO_2 and pyrazole(C)·(O)CO_2 interactions. Shengqian Ma and coworkers reported two rht-type MOFs based on tetrazolate and pyrazolate motifs [1]. Their investigation in CO_2 adsorption and CO_2/CH_4 selectivity shows that the rht-MOF featuring the pyrazolate moiety demonstrates superior performances compared to the rht-MOF on the basis of the tetrazolate moiety. Computational studies reveal that the local electric field favors more than the richness of exposed nitrogen atoms for the interactions with CO_2 molecules.

Moreover than these common chemical properties, each function of heterocyclic azoles has its unique chemical properties which give special host-guest chemistry and chemical behavior to the heterocyclic azole.

Jeffrey R. Long and coworker applied a series of tritopic ligands with pyrazole, triazole and tetrazole coordinating site for selective O_2 capture. With this aim they applied a series of MOFs isostructure with Cr(III) and Co(II) open metal sites. They find that rational selection of heterocyclic azole binding site has a remarkable effect on ligands field of metal centers in a way that can lead to large increases in the O_2 heat of adsorption and O_2/N_2 selectivity as well as stability of the framework.

In the first achievement, they applied $Cr_3(BTC)_2$ (where BTC^{3-} is 1,3,5-benzenetri-carboxylate) in selective O_2 binding with the aim that exposed Cr(II) centers might engage in charge transfer interactions with O_2 (but not N_2) and find that $Cr_3(BTC)_2$ has high O_2 loading capacity and strong selectivity for binding O_2 over N_2 at 298 K (Figure 6.1) [40]. In this condition, the O_2 isotherm rises sharply, reaching 11 wt% at just 2 mbar while the N_2 adsorption isotherm climbs gradually to a capacity of 0.58 wt% at 1 bar. $Cr_3(BTC)_2$ reaches O_2/N_2 selectivity factor of 22 based on interpolated

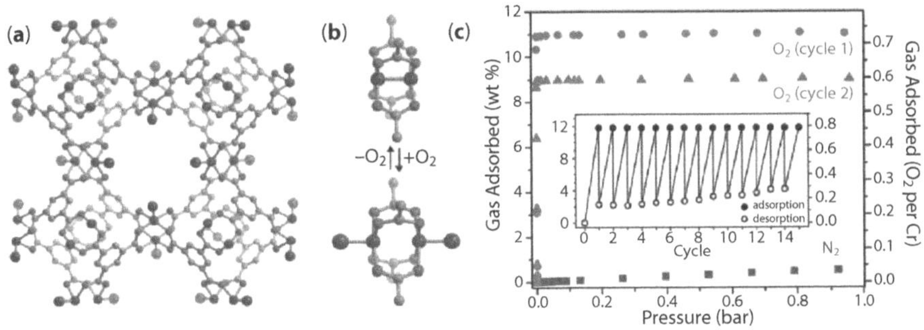

Figure 6.1 Application of $Cr_3(BTC)_2$ in selective O_2 capture and release. (a) Crystal structure of the material. (b) Possible O_2 separation mechanism. (c) Adsorption isotherms and recyclability [40].

uptakes of 0.73 $mmol{\cdot}g^{-1}$ O_2 at 0.21 bar and 0.033 $mmol{\cdot}g^{-1}$ N_2 at 0.78 bar which is correspond to partial pressures in air. $Cr_3(BTC)_2$ readsorbs 9.1 wt % O_2 at 0.21 bar under 50 °C and 48 hour dynamic vacuum in next cycle. This gradual reduction is due to the incomplete release of bound O_2 under the regeneration conditions and or partial decomposition of the material as a result of the highly exothermic reaction with O_2. Experimental analyses reveal that there is partial charge transfer from the Cr(II) center to the bound O_2 molecule but not necessarily complete charge transfer to give a Cr(III)-superoxide adduct.

In next work by same group, $Cr_3[(Cr_4Cl)_3(BTT)_8]_2$ (Cr-BTT; BTT^{3-}= 1,3,5-benzenetristetrazolate) synthesized and applied for selective O_2 capture and release (Figure 6.2) [41]. This MOF features coordinatively unsaturated, electron donating, redox-active Cr(II) open metal sites. O_2 and N_2 adsorption isotherms were measured at 298 K. N_2 adsorption is minimal and climbs very gradually to just 0.69 wt% at 1 bar while O_2 adsorption isotherm is extraordinarily steep and reaches 7.01 wt% at 0.20 bar before saturating at 7.59 wt% at 1 bar. This adsorption capacity is lower that $Cr_3(BTC)_2$ which indicate that a portion of the framework cations are somehow incapable of strongly binding oxygen. Cr-BTT framework shows drastically larger O_2 adsorption enthalpy of 65 $kJ{\cdot}mol^{-1}$ which is consistent with steeper nature of the isotherm. Selectivity calculations show that for the separation of a 1:4 O_2:N_2 mixture at 298 K and 1 bar, Cr-BTT displays a selectivity in excess of 2,570 corresponds to 99.3% pure O_2 in the adsorbed phase at concentrations relevant to separation of air. After fifteen adsorption–desorption cycles, Cr-BTT reaches 4.6 wt%. Such impressive results are

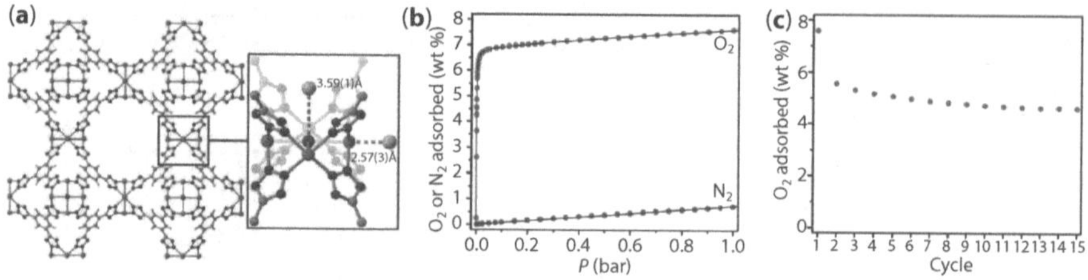

Figure 6.2 Application of Cr-BTT in selective O_2 capture and release. (a) Crystal structure of Cr-BTT. Atom colors: Cr dark green, Cl purple, C gray, N blue, H white. D_2 is shown as a single yellow sphere. Charge-balancing Cr atoms have been omitted for clarity. Right: Expanded portion of the structure of Cr-BTT dosed with D_2. (b) Excess O_2 and N_2 adsorption isotherms collected for Cr-BTT at 298 K. (c) Uptake of O_2 at 200 mbar in Cr-BTT over 15 cycles at 298 K. Adsorption experiments were performed over 30 min and desorption was carried out by placing the sample under a dynamic vacuum at 423 K for 30 min [41].

gained due to the selective electron transfer from Cr(II) to O_2 to form Cr(III) superoxide moieties.

In next work by same group, Co-BTTri (with formula $Co_3[(Co_4Cl)_3(BTTri)_8]_2$·DMF, H_3BTTri = 1,3,5-tri(1H-1,2,3-triazol-5-yl)benzene) and Co-BDTriP (with formula $Co_3[(Co_4Cl)_3(BDTriP)_8]_2$·DMF, H_3BDTriP = 5,5′-(5-(1*H*pyrazol-4-yl)-1,3-phenylene)bis(1*H*-1,2,3-triazole))) applied in selective O_2 separation (Figure 6.3) [42] At 195 K, the O_2 loading rapidly rises to 3.3 mmol·g^{-1} at 0.21 bar and reaches a maximum of 4.8 mmol·g^{-1} at 1 bar, while the N_2 isotherm is much flatter, achieving only 2.0 mmol·g^{-1} at 1 bar. At low loadings, the isosteric heat (Q_{st}) of O_2 adsorption is −34 kJ·mol^{-1} which decreases to −5 kJ·mol^{-1}. The large isosteric heat at low loadings is due to interactions between O_2 and the coordinatively unsaturated cobalt(II) centers, while the lower isosteric heat at higher loadings is a result of weak physisorption to the framework surface. Calculated selectivity for a mixture of 0.21 bar O_2 and 0.79 bar N_2, reach 41 at 195 K and gradually drop to 13 by 243 K corresponding to O_2 purities of 92 and 77%, respectively. Structural analysis and calculations show that the Co–O_2 adduct in Co-BTTri has more Co(II)-dioxygen character than Co(III)–superoxo (O^{2-}) character. Application of Co-BDTriP shows that Co-BDTriP has higher affinity toward O_2 molecules (−47 kJ·mol^{-1}) rather Co-BTTri and selectivity improved more than doubles to 105 (97% purity) at 195 K and drops only to 40 at 243 K (91% purity). Despite the higher affinity between Co-BDTriP and O_2 molecules, no loss in stability is observed. Electronic structure calculations suggest that stronger binding sites in Co-BDTriP form cobalt(III)–superoxo moieties.

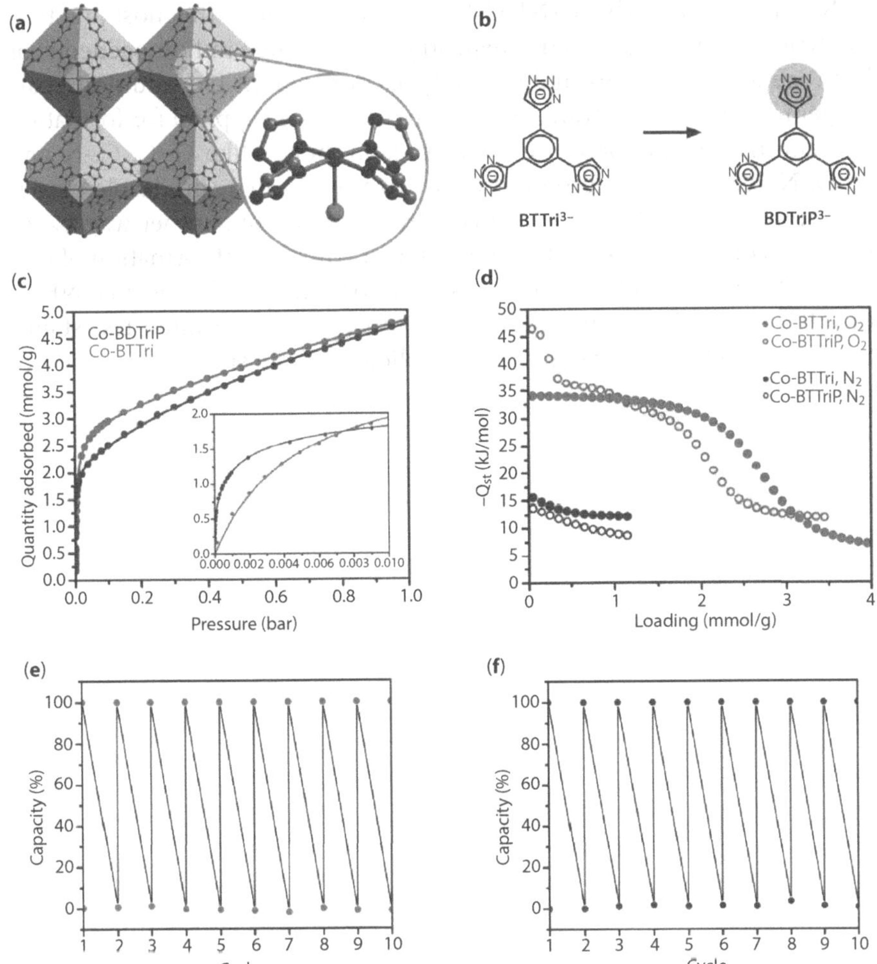

Figure 6.3 Application of Co-BTTri and Co-BDTriP in selective O_2 capture. (a) Structure of Co-BTTri. Expanded view: Local coordination environment of the framework cobalt(II) centers. (b) Structure of the ligands used in this work, $BTTri^{3-}$ and $BDTriP^{3-}$. In $BDTriP^{3-}$, one of the bridging triazolates is replaced with a more electron-donating pyrazolate. (c) Comparison of O_2 adsorption isotherms collected for Co-BTTri (red) and Co-BDTriP (purple) at 195 K. Inset: Low-pressure region of the 195 K O_2 isotherms. (d) Comparison of O_2 and N_2 isosteric heats of adsorption. (e) Cycling experiments for Co-BTTri, with O_2 adsorption occurring at 195 K and 0.21 bar. Desorption was performed by quickly heating the sample to 323 K for 5 min. Note that the before the last cycle, the sample was exposed to humid air for 24 h and then reactivated at 423 K for 6 h, demonstrating the excellent air-stability of Co-BTTri. (f) Cycling experiments for Co-BDTriP, with O_2 adsorption occurring at 195 K and 0.21 bar. Desorption was performed by quickly heating the sample to 323 K for 5 min [42].

N-heterocyclic carbene (NHCs) are known as one of the most practical catalytic sites in the synthetic chemistry. They applied in homogeneous or heterogeneous reactions in different types of reactions like addition, rearrangement, and group-transfer reactions. Also, they applied for formation and elimination of some important organic bonds like C–C, C–N and C–O. NHCs can be deprotonated in two way. In first, imidazole ring is deprotonated on C2 atom which is more common. In another way, C4 or C5 atoms of carbene could be deprotonated. This way of formation of carbenic site on C4 and C5 atom is less common and called abnormal NHCs (aNHCs). aNHC sites are significantly more basic, and often less stable. Since the imidazolium ring is most applied catalytic site for formation of

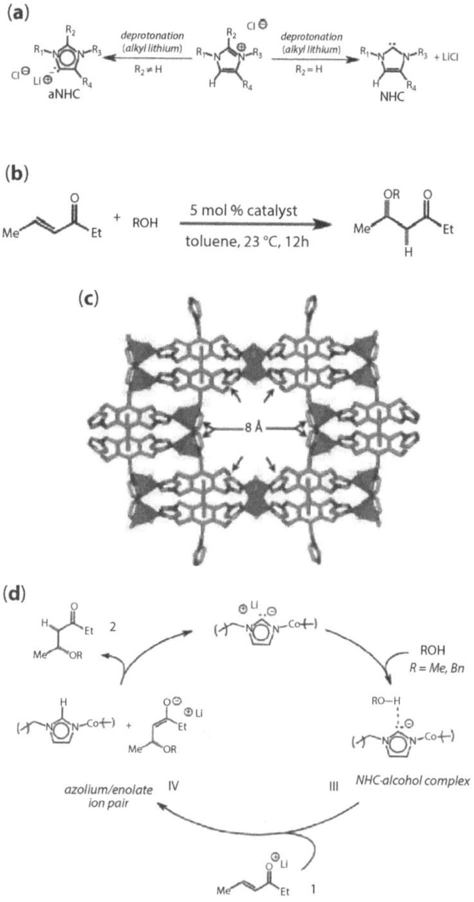

Figure 6.4 Application of TIF-1 in conjugate addition reaction. (a) Formation of Free NHCs versus Free aNHCs. (b) conjugate addition of alcohol. (c) Structure of TIF-1. (d) Proposed mechanism for catalysis of conjugate addition reaction by lithiated TIF-1 [43].

NHCs and aNHCs, synthesis of imidazolium functionalized MOFs is an ideal candidate to develop heterogeneous NHC-based catalysts.

With such idea, Hupp and coworkers proved that n-butyl lithium could deprotonate tripodal imidazolate framework-1 (TIF-1) so that produced N-heterocyclic carbene (NHC) sites can be accessible which are remarkable competent as Brønsted-base-type NHC catalysts (Figure 6.4) [43]. As results, an α,β-unsaturated ketone was converted to the corresponding benzyl ether with benzyl alcohol at a faster rate and in higher yield than with a traditional homogeneous NHC catalyst. In practical reaction, conjugate addition of benzyl alcohol to (E)-hex-4-ene-3-one to form the corresponding ethers show 72 and 83% conversion for TIF-1 and homogeneous NHC catalyst respectively.

Imidazole is incorporated into many important biological molecules especially amino acid histidine, which has an imidazole side-chain. Histidine is present in many proteins and enzymes and plays a vital part in the structure and binding functions of hemoglobin. Imidazole-based histidine compounds play a very important role in intracellular buffering. Thus, in biological systems histidine as imidazole based amino acid can be both proton acceptor and proton donor. Such ability of imidazole ring in pH dependent abstractable protons leads to designing material or membranes for pH sensing, proton conductivity and molecular buffers.

Similar to imidazolium group, imidazole is source of proton in intramolecular or intermolecular proton transfer materials. This proton transfer could be observed as a response to external stimuli. Cheng-Yong Su and coworkers reported a Sm-based MOF with emission in visible and near-infrared region which is sensitized by excited-state intramolecular proton transfer (ESIPT) of imidazole containing ligand. H_3PI2C ligand (5-(2-(2-hydroxyphenyl)-4,5-diphenyl-1H-imidazol-1-yl)isophthalic acid) and Sm^{3+} applied together for construction of $[Sm_2(HPI2C)_3(DMF)_2(H_2O)_2]\cdot 2DMF$ (Sm-MOF) (Figure 6.5) [44]. In ESIPT mechanism, the S_0 ground state of $HPI2C^{2-}$ ligand in enol from (E) absorbs UV light and then excited to $[S_1(E^*)]$ state. Then, the E* form of the ligand transform to excited keto form $[S_1(K^*)]$ after light irradiation in photoinduced ESIPT process. In a non-radiative process, the $[S_1(K^*)]$ state alters to $[T_1(K^*)]$ through intersystem crossing (isc) and in a radiative process, $[S_1(K^*)]$ state switches to its ground state (K) with keto-based emission with longer wavelength visible light. Both $[S_1(K^*)]$ and $[T_1(K^*)]$ states have appropriate energy levels to undergo intermolecular energy transfer (ET) to the nearest emitting level ($^4G_{5/2}$) of Sm^{3+} ion, after which radiative transitions in the visible or NIR domains characterized by the specific emissions of Sm^{3+} ions are produced.

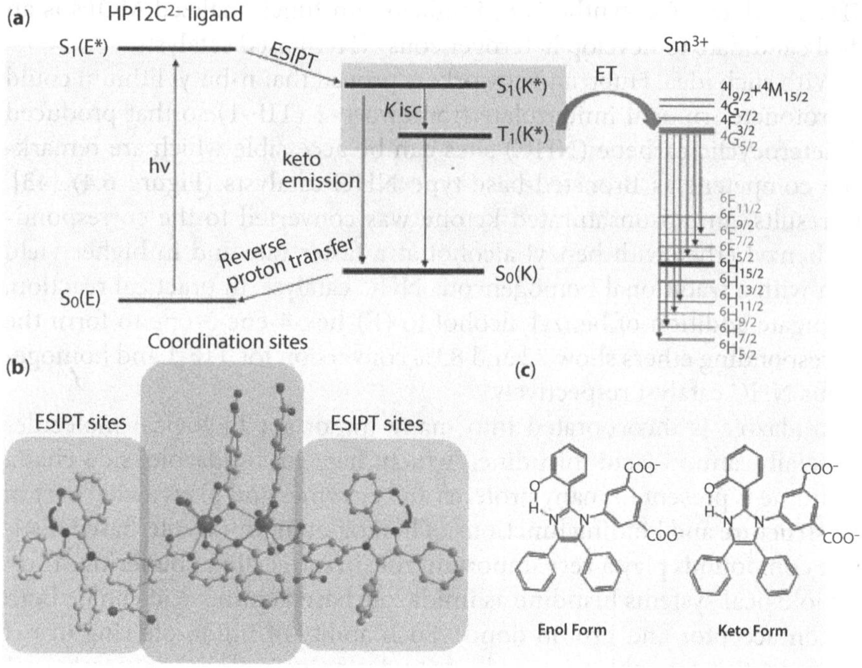

Figure 6.5 Application of Sm-MOF in excited-state intramolecular proton transfer. (a) Schematic representation of ESIPT phenomenon and energy absorption, migration, emission (plain arrows), and dissipation (dotted arrows) processes in Sm-MOF. S = singlet state; T = triplet state; ET = energy transfer. (b) Illustration of the linkage between coordination sites and ESIPT sites (solvent molecules and H atoms are omitted for clarity). (c) Representation of excited-state intramolecular proton transfer [44].

Energetic materials containing high detonation performance (gathering the maximum energetic groups in the smallest volume) and low sensitivities are of special interests. To provide energetic materials with mentioned characters, two main types of materials had been applied: (I) explosive and hard to synthesize energetic motifs and their nitrogen-rich salts and (II) coordination polymers constructed by specific metal ions and nitrogen rich ligands. In comparison of 1D, 2D and 3D coordination polymers, due to low rigidity of 1D and 2D coordination polymers they are highly sensitive to impact. But 3D structure of MOFs constructed based on complicated modes and rigid building blocks so that provide highly thermal stability, perfect insensitivity and attractive detonation performance. About energetic MOFs, thermostability and energetic performance directly related to the energetic organic-inorganic building blocks. As energetic building blocks, nitrogen-rich heterocycles especially azole-based compounds like triazole and tetrazole are generally highly endothermic with high densities and low sensitivities towards outer stimuli that provide many requirements in the challenging field of energetic-MOFs. Tetrazole and triazole rings with energetic substituents like nitro, amine, azide, furazan and furoxan have been greatly concerned because of their high density, high nitrogen content, high heat of formation, popular insensitivity. Mentioned detonation properties of triazole and tetrazole ring is directly attributable to the large number of inherently energetic N–N, N=N, C–N, C=N, N–O, and N=O bonds. Combinations of these energetic bonds in the structure of organic linker with diverse coordination modes of these energetic ligands with metal ions in the rigid structure of MOFs provide energetic materials with good insensitivity and high energetic performance [45–56]. 1,2,4-triazoles show a perfect balance between thermal stability and high positive heats of formation, high density, high heat of detonation and outstanding detonation performance by gathering the maximum energetic groups in the smallest volume required for applications as prospective HEDMs. Additionally, to design energetic MOFs with characters like high energy, safety and green, thermal stability, good insensitivity and excellent detonation properties, moreover that energetic organic linker, rational selection of metal ions should be taken into account. As good example, Cu(I)/Cu(II) ions possess strong coordination ability with N atoms and O atoms and the Cu(I)/Cu(II)-based energetic compounds are converted to their oxidation states/reduction states in the detonation reaction, which are more environmentally friendly choices.

Siping Pang and coworkers synthesized $[Cu(atrz)_3(NO_3)_2]_n$ (Cu-atrz or **1**) and $[Ag(atrz)_{1.5}(NO_3)]_n$ (Ag-atrz or **2**) as highly energetic and eco-friendly energetic materials (Figure 6.6) [46]. They selected atrz as a ligand because it has a nitrogen-rich heterocyclic backbone with high nitrogen content (N% = 68.3), high heat of formation (878 kJ·mol^{-1}) and good thermal stability with a decomposition temperature of up to 313 °C which are suitable characters for improving the energetic performances of the target

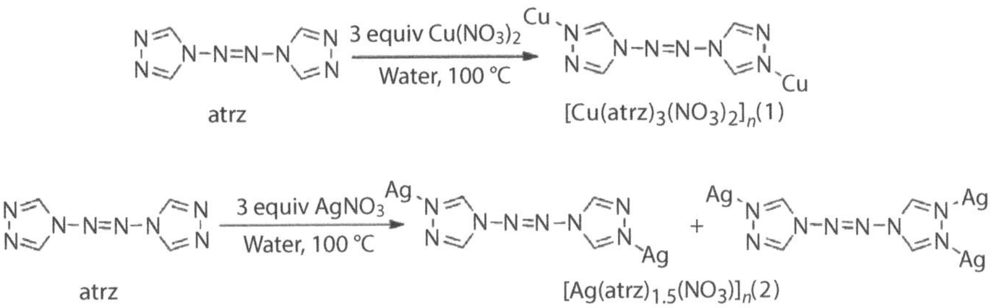

Figure 6.6 Synthesis, coordination mode, and formulae for 3D energetic $[Cu(atrz)_3(NO_3)_2]_n$ (Cu-atrz or **1**) and $[Ag(atrz)_{1.5}(NO_3)]_n$ (Ag-atrz or **2**) [46].

polymers. To estimate the heat of detonation of these two new explosives, they mention that possible detonation reaction is as follows:

$$\mathbf{1}(CuC_{12}H_{12}N_{26}O_6) \rightarrow CuO + 5\ H_2O + 2/3\ NH_3 + 38/3\ N_2 + 12\ C$$

$$\mathbf{2}(AgC_6H_6N_{13}O_3) \rightarrow 1/2\ Ag_2O + 5/2\ H_2O + 1/3\ NH_3 + 19/3\ N_2 + 12\ C$$

For both frameworks, water, nitrogen, carbon, and ammonia were assumed to be the final products of decomposition of the organic part of the framework and the formation of metal oxides was assumed to be governed by the deficiency of oxygen. Then, they applied density functional theory to compute the energy of detonation ($\Delta E_{Detonation}$) from ($\Delta H_{Detonation}$) of mentioned reactions. The results show that despite the low sensitivity, Cu-atr (**1**) shows exceptional detonation heat of 3.62 kcal·g^{-1} (6.08 kcal·cm^{-3}). Same results are about 1.4 kcal·g^{-1} (3 kcal·cm^{-3}) for Ag-atrz (**2**). Overall, these frameworks exhibit excellent physical and detonation properties, such as high thermal stabilities, low sensitivities, and high heats of detonation especially detonation heat of Cu-atr (**1**) is high.

In other work, Guo-Cong Guo and coworkers synthesized two coordination polymers, $[M(N_3)_2(atrz)]_n$ (M = Co **1**, Cd **2**; atrz = 4,4'-azo-1,2,4-triazole), with aim of development of low sensitive materials with high detonation heat (Figure 6.7) [57]. Thermogravimetric analyses show

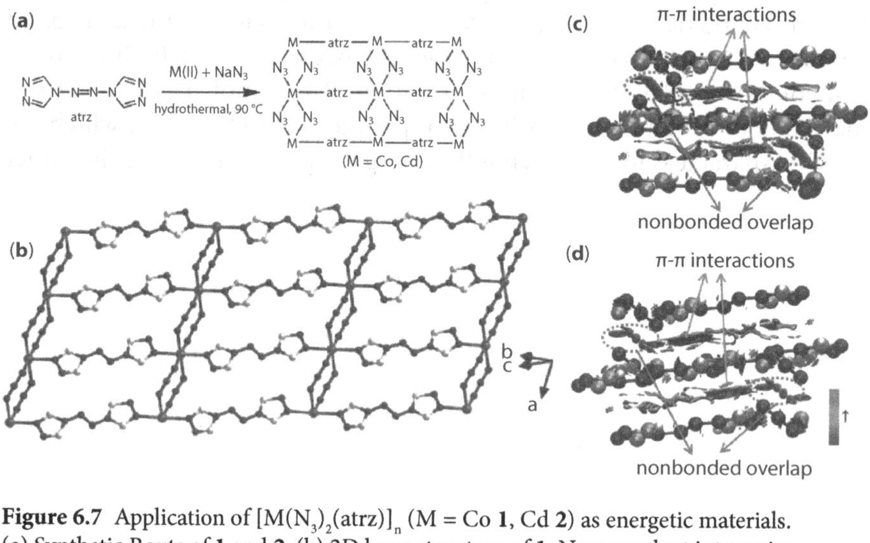

Figure 6.7 Application of $[M(N_3)_2(atrz)]_n$ (M = Co **1**, Cd **2**) as energetic materials. (a) Synthetic Route of **1** and **2**. (b) 2D layer structure of **1**. Non-covalent interaction plots of gradient isosurfaces for **1** (c) and **2** (d) [57].

that the thermal decomposition temperatures of **1** and **2** are 246 and 273 °C, respectively. Since these values are higher than 200 °C, these coordination polymers meet the requirements for practical applications. Since compounds **1** and **2** have high nitrogen contents with 63.81% for **1** and 54.39% for **2**, they can apply as energetic materials. Enthalpies of formation ($\Delta H_{formation}$) and detonation ($\Delta H_{detonation}$) are important energetic properties of an energetic material. The highly positive enthalpies of formation ($\Delta H_{detonation}$) are the main energy output source of nitrogen-rich energetic compounds. $\Delta H_{formation}$ values calculated based on Hess's low and combustion reaction of **1** and **2** compounds by the known enthalpies of CoO (s, −237.94 kJ·mol^{-1}), CdO (s, −258.35 kJ·mol^{-1}), CO_2 (g, −393.51 kJ·mol^{-1}), and H_2O (l, −285.83 kJ·mol^{-1}).

For **1**, combustion reaction and Hess low are:

$$CoC_4H_4N_{14}(s) + 11/2\,O_2(g) + CoO(s) + 4CO_2(g) + 2H_2O(l) + 7N_2\,(g)$$

$$\Delta H_f^\circ[\mathbf{1},s] = \Delta H_f^\circ[CoO,s] + 4\,\Delta H_f^\circ[CO_2,g] + 2\,\Delta H_f^\circ[H_2O,l] - \Delta H_c[\mathbf{1},s]$$

For **2**, combustion reaction and Hess low are:

$$CdC_4H_4N_{14}(s) + 11/2\,O_2(g) + CdO(s) + 4CO_2(g) + 2H_2O(l) + 7N_2\,(g)$$

$$\Delta H_f^\circ[\mathbf{2},s] = \Delta H_f^\circ[CdO,s] + 4\,\Delta H_f^\circ[CO_2,g] + 2\,\Delta H_f^\circ[H_2O,l] - \Delta H_c[\mathbf{2},s]$$

where ΔH_c[**1**,s] and ΔH_c[**2**,s] are the enthalpies of combustion for **1** and **2**. ΔH_c is calculated based on $\Delta H_c = \Delta U_c + \Delta nRT$; where Δn is the change of the number of gas constituents in the reaction process, R = 8.314 J·mol^{-1}.K^{-1}, T = 298.15 K and ΔU_c is constant-volume combustion heat which could be calculated by oxygen bomb calorimetry (12.013 kJ·g^{-1} for **1** and 10.783 kJ·g^{-1} for **2**). So, final value of $\Delta H_f^\circ[\mathbf{1},s]$ and $\Delta H_f^\circ[\mathbf{2},s]$ are 4.21 kJ·g^{-1} and 4.08 kJ·g^{-1}, respectively. These values are higher than traditional primary explosives.

$\Delta H_{detonation}$ values for **1** and **2** calculated by the assumption that metal, ammonia, nitrogen gas, and carbon are explosion products.

$$CuC_4H_4N_{14}(s) \rightarrow Cu(s) + 4/3\,NH_3 + 19/3\,N_2 + 4\,C$$

$$CdC_4H_4N_{14}(s) \rightarrow Cd(s) + 4/3\,NH_3 + 19/3\,N_2 + 4\,C$$

Finally, the $\Delta H_{detonation}$ could be calculated using following equation:

$$\Delta H_{detonation} = -\left(\left(\Delta H^{\circ}_{f(detonation\ products)} - \Delta H^{\circ}_{f(explosive)}\right) / (\text{formula weight of explosive})\right)$$

Final values of $\Delta H_{detonation}$ for **1** and **2** are 4.407 and 4.248 kJ·g^{-1}, respectively, with the known ΔH°_{f} values of Co, Cd, NH_3, N_2, and C, and the above experimentally determined ΔH°_{f} of **1** and **2**. These remarkable data are much larger than commercial primary explosives and many coordination polymers. Evaluation of safety parameters of **1** and **2** like the impact, friction, and electrostatic discharge sensitivities reveals that these compounds are acceptable for use as primary explosives.

In structural view it is proved that the face-to-face π–π interactions can effectually buffer against mechanical actions to remarkably reduce the sensitivity. So, owing to the presence of interlayer π–π interactions in the structure of **1** and **2**, these two compounds show low sensitivity.

As guest interactive site, tetrazole function is of high number of free N sites. As a result tetrazole function is a good guest-interactive site with localized high charge density to polarize some of non-polar gas molecules like H_2 and Xe. On the other tetrazole can interact with acidic gas like H_2S and CO_2 through both its acidic and basic characters and (–NH–) site.

6.3 Function–Structure Properties

General chemical properties of heterocyclic azoles (Figure 6.8) and their coordination chemistry (Figure 6.9) depend on the number of N atom in five member ring. These properties will be discussed member by member.

Pyrazole (1H-pyrazole; 1,2-diazole) has been drawn very considerable attentions due to broad variety of properties in supramolecular chemistry as molecular tectons and coordination chemistry as coordinating site [59]. Two pyridinic (N2) and pyrrolic (N–H at N1) nitrogen atoms in pyrazole ring are adjacent.

As mentioned earlier, pyrazole has two types of nitrogen atoms, N1 site which is pyrrolic NH site and N2 site which is N2 pyridinic site. Typically, in coordination chemistry pyrazole coordinate to metal ions with pyridinic N2 atom. In this condition N1 pyrrolic nitrogen atom remains uncoordinated. On one hand, this free NH site can act as a guest-interactive site to interact with anions and other electron-rich analytes. On the other hand, this free NH site can generate pyrazole-based frameworks with higher dimensionality through hydrogen bonding with other parts of the

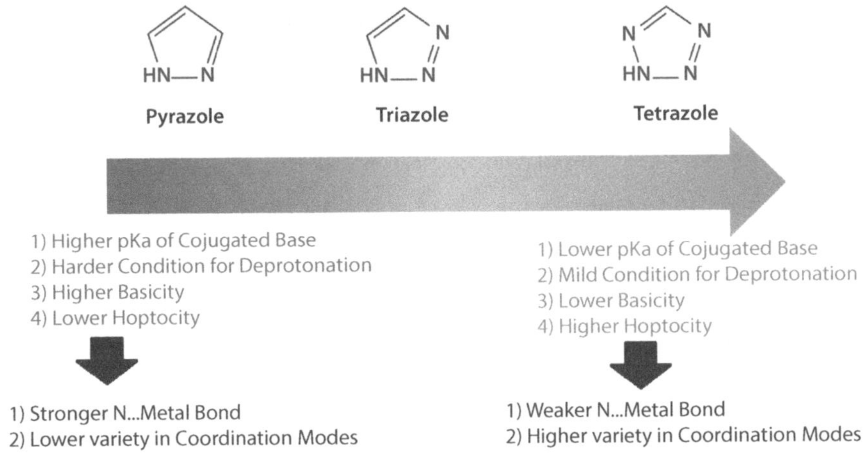

Figure 6.8 Comparison of general properties of heterocyclic azoles [58].

structure. This means pyrazole ring can act as structure-directing function through N2 site by coordination interaction and NH site by hydrogen bond while it can role as guest-interactive site by its free NH group to develop the structure of coordination complexes.

Pyrazole is a weak acid, comparable to aliphatic alcohols, and is also a weaker base than pyridine or simple aromatic amines. Pyrazoles with free N–H groups are amphoteric though more notably basic than acidic. Deprotonation of pyrazole increases its hapticity and basicity.

Pyrazoles are particularly thermally and hydrolytically very stable compounds [60]. The ditopic nature of pyrazole and the coordinative flexibility of pyrazolate, mean that the pyrazole ring can be considered to be a flexible ligand group in coordination chemistry [61]. Investigations show pyrazole ring is one of the easiest and most flexible N-donor heterocycles to incorporate into larger polydentate ligand structures, even outside a chelating ligand framework. However, an isolated pyrazole ring is a remarkably flexible ligand in its own right.

Considering these structural and interactive features of pyrazole ring like stability, flexibility, structure-directing effect, different coordination modes, hydrogen bond participating character and Lewis basicity, it has gained lots of attention and still is one of the most studied heterocyclic compounds in coordination chemistry of complexes [62]. So, it seems that introduction of pyrazole function is a suitable idea to construct MOFs for different structural and practical targets.

Pyrazole function usually applied as coordinating site rather guest-interactive site inside the structure of MOFs because it is flexible, stable

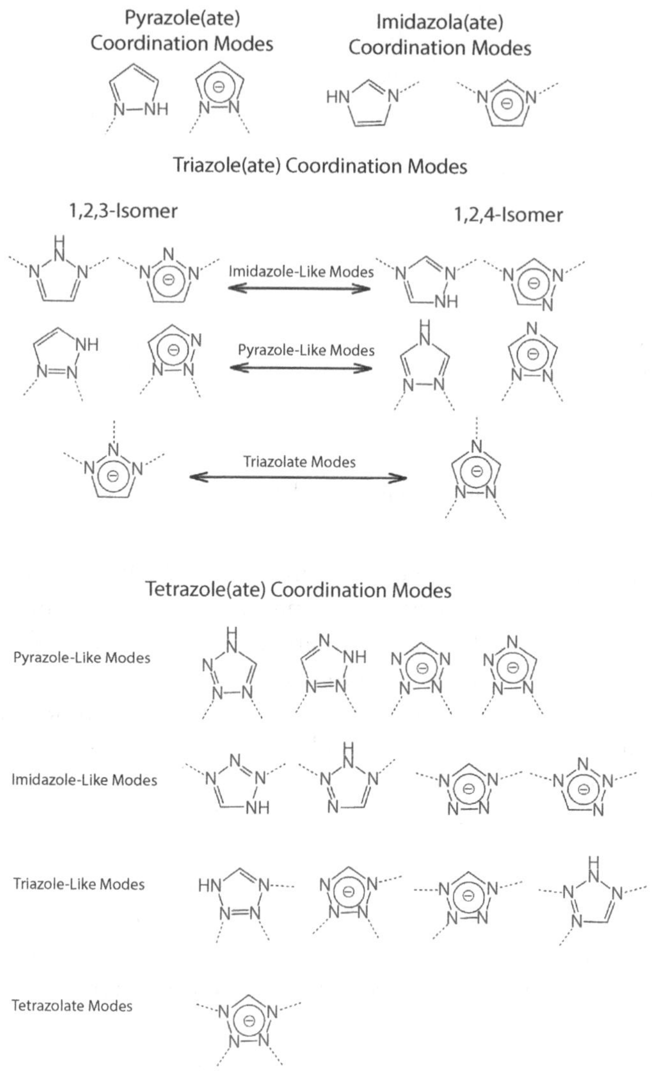

Figure 6.9 Coordination modes of heterocyclic azole N-based functions.

and has ability to show structure-directing character through hydrogen bonding and π–π stacking. Burrows and coworkers reported a series of pyrazole functionalized MOFs [63]. In some of them, the pyrazole NH groups are involved in hydrogen bonding to develop the dimensionality of the framework through establishing connection between interpenetrated networks or neighboring sheets. In others, the pyrrolic NH groups remain free and directed into the pores to interact with guest species.

Great number of MOFs based on ligands with pyrazole binding sites exhibit exceptional stability against temperature, high acidic and basic solutions, boiling water and water vapors. The pyrazole (pK_a = 19.8) ligand is more likely to form the relatively stable MOFs compared to the analogues with triazole and tetrazole ligands [64].

MOFs based on pyrazole function as coordinating site show two different structural features which are flexibility and high thermal and chemical stability. For example, BUT-31 (with formula $[Ni_8(OH)_4(H_2O)_2(BDP\text{-}X)_6]$ where H_2BDP is 1,4-bis(4-pyrazolyl)benzene and X = –CHO, –CN, –COOH) is highly stable in boiling water and highly basic aqueous solution of 4 m sodium hydroxide [66]. Kitagawa and coworkers synthesized an MOF based on Cu(I) metal ions and a pyrazole ligand (3,3',5,5'-tetramethyl-4,4'-bipyrazole) with ultrahigh thermal stability up to 500 °C in nitrogen atmosphere [70]. FJU-66 (with formula $[Cu_6(NDI)_3]$ where H_2NDI = 2,7-bis(3,5-dimethyl)dipyrazol-1,4,5,8-naphthalene-tetracarboxydiimide) structurally remain intact up to 803 K [65]. Also, FJU-66 saves its crystallinity in aqueous solution in pH range of 3–14 as well as 10 M sodium hydroxide solution over 12 h. Other authors mentioned that polyazolate-based MOFs are highly stable under humid and basic conditions [66–69]. Such high capacity of pyrazole-based MOFs is due to the high pK_a of pyrazole so that pyrazolate-based ligands can form much stronger M–N bonds, giving rise in a high thermally and chemically stable MOFs.

As mentioned, flexibility as well as stability, is another feature of MOFs based on pyrazole coordinating sites. Two nitrogen atoms of the pyrazolate ligand can provide the bridging mode to connect two metal ions. The bridging nitrogen atoms of the pyrazolate rings act as a "kneecap" around the M·M axis, which commonly leads to the rotation of ligands around the M·M axis in response to external stimulus and then gives rise to the reversible flexibility and breathing adsorption behaviors of the framework or under heat [61, 71–75]. Moreover, Jian Zhang and coworkers synthesized flexible frameworks which during adsorption of the C1 to C3 light hydrocarbons exhibit the gate-opening and hysteretic adsorption behaviors in response to light hydrocarbons with high capacity [73].

Coordination geometry of pyrazolate is somewhat similar to the syn,syn-μ-η^1,η^1-bidentate mode of carboxylate, which is the basis of paddle-wheel $M_2(RCOO)_4$, trigonal-prismatic $M_3O(RCOO)_6$, and octahedral $M_4O(RCOO)_6$ SBUs commonly used in porous metal carboxylate frameworks [64]. For example some polyazolate ligand like $H_3BDTriP$ (5,5′-(5-(1H-pyrazol-4-yl)-1,3-phenylene)bis(1H-1,2,3-triazole))), H_2bpz (3,3′,5,5′-tetramethyl-4,4′-bipyrazole) and H_2pzca=(1H-pyrazole-4-carboxylic acid) applied in constructing isostructural MOFs with

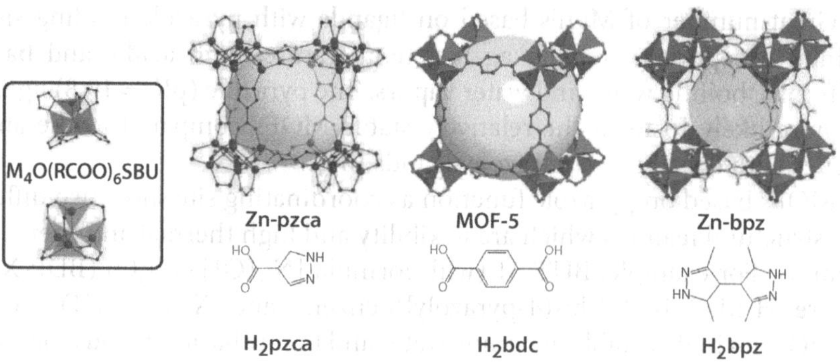

Figure 6.10 Similarity between carboxylate and pyrazolate coordination modes in octahedral $M_4O(RCOO)_6$ SBUs. Zn-pzca is $[Zn_4(\mu_4\text{-}O)(\mu_4\text{-}pzca)_3]_n$ where H_2pzca = 1*H*-pyrazole-4-carboxylic acid. Zn-bpz is $[Zn_4O(bdc)(bpz)_2]$ where H_2bdc = 1,4-benzenedicarboxylate and H_2bpz = 3,3′,5,5′-tetramethyl-4,4′-bipyrazole.

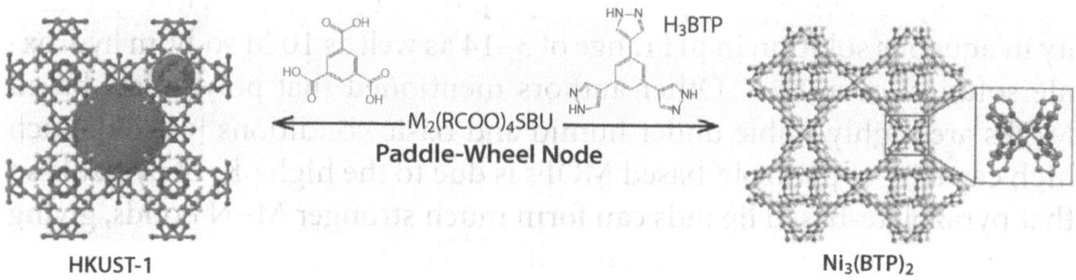

Figure 6.11 Similarity between carboxylate and pyrazolate coordination modes in paddle-wheel $M_2(RCOO)_4$ units.

HKUST-1 and IRMOF-5. In prototype MOF-5 ($[Zn_4O(bdc)_3]$ where H_2bdc = 1,4-benzenedicarboxylic acid) by partial or complete replacing dicarboxylate bdc linker with bipyrazolate linkers, isostructural MOFs with MOF-5 are generated. While the coordination mode of pyrazolate is very similar to that of the bidentate bridging carboxylate groups (Figures 6.10–6.11), there are still differences between them. First, polypyrazolates ligands can extend novel structures that could not be developed by polycarboxylates ligands. Also, polypyrazolates ligands have very stronger affinity to common late 3d metal ion rather carboxylate ligands. Consequently, metal pyrazolate frameworks based on 3d metal ions are more stable than metal carboxylate frameworks.

Owing to the simple coordination mode, small bridging angle, and short bridging distances of pyrazolates, their coordination polymers have quite limited structure types. Meanwhile, polypyrazolates adopt the advantages of the adjustable length and geometry of the donor groups, just as polycarboxylates and polypyridyls [76]. Moreover than binding through pyridinic nitrogen site, pyrazole motif can binds to metal sites through two nitrogen atoms simultaneously. Deprotonation of pyrazole increases its hapticity and basicity. The position of two donor atoms makes pyrazolate an *exo*-bidentate ligand; As a matter of fact, twenty different terminal or bridging coordination modes have been identified so far for pyrazolates. With these attractive features, it is not surprising that pyrazolates have received and still receive a massive attention for the construction of coordination complexes and polymers.

Imidazole (1H-Imidazole; 1,3-diazole) is another five member heterocycles ring containing two nitrogen possesses the two types of heterocyclic nitrogens and illustrates several aspects of amine acidity-basicity. Nitrogen number 1 of imidazole ring which has hydrogen on it contributes its electron pair to complete the aromatic sextet and is "pyrrole-like" nitrogen. It is weakly acidic rather than basic. The nitrogen in position 3 has an electron pair available and is "pyridine-like" nitrogen. However imidazole is both somewhat more acidic than pyrrole and somewhat more basic than pyridine, probably because six electrons are delocalized on five atoms, resulting in higher electron density. So, as well as Lewis basicity, imidazole ring contains of different type of hydrogen bond participating sites; Hydrogen bond acceptor from pyridinic nitrogen and hydrogen bond donor from pyrrolic nitrogen [77].

Substitution on nitrogen of imidazole ring classified in three types: (I) alkyl–alkyl substitution which results in construction of cationic imidazolium ring; (II) metal–metal substitution which results in construction of anionic imidazolate ring and (III) alkyl–metal substitution which results in construction of neutral substituted imidazole ring (Figure 6.12) [43]. So, imidazole ring applied as supramolecular building block in designing of ionic and neutral functional materials.

Imidazole function of organic linkers can act as both guest interactive site and binding site. If nitrogen atoms of imidazole rings remain uncoordinated, they can be accessible in the pore surface of the framework to

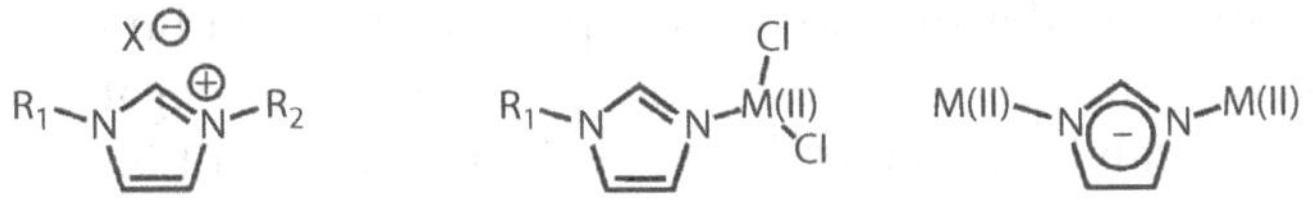

Figure 6.12 From left to right: an alkyl–alkyl imidazolium cation; an imidazole substituted with one alkyl group and one metal; and finally, an imidazolate (anion) substituted with two metal cations. (Note that each M(II) in the structure at right is coordinated to three additional imidazolate ions when the structure constitutes part of an overall neutral metal–organic framework) [43].

interact with guests. In first binding form, imidazole can bind to metal centers from N2 atom to construct MOFs with neutral imidazole ring to construct metal imidazole frameworks. The free NH site imidazole ring can interact with guests again or participate in self-assembly process to extend dimensionality of the framework form 2D metal–organic nanosheets to 3D supramolecular organic-inorganic frameworks. In second binding form, deprotonation of imidazole ring on NH site leads to coordination of two nitrogen atoms to metal centers and construction of metal-imidazolate frameworks. Deprotonation not only allows all N atoms to coordinate with metal ions but also further increases the basicity of these donors.

As an important class of metal-imidazolate frameworks, zeolite imidazolate frameworks (ZIFs) are developed by simple imidazole and functionalized imidazolate ligands which have emerged as a novel type of porous zeolite-like crystalline material based on the similarity of the structural units between metal(II)–Im^- (Im^- = imidazolate) and tetrahedral Si(Al)O_4 moieties. In case of simple imidazole, it predominantly forms simple zigzag chain structures in anionic imidazolate form. The two N-donors in imidazolate point outward from the five membered rings with an angle of ca. 140°, which is essential for the generation of novel framework structures. Most reported metal–imidazolate frameworks are based on univalent coinage metal ions and/or divalent first-row transition-metal ions. Note that in the structure of ZIFs each M(II) in the structure is coordinated to three additional imidazolate ions when the structure constitutes part of an overall neutral metal–organic framework.

Triazole isomers (1H-1,2,3-triazoles, 1H-1,2,4-triazoles) considered as one of nitrogen rich azole heterocycle ring. Triazole diverse supramolecular interactions, variable conformation and coordination modes make it enable to apply as binding site in coordination chemistry and supramolecular building blocks [78]. The nitrogen-rich property of triazole ring features good chemical characters like polarized carbon atoms in the ring which allow the complexation of anions by hydrogen and halogen bonding for ion-pair recognition, different Lewis basic sites for dipole or hydrogen bond donation and acceptance and π–π interactions. Several N-coordination modes of triazole ring provide different binding forms to metals in coordination complexes and polymers. Even negatively charged triazolate ring can bind to metal sites by deprotonation of N–H bond.

Tetrazole is the nitrogen-richest member of heterocyclic azole N-based functions possessing both a basic character and acidic hydrogen. Tetrazole derivatives are well-known for their instability, decomposing with release of energy. As potential energetic materials, various alkali-metal tetrazolate salts have been synthesized and structurally characterized.

In comparison with azole rings, tetrazole, hold an additional nitrogen atoms and consequently, compared to the triazole, imidazole or pyrazole rings, the tetrazole unit exhibits more bridging coordination possibilities. In structural coordination chemistry, neutral tetrazole may function both as a dinucleating ligand and a trinucleating ligand but deprotonated tetrazolate can bridge two, three and up to four metal ions. Anyway, at same time the uncoordinated N-heteroatom sites of tetrazole ring can act as interactive site and enhance the interaction between tetrazole functionalized host and guests molecules.

The numerous bridging coordination modes of tetrazole ring using its four nitrogen donor atoms afford great synthetic possibilities for the design and preparation of hybrid inorganic/organic architectures with diverse SBUs of varying connectivity and symmetry. Also the tetrazole-based ligands have strong coordination ability with metal because of the aromaticity and multiple N-donor atoms and have been utilized as multifunctional organic linkers for the generating coordination polymers [78].

Tetrazole and carboxy functional group are applied extensively as coordinating site in coordination chemistry for synthesis of coordination complexes and polymers. There are some properties about tetrazole and carboxy functions which must be compared together [58]. The pH and acidity of carboxylate and tetrazolate functions are closely similar and owing to this similarity, similar experimental conditions can be used for the synthesis of MOFs based on carboxylate–tetrazolate linkers. The utility of tetrazoles for producing robust MOFs with permanent porosity, and with topologies and gas adsorption properties is comparable to those of the pioneering, carboxylate-based MOF materials. Owing to the N-donor nature of tetrazolate ring, coordination bonds between 3d late metal ions and nitrogen sites of tetrazole are strong while owing to hard nature of carboxylate and soft nature of these metal ions, coordination bonds between 3d late metal ions and carboxylate are weak. Carboxylate function is a bidentate ligand while coordination modes of tetrazole are versatile from monodentate up to tetradentate which results in expanding in the range of metal–tetrazolate frameworks rather metal–carboxylate frameworks.

References

1. Gao, W.-Y., Pham, T., Forrest, K.A., Space, B., Wojtas, L., Chen, Y.-S., Ma, S., The local electric field favours more than exposed nitrogen atoms on CO2

capture: A case study on the rht-type MOF platform. *Chem. Commun.*, 51, 9636–9639, 2015.
2. He, T., Zhang, Y.Z., Wang, B., Lv, X.L., Xie, L.H., Li, J.R.J.C., A Base-Resistant ZnII-Based Metal–Organic Framework: Synthesis, Structure, Postsynthetic Modification, and Gas Adsorption. *ChemPlusChem*, 81, 864–871, 2016.
3. Baima, J., Macchieraldo, R., Pettinari, C., Casassa, S., Ab initio investigation of the affinity of novel bipyrazolate-based MOFs towards H2 and CO2. *CrystEngComm*, 17, 448–455, 2015.
4. Li, G.-P., Liu, G., Li, Y.-Z., Hou, L., Wang, Y.-Y., Zhu, Z., Uncommon pyrazoyl-carboxyl bifunctional ligand-based microporous lanthanide systems: Sorption and luminescent sensing properties. *Inorg. Chem.*, 55, 3952–3959, 2016.
5. Wang, H.-H., Jia, L.-N., Hou, L., Shi, W.-j., Zhu, Z., Wang, Y.-Y., A new porous MOF with two uncommon metal–carboxylate–pyrazolate clusters and high CO2/N2 selectivity. *Inorg. Chem.*, 54, 1841–1846, 2015.
6. Tomar, K., Gupta, A.K., Gupta, M., Change in synthetic strategy for MOF fabrication: From 2D non-porous to 3D porous architecture and its sorption and emission property studies. *New J. Chem.*, 40, 1953–1956, 2016.
7. Xu, G., Peng, Y., Hu, Z., Yuan, D., Donnadieu, B., Zhao, D., Cheng, H., A 2D metal–organic framework composed of a bi-functional ligand with ultra-micropores for post-combustion CO2 capture. *RSC Adv.*, 5, 47384–47389, 2015.
8. Willans, C.E., French, S., Barbour, L.J., Gertenbach, J.-A., Junk, P.C., Lloyd, G.O., Steed, J.W., A catenated imidazole-based coordination polymer exhibiting significant CO2 sorption at low pressure. *Dalton Trans.*, 33, 6480–6482, 2009.
9. Nandi, S., Maity, R., Chakraborty, D., Ballav, H., Vaidhyanathan, R., Preferential Adsorption of CO2 in an Ultramicroporous MOF with Cavities Lined by Basic Groups and Open-Metal Sites. *Inorg. Chem.*, 57, 5267–5272, 2018.
10. Chen, S.-S., Wang, P., Takamizawa, S., Okamura, T.-a., Chen, M., Sun, W.-Y., Zinc (II) and cadmium (II) metal–organic frameworks with 4-imidazole containing tripodal ligand: Sorption and anion exchange properties. *Dalton Trans.*, 43, 6012–6020, 2014.
11. Dey, A., Bairagi, D., Biradha, K., MOFs with PCU Topology for the Inclusion of One-Dimensional Water Cages: Selective Sorption of Water Vapor, CO2, and Dyes and Luminescence Properties. *Cryst. Growth Des.*, 17, 3885–3892, 2017.
12. Lv, X., Li, L., Tang, S., Wang, C., Zhao, X., High CO2/N2 and CO2/CH4 selectivity in a chiral metal–organic framework with contracted pores and multiple functionalities. *Chem. Commun.*, 50, 6886–6889, 2014.
13. Jayaramulu, K., Reddy, S.K., Hazra, A., Balasubramanian, S., Maji, T.K., Three-Dimensional Metal–Organic Framework with Highly Polar Pore Surface: H2 and CO2 Storage Characteristics. *Inorg. Chem.*, 51, 7103–7111, 2012.

14. Meng, L., Niu, Z., Liang, C., Dong, X., Liu, K., Li, G., Li, C., Han, Y., Shi, Z., Feng, S., Integration of Open Metal Sites and Lewis Basic Sites for Construction of a Cu MOF with a Rare Chiral Oh-type cage for high performance in methane purification. *Chem.—Eur. J.*, 24, 50, 13181–13187, 2018.
15. Wang, H.H., Shi, W.J., Hou, L., Li, G.P., Zhu, Z., Wang, Y.Y., A Cationic MOF with High Uptake and Selectivity for CO2 due to Multiple CO2-Philic Sites. *Chem.—Eur. J.*, 21, 16525–16531, 2015.
16. Lin, J.-B., Zhang, J.-P., Chen, X.-M., Nonclassical active site for enhanced gas sorption in porous coordination polymer. *J. Am. Chem. Soc.*, 132, 6654–6656, 2010.
17. Liu, B., Yao, S., Shi, C., Li, G., Huo, Q., Liu, Y., Significant enhancement of gas uptake capacity and selectivity *via the* judicious increase of open metal sites and Lewis basic sites within two polyhedron-based metal–organic frameworks. *Chem. Commun.*, 52, 3223–3226, 2016.
18. Ye, Y., Xiong, S., Wu, X., Zhang, L., Li, Z., Wang, L., Ma, X., Chen, Q.-H., Zhang, Z., Xiang, S., Microporous Metal–Organic Framework Stabilized by Balanced Multiple Host–Couteranion Hydrogen-Bonding Interactions for High-Density CO2 Capture at Ambient Conditions. *Inorg. Chem.*, 55, 292–299, 2015.
19. Li, Y.-W., He, K.-H., Bu, X.-H., Bottom-up assembly of a porous MOF based on nanosized nonanuclear zinc precursors for highly selective gas adsorption. *J. Mater. Chem. A*, 1, 4186–4189, 2013.
20. Li, P.-Z., Wang, X.-J., Liu, J., Lim, J.S., Zou, R., Zhao, Y., A triazole-containing metal–organic framework as a highly effective and substrate size-dependent catalyst for CO2 conversion. *J. Am. Chem. Soc.*, 138, 2142–2145, 2016.
21. Seth, S., Savitha, G., Moorthy, J.N., Carbon Dioxide Capture by a Metal–Organic Framework with Nitrogen-Rich Channels Based on Rationally Designed Triazole-Functionalized Tetraacid Organic Linker. *Inorg. Chem.*, 54, 6829–6835, 2015.
22. Chen, D.-M., Tian, J.-Y., Chen, M., Liu, C.-S., Du, M., Moisture-Stable Zn(II) Metal–Organic Framework as a Multifunctional Platform for Highly Efficient CO2 Capture and Nitro Pollutant Vapor Detection. *ACS Appl. Mater. Interfaces*, 8, 18043–18050, 2016.
23. Mohideen, M.I.H., Pillai, R.S., Adil, K., Bhatt, P.M., Belmabkhout, Y., Shkurenko, A., Maurin, G., Eddaoudi, M., A Fine-Tuned MOF for Gas and Vapor Separation: A Multipurpose Adsorbent for Acid Gas Removal, Dehydration, and BTX Sieving. *Chem*, 3, 822–833, 2017.
24. Wang, F., Fu, H.-R., Kang, Y., Zhang, J., A new approach towards zeolitic tetrazolate–imidazolate frameworks (ZTIFs) with uncoordinated N-heteroatom sites for high CO2 uptake. *Chem. Commun.*, 50, 12065–12068, 2014.
25. Dong, B.X., Zhang, S.Y., Liu, W.L., Wu, Y.C., Ge, J., Song, L., Teng, Y.L., Gas storage and separation in a water-stable [CuI5BTT3]4– anion framework

comprising a giant multi-prismatic nanoscale cage. *Chem. Commun.*, 51, 5691–5694, 2015.

26. Zhao, Y.-P., Li, Y., Cui, C.-Y., Xiao, Y., Li, R., Wang, S.-H., Zheng, F.-K., Guo, G.-C., Tetrazole–Viologen-based Flexible Microporous Metal–Organic Framework with High CO2 Selective Uptake. *Inorg. Chem.*, 55, 7335–7340, 2016.
27. Poloni, R., Smit, B., Neaton, J.B., Ligand-Assisted Enhancement of CO2 Capture in Metal–Organic Frameworks. *J. Am. Chem. Soc.*, 134, 6714–6719, 2012.
28. Cui, P., Ma, Y.-G., Li, H.-H., Zhao, B., Li, J.-R., Cheng, P., Balbuena, P.B., Zhou, H.-C., Multipoint Interactions Enhanced CO2 Uptake: A Zeolite-like Zinc–Tetrazole Framework with 24-Nuclear Zinc Cages. *J. Am. Chem. Soc.*, 134, 18892–18895, 2012.
29. Demessence, A., D'Alessandro, D.M., Foo, M.L., Long, J.R., Strong CO2 Binding in a Water-Stable, Triazolate-Bridged Metal–Organic Framework Functionalized with Ethylenediamine. *J. Am. Chem. Soc.*, 131, 8784–8786, 2009.
30. Ye, J. and Johnson, J.K., Design of Lewis pair-functionalized metal organic frameworks for CO2 hydrogenation. *ACS Catal.*, 5, 2921–2928, 2015.
31. Liu, L.-H., Qiu, X.-T., Wang, Y.-J., Shi, Q., Sun, Y.-Q., Chen, Y.-P., NIR emission and luminescent sensing of a lanthanide–organic framework with Lewis basic imidazole and pyridyl sites. *Dalton Trans.*, 46, 12106–12113, 2017.
32. Zhang, X., Xia, T., Jiang, K., Cui, Y., Yang, Y., Qian, G., Highly sensitive and selective detection of mercury (II) based on a zirconium metal–organic framework in aqueous media. *J. Solid State Chem.*, 253, 277–281, 2017.
33. Liu, J.Q., Li, G.P., Liu, W.C., Li, Q.L., Li, B.H., Gable, R.W., Hou, L., Batten, S.R., Two Unusual Nanocage-Based Ln-MOFs with Triazole Sites: Highly Fluorescent Sensing for Fe3+ and Cr2O72–, and Selective CO2 Capture. *ChemPlusChem*, 81, 1299–1304, 2016.
34. Yang, Y., Jiang, F., Chen, L., Pang, J., Wu, M., Wan, X., Pan, J., Qian, J., Hong, M., An unusual bifunctional Tb-MOF for highly sensitive sensing of Ba2+ ions and with remarkable selectivities for CO2–N2 and CO2–CH4. *J. Mater. Chem. A*, 3, 13526–13532, 2015.
35. Singha, D.K. and Mahata, P., Highly Selective and Sensitive Luminescence Turn-On-Based Sensing of Al3+ Ions in Aqueous Medium Using a MOF with Free Functional Sites. *Inorg. Chem.*, 54, 6373–6379, 2015.
36. Li, Q., Xue, D.-X., Zhang, Y.-F., Zhang, Z.-H., Gao, Z., Bai, J., A dual-functional indium–organic framework towards organic pollutant decontamination *via* physically selective adsorption and chemical photodegradation. *J. Mater. Chem. A*, 5, 14182–14189, 2017.
37. Yu, Y., Zhang, X.-M., Ma, J.-P., Liu, Q.-K., Wang, P., Dong, Y.-B., Cu (I)-MOF: Naked-eye colorimetric sensor for humidity and formaldehyde in single-crystal-to-single-crystal fashion. *Chem. Commun.*, 50, 1444–1446, 2014.

38. Han, Y.-H., Tian, C.-B., Li, Q.-H., Du, S.-W., Highly chemical and thermally stable luminescent EuxTb1−x MOF materials for broad-range pH and temperature sensors. *J. Mater. Chem. C*, 2, 8065–8070, 2014.
39. Hu, Q., Qin, L., Lei, J., Tan, X.-X., Ni, G., Wang, Y.-Q., Li, J., Cao, F.-H., Nitrogen-rich metal–organic framework: Dye adsorptions and electrochemical performance. *Polyhedron*, 151, 33–36, 2018.
40. Murray, L.J., Dinca, M., Yano, J., Chavan, S., Bordiga, S., Brown, C.M., Long, J.R., Highly-selective and reversible O2 binding in Cr3 (1,3,5-benzenetricarboxylate) 2. *J. Am. Chem. Soc.*, 132, 7856–7857, 2010.
41. Bloch, E.D., Queen, W.L., Hudson, M.R., Mason, J.A., Xiao, D.J., Murray, L.J., Flacau, R., Brown, C.M., Long, J.R., Hydrogen Storage and Selective, Reversible O2 Adsorption in a Metal–Organic Framework with Open Chromium (II) Sites. *Angew. Chem. Int. Ed.*, 55, 8605–8609, 2016.
42. Xiao, D.J., Gonzalez, M.I., Darago, L.E., Vogiatzis, K.D., Haldoupis, E., Gagliardi, L., Long, J.R., Selective, tunable O2 binding in cobalt (II)–triazolate/pyrazolate metal–organic frameworks. *J. Am. Chem. Soc.*, 138, 7161–7170, 2016.
43. Lalonde, M.B., Farha, O.K., Scheidt, K.A., Hupp, J.T., N-Heterocyclic Carbene-Like Catalysis by a Metal–Organic Framework Material. *ACS Catal.*, 2, 1550–1554, 2012.
44. Chen, L., Zhang, H., Pan, M., Wei, Z.W., Wang, H.P., Fan, Y.N., Su, C.Y., An Efficient Visible and Near-Infrared (NIR) Emitting SmIII Metal–Organic Framework (Sm-MOF) Sensitized by Excited-State Intramolecular Proton Transfer (ESIPT) Ligand. *Chem.—Asian J.*, 11, 1765–1769, 2016.
45. Dippold, A.A. and Klapötke, T.M., Nitrogen-Rich Bis-1, 2, 4-triazoles—A Comparative Study of Structural and Energetic Properties. *Chem.—Eur. J.*, 18, 16742–16753, 2012.
46. Li, S., Wang, Y., Qi, C., Zhao, X., Zhang, J., Zhang, S., Pang, S., 3D energetic metal–organic frameworks: Synthesis and properties of high energy materials. *Angew. Chem.*, 125, 14281–14285, 2013.
47. Dong, Y., Peng, P., Hu, B., Su, H., Li, S., Pang, S., High-Density Energetic Metal–Organic Frameworks Based on the 5,5′-Dinitro-2H, 2′H-3,3′-bi-1, 2, 4-triazole. *Molecules*, 22, 1068, 2017.
48. Liu, X., Gao, W., Sun, P., Su, Z., Chen, S., Wei, Q., Xie, G., Gao, S., Environmentally friendly high-energy MOFs: Crystal structures, thermostability, insensitivity and remarkable detonation performances. *Green Chem.*, 17, 831–836, 2015.
49. Qin, J.-S., Zhang, J.-C., Zhang, M., Du, D.-Y., Li, J., Su, Z.-M., Wang, Y.-Y., Pang, S.-P., Li, S.-H., Lan, Y.-Q., A Highly Energetic N-Rich Zeolite-Like Metal-Organic Framework with Excellent Air Stability and Insensitivity. *Adv. Sci.*, 2, 1500150, 2015.
50. Zhang, Q. and Shreeve, J.N.M., Metal-organic frameworks as high explosives: A new concept for energetic materials. *Angew. Chem. Int. Ed.*, 53, 2540–2542, 2014.

51. Yang, Q., Yang, G., Zhang, W., Zhang, S., Yang, Z., Xie, G., Wei, Q., Chen, S., Gao, S., Superior Thermostability, Good Detonation Properties, Insensitivity, and the Effect on the Thermal Decomposition of Ammonium Perchlorate for a New Solvent-Free 3D Energetic PbII-MOF. *Chem.—Eur. J.*, 23, 9149–9155, 2017.
52. Zhang, S., Liu, X., Yang, Q., Su, Z., Gao, W., Wei, Q., Xie, G., Chen, S., Gao, S., A New Strategy for Storage and Transportation of Sensitive High-Energy Materials: Guest-Dependent Energy and Sensitivity of 3D Metal–Organic-Framework-Based Energetic Compounds. *Chem.—Eur. J.*, 20, 7906–7910, 2014.
53. Feng, Y., Liu, X., Duan, L., Yang, Q., Wei, Q., Xie, G., Chen, S., Yang, X., Gao, S., In situ synthesized 3D heterometallic metal–organic framework (MOF) as a high-energy-density material shows high heat of detonation, good thermostability and insensitivity. *Dalton Trans.*, 44, 2333–2339, 2015.
54. Zhang, J. and Jean'ne, M.S., 3D Nitrogen-rich metal–organic frameworks: Opportunities for safer energetics. *Dalton Trans.*, 45, 2363–2368, 2016.
55. Xu, Y., Liu, W., Li, D., Chen, H., Lu, M., In situ synthesized 3D metal–organic frameworks (MOFs) constructed from transition metal cations and tetrazole derivatives: A family of insensitive energetic materials. *Dalton Trans.*, 46, 11046–11052, 2017.
56. Feng, Y., Bi, Y., Zhao, W., Zhang, T., Anionic metal–organic frameworks lead the way to eco-friendly high-energy-density materials. *J. Mater. Chem. A*, 4, 7596–7600, 2016.
57. Xu, J.-G., Sun, C., Zhang, M.-J., Liu, B.-W., Li, X.-Z., Lu, J., Wang, S.-H., Zheng, F.-K., Guo, G.-C., Coordination Polymerization of Metal Azides and Powerful Nitrogen-Rich Ligand toward Primary Explosives with Excellent Energetic Performances. *Chem. Mater.*, 29, 9725–9733, 2017.
58. Ali Akbar Razavi, S. and Morsali, A., Linker functionalized metal–organic frameworks. *Coord. Chem. Rev.*, 399, 213023, 2019.
59. Schmidt, A. and Dreger, A., Recent advances in the chemistry of pyrazoles. Properties, biological activities, and syntheses. *Curr. Org. Chem.*, 15, 1423–1463, 2011.
60. Li, N., Xu, J., Feng, R., Hu, T.-L., Bu, X.-H., Governing metal–organic frameworks towards high stability. *Chem. Commun.*, 52, 8501–8513, 2016.
61. Halcrow, M.A., Pyrazoles and pyrazolides—Flexible synthons in self-assembly. *Dalton Trans.*, 52.55, 2059–2073, 2009.
62. Pettinari, C., Tăbăcaru, A., Galli, S., Coordination polymers and metal–organic frameworks based on poly (pyrazole)-containing ligands. *Coord. Chem. Rev.*, 307, 1–31, 2016.
63. Burrows, A.D., Chan, S., Gee, W.J., Mahon, M.F., Richardson, C., Sebestyen, V.M., Turski, D., Warren, M.R., The impact of N, N′-ditopic ligand length and geometry on the structures of zinc-based mixed-linker metal–organic frameworks. *CrystEngComm*, 19, 5549–5557, 2017.

64. Zhang, J.-P., Zhang, Y.-B., Lin, J.-B., Chen, X.-M., Metal Azolate Frameworks: From Crystal Engineering to Functional Materials. *Chem. Rev.*, 112, 1001–1033, 2012.
65. Li, Z., Zhang, Z., Ye, Y., Cai, K., Du, F., Zeng, H., Tao, J., Lin, Q., Zheng, Y., Xiang, S., Rationally tuning host-guest interactions to free hydroxide ions within intertrimerically cuprophilic metal-organic frameworks for high OH- conductivity. *J. Mater. Chem. A*, 5, 7816–7824, 2017.
66. He, T., Zhang, Y.-Z., Wu, H., Kong, X.-J., Liu, X.-M., Xie, L.-H., Dou, Y., Li, J.-R., Functionalized Base-Stable Metal-Organic Frameworks for Selective CO2 Adsorption and Proton Conduction. *ChemPhysChem*, 18, 3245–3252, 2017.
67. Padial, N.M., Quartapelle Procopio, E., Montoro, C., López, E., Oltra, J.E., Colombo, V., Maspero, A., Masciocchi, N., Galli, S., Senkovska, I., Kaskel, S., Barea, E., Navarro, J.A.R., Highly Hydrophobic Isoreticular Porous Metal-Organic Frameworks for the Capture of Harmful Volatile Organic Compounds. *Angew. Chem.*, 125, 8448–8452, 2013.
68. Hitzbleck, J., O'Brien, A.Y., Forsyth, C.M., Deacon, G.B., Ruhlandt-Senge, K., Heavy Alkaline Earth Metal Pyrazolates: Synthetic Pathways, Structural Trends, and Comparison with Divalent Lanthanoids. *Chem.—Eur. J.*, 10, 3315–3323, 2004.
69. Yang, H., Zhang, X., Zhang, G., Fei, H., An alkaline-resistant Ag(i)-anchored pyrazolate-based metal-organic framework for chemical fixation of CO2. *Chem. Commun.*, 54, 4469–4472, 2018.
70. Zhang, J.-P. and Kitagawa, S., Supramolecular Isomerism, Framework Flexibility, Unsaturated Metal Center, and Porous Property of Ag(I)/Cu(I) 3,3',5,5'-Tetrametyl-4,4'-Bipyrazolate. *J. Am. Chem. Soc.*, 130, 907–917, 2008.
71. Colombo, V., Galli, S., Choi, H.J., Han, G.D., Maspero, A., Palmisano, G., Masciocchi, N., Long, J.R., High thermal and chemical stability in pyrazolate-bridged metal-organic frameworks with exposed metal sites. *Chem. Sci.*, 2, 1311–1319, 2011.
72. Quartapelle Procopio, E., Fukushima, T., Barea, E., Navarro, J.A., Horike, S., Kitagawa, S., A soft copper (II) porous coordination polymer with unprecedented aqua bridge and selective adsorption properties. *Chem.—Eur. J.*, 18, 13117–13125, 2012.
73. Fu, H.R. and Zhang, J., Structural Transformation and Hysteretic Sorption of Light Hydrocarbons in a Flexible Zn–Pyrazole–Adenine Framework. *Chem.—Eur. J.*, 21, 5700–5703, 2015.
74. Choi, H.J., Dinca, M., Long, J.R., Broadly hysteretic H2 adsorption in the microporous metal-organic framework Co (1, 4-benzenedipyrazolate). *J. Am. Chem. Soc.*, 130, 7848–7850, 2008.
75. Fu, H.-R. and Zhang, J., Flexible porous zinc-pyrazole-adenine framework for hysteretic sorption of light hydrocarbons. *Cryst. Growth Des.*, 15, 1210–1213, 2015.

76. Pettinari, C., Tăbăcaru, A., Galli, S., Coordination polymers and metal–organic frameworks based on poly(pyrazole)-containing ligands. *Coord. Chem. Rev.*, 307, 1–31, 2016.
77. Matuszak, C. and Matuszak, A., Imidazole-versatile today, prominent tomorrow. *J. Chem. Educ.*, 53, 280, 1976.
78. Aromí, G., Barrios, L.A., Roubeau, O., Gamez, P., Triazoles and tetrazoles: Prime ligands to generate remarkable coordination materials. *Coord. Chem. Rev.*, 255, 485–546, 2011.

7

Functional Metal–Organic Frameworks by Oxygen and Sulfur Based Functions

Abstract
The content of this chapter is about functional groups based on O and S atoms. Most important functions in these groups are hydroxy, ether, enoxide, thiol, sulfide and sulfonate. Application of these functions in gas adsorption, separation, conductivity and detection is discussed. Also, a succinct discussion about metal sulfonate/carboxylate frameworks is presented.

Keywords: Thiol, sulfide, thiourea, hydroxyl, ether, sulfonate, proton conductivity, metal-sulfonate frameworks

7.1 Functionalized Metal–Organic Frameworks by Oxygen Based Functions

7.1.1 Function–Application Properties

O-atom is highly electronegative with high density of electrons on it. In this regard functional groups based on O-atom like hydroxy, ether, azoxy, enoxide and oxadiazole are electron-rich, Lewis base and hydrogen bond acceptor. So, they can interact with different types of guests like dipolar or quadrupolar molecules through polar/quadrupolar interactions, Lewis acids through donor–acceptor interactions and H-bond donor molecules. These chemical properties are effective in construction of functional MOFs.

As mentioned, O-based functions could commonly interact with electron deficient species. In this regard, hydroxy, ether, azoxy and oxadizole decorated MOFs applied for CO_2 separation through functional group (O)·(C)CO_2 donor–acceptor or polar/quadrupolar interactions [1–17], H_2 adsorption [18, 19] through dipole(O center of functional group)

Ali Morsali and Sayed Ali Akbar Razavi. Functional Metal-Organic Frameworks: Structure, Properties and Applications, (133–164) © 2021 Scrivener Publishing LLC

quadrupole (H_2) interaction as well as detection of metal ions [8, 20–24] through oxygen based donor–acceptor interactions.

Hongwei Hou and coworkers synthesized an azoxy functionalized MOF with formula $[Zn_3L_3(BPE)_{1.5}]_n$ (H_2L = 4,4'-azoxydibenzoic acid, BPE = bis(4-pyridyl)ethylene) and applied it for selective high capacity Pb(II) removal (616.64 mg·g^{-1}) from aqueous solution (Figure 7.1) [25] XPS analysis reveals that strong electrostatic attraction and coordination interactions between the highly accessible (O^-) of azoxy groups and Pb^{2+} ions is responsible for the adsorption process.

Yunwu Li and coworkers applied two O-rich MOFs ($[Ln(L2)(H_2O)(DMF)]_n$ where Ln is Eu in Eu-L2 and Tb in Tb-L2) for luminescent detection of metal ions. H_3L2 at room temperature exhibits an emission peak at 454 nm (λ_{ex} = 324 nm) (Figure 7.2) [26]. Eu-L2 and Tb-L2 represent characteristic fluorescent peaks of Eu at 592, 620, 653, and 702 nm and Tb at 490, 546, 585, and 622 nm, respectively. Except for the characteristic emission peaks of Eu(III) and Tb(III) ions, Eu-L2 and Tb-L2 present additional broad emission peaks at about 450 nm, respectively which are similar to the ligand of H_3L2. These emissions are assigned to the blue emissions of the n–π or π–π* intraligand electronic transition, indicating that the energy of the ligand only transfers partly to the Ln(III) ions within these complexes. Application of these MOFs in metal ions detection represent that Eu-L2 and Tb-L2 are highly sensitive to Fe(III) metal ions. The K_{sv} values of Eu-L2 and Tb-L2 were 3.10 and 2.89 × 10^{+4} M^{-1}, respectively. The detection limits of 1.57 μM (for Eu-L2) and 0.91 μM (for Tb-L2) proved high sensitivity of these complexes. Based on X-ray photoelectron

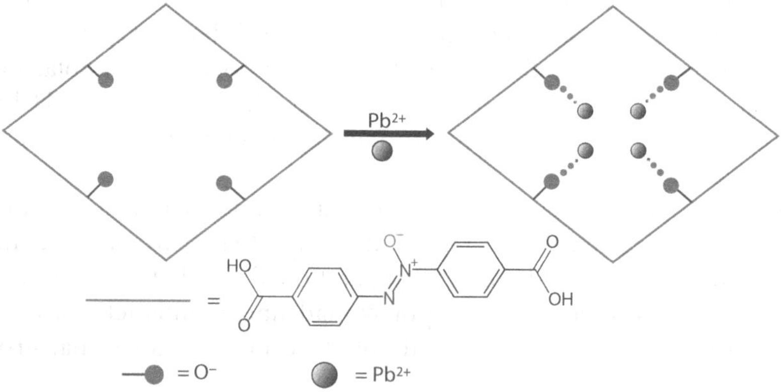

Figure 7.1 Possible interaction between azoxy functionalized framework and Pb(II) ions [25].

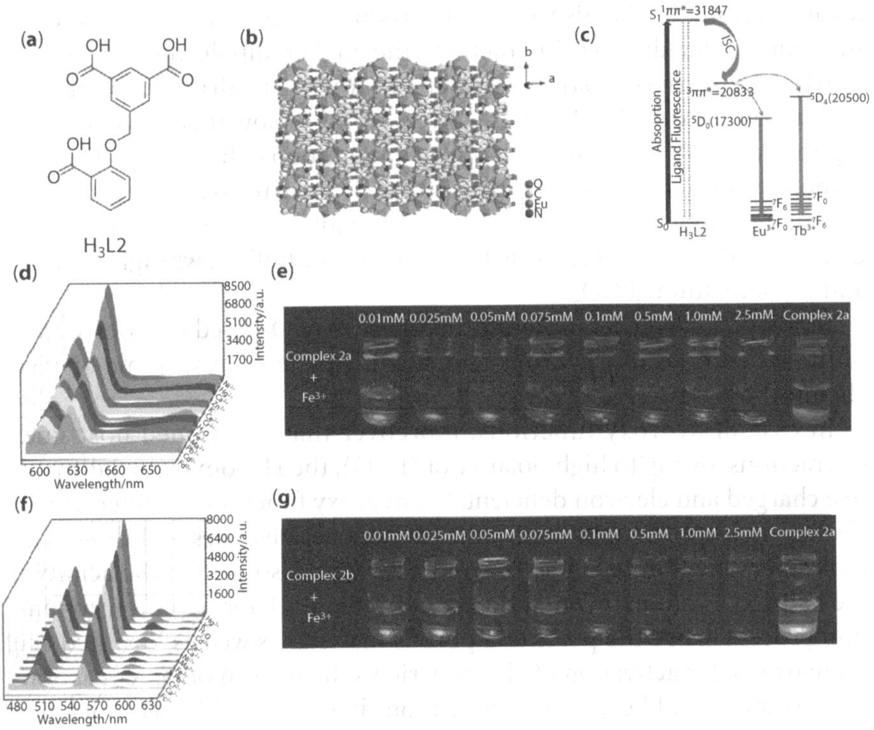

Figure 7.2 Application of Eu-L2 and Tb-L2 in metal ion detection. (a) structure of H_3L2 ligand. (b) 3D structure of Eu-L2 from the c axis. (c) Schematic of the energy-transfer process of Eu-L2. (d) Luminescence spectra and intensity of $^5D_0 \rightarrow {}^7F_2$ transitions (620 nm) of Eu-L2 dispersed in aqueous solutions with different cations (10^{-3} M). (f) Luminescence spectra and intensity of $^5D_4 \rightarrow {}^7F_5$ transitions (546 nm) of Tb-L2 dispersed in aqueous solutions with different cations (10^{-3} M). (e, g) Photos of luminescence quenching effects of complexes Eu-L2 and Tb-L2 with the presence of Fe(III) at different concentrations in aqueous solutions under UV light irradiation at 365 nm [26].

spectroscopy (XPS), the O(1s) peaks display a slight shift from 531.78 to 531.83 eV for Eu-L2 and from 531.90 to 532.07 eV for Tb-L2, suggesting the probable weak binding of Fe(III) ions to uncoordinated oxygen atoms within the frameworks.

The nature of interaction between CO_2 molecules and organic functional groups can be donor–acceptor interaction or dipole–quadrupole interaction. In donor–acceptor interaction, with increasing the nucleophilicity or Lewis basicity of the donor atom stronger interaction is established with C-atom of carbon dioxide. The ultimate interaction in this form can result in chemisorption of carbon dioxide on functional groups. In a dipole–quadrupole interaction, the higher dipole moment of the functions give rises in stronger interaction with quadrupole moment of the carbon dioxide. De-Li Chen and coworkers show that up on increasing in the dipole moment of the functional groups, the higher affinity to carbon dioxide is observed (Figure 7.3) [27]. Since the oxadiazole (6.23 D) has higher dipole moment compared to thiadiazole (4.80 D) and selenadiazole (2.61 D), the oxadiazole functionalized MOF shows higher affinity and capacity toward CO_2.

Moreover than general chemical properties of O-based functions, some of these functions like hydroxy and enoxide benefits from some of other chemical properties.

In case of hydroxy functions, moreover that mentioned host-guest interactions, owing to high polarity of (H–O), the H atom is partially positive charged and electron deficient. So, hydroxy function can interact with electron-rich species, molecules with partially negative sites and those guest molecules able to interact as H-bond acceptor. Also, due to the acidity of the hydroxy H atom, this function could act as initiator of proton conduction process inside the pores of a porous material as well as Brønsted acid catalytic site for activation of electron-rich substrates in organic reactions. Also, H atom could exchange with cations like Li(I) and Na(I) which help to insert ionic ($-O^-Li^+$) sites in the pores of functional MOFs. These features are beneficial for improvement of host–guest interaction of hydroxy functional MOFs.

In comparison of –OH and $-NH_2$ we can state that both functions can interact with C and O atoms of CO_2 molecules. But it is necessary to mention that owing to higher polarity of O–H bond rather N–H bond, there is stronger interaction between O atom of hydroxy and C atom of CO_2 as

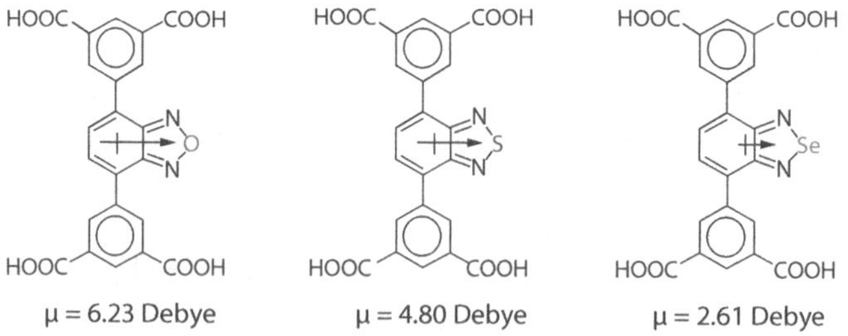

Figure 7.3 Representation of oxadiazole, thiadiazole, and selenadiazole ligands and their related dipole moment [27].

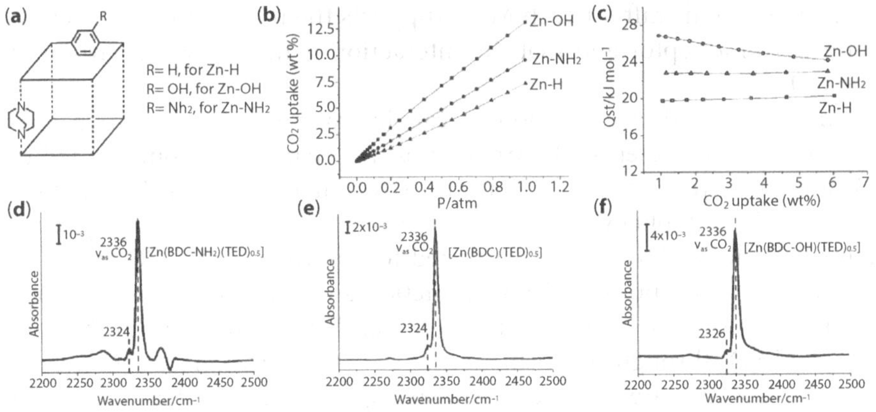

Figure 7.4 Application of Zn–H, Zn–OH and Zn–NH_2 in CO_2 adsorption. (a) Schematic representation of the modification on the MOF pore surface. (b) CO_2 adsorption-desorption isotherms at 298 K. (c) Isosteric heats of CO_2 adsorption (Q_{st}) values. (d) Room temperature IR absorption spectra of CO_2 in Zn–NH_2 (d), Zn–H (e) and Zn–OH (f) [28].

well as stronger interaction between H atom of hydroxy and O atom of CO_2. These interactions could be reinforced by higher polarity of hydroxy O-H bond.

Li and coworkers synthesized three isostructure MOFs and applied them in CO_2 separation (Figure 7.4) [28]. These MOFs are non-functional $[Zn(BDC)(TED)_{0.5}]\cdot2DMF\cdot0.2H_2O$ (Zn–H) parent framework, hydroxy functionalized $[Zn(BDC\text{-}OH)(TED)_{0.5}]\cdot1.5DMF\cdot0.3H_2O$ (Zn–OH) framework and amine functionalized $[Zn(BDC\text{-}NH_2)(TED)_{0.5}]\cdot xDMF\cdot yH_2O$ (Zn–NH_2) framework, where H_2BDC = terephthalic acid, TED = triethylenediaminc, H_2BDC-OH = 2-hydroxylterephthalic acid, H_2BDC-NH_2=2-aminoterephthalic acid. Brunauer-Emett–Teller (BET) surface areas of these MOFs are 1,937 $m^2\cdot g^{-1}$ for Zn-H, 1023 $m^2\cdot g^{-1}$ for Zn–OH and 1,081 $m^2\cdot g^{-1}$ for Zn–NH_2. CO_2 adsorption measurements show that Zn–OH could adsorb higher amount of CO_2 rather other two frameworks which clarify that the role of OH groups in Zn–OH are more dominant rather higher surface area of parent Zn–H framework and amine decorated Zn–NH_2 framework. Calculation of adsorption enthalpy at low coverage shows that Zn–OH framework has higher affinity to CO_2 molecules. The IR absorption spectroscopy measurements of CO_2 adsorption were also carried out on all three compounds. In comparison with unperturbed $\nu_{as}(CO_2)$ in gas phase at 2,349 cm^{-1}, a red-shift in $\nu_{as}(CO_2)$ for all three component is observed (Δ = 13 cm^{-1} for Zn-H and Zn–NH_2 with $\nu_{as}(CO_2)$ at 2,336 cm^{-1} and Δ = 11 cm^{-1} for Zn–OH with $\nu_{as}(CO_2)$ at 2,338 cm^{-1}).

The value of red shift for each MOF supports the finding that CO_2 acts as an electron acceptor and that the interaction is involving the carbon of the CO_2.

In case of hydroxy function, it can interact with electron rich species because of presence of partially positive charged H atom bonded to O atom of hydroxy. NOTT-300 (with formula $[Al_2(OH)_2(L1)](H_2O)_6)$ where H_4L1 is biphenyl-3,3',5,5'-tetracarboxylic acid) as a hydroxyl containing framework has been developed to remove harmful CO_2 and SO_2 gases [29]. *In situ* powder X-ray diffraction and inelastic neutron scattering studies combined with modeling reveal that hydroxyl groups bind to CO_2 and SO_2 through the formation of O=C(S)=O(δ^-)...H(δ^+)–O hydrogen bonds, which are reinforced by weak supramolecular interactions with C–H atoms on the aromatic rings of the framework. Moreover, hydroxy decorated MOFs are able to interact with analytes like cathecol containing H-bond donor/acceptor site in their molecular structure. In addition, owing to acidity of hydroxy group, this function could apply as originator in proton conduction process. Possible host–guest interactions between CO_2 molecules and hydroxy function are depicted in Figure 7.5.

Considering the ability of hydroxy function as H-bond donor or Brønsted acidic site, this function is applied to activate epoxy compounds in different reactions. Recently, we stabilized L-(+)-tartrate anion with two chiral hydroxy functions on the metal nodes of MIL-101-Cr(III) through post-synthesis anion exchange and applied for asymmetric alkene epoxidation [30]. In another work, En-Qing Gao and coworkers had proved that introduction of hydroxy sites is an effective strategy in efficient conversion of CO_2 through activation of epoxide substrates by hydroxy(-OH)·(-O-) epoxy hydrogen bonding (Figure 7.6) [31].

Owing to acidity of hydroxy function and its ability to exchange the H^+ ions with cations like Li(I), cation exchange process via solution method conducted to improve the affinity between guest molecules and MOF adsorbent. After exchange, (-OH) is changed into ionic (-O-Li^+) sites which are effective in interaction with polar or quadrupolar molecules

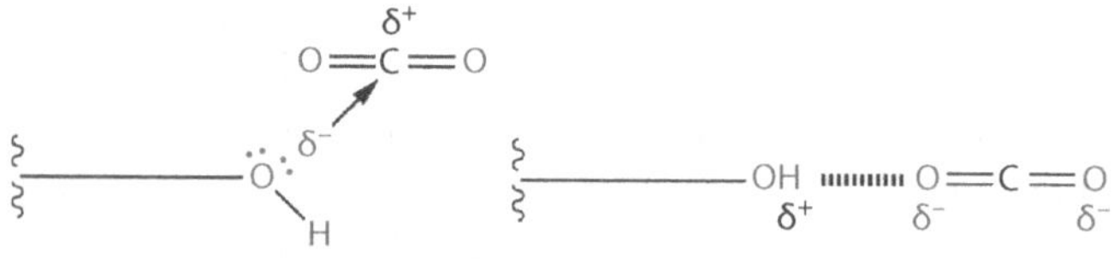

Figure 7.5 Possible host–guest interactions between CO_2 molecules and hydroxy function.

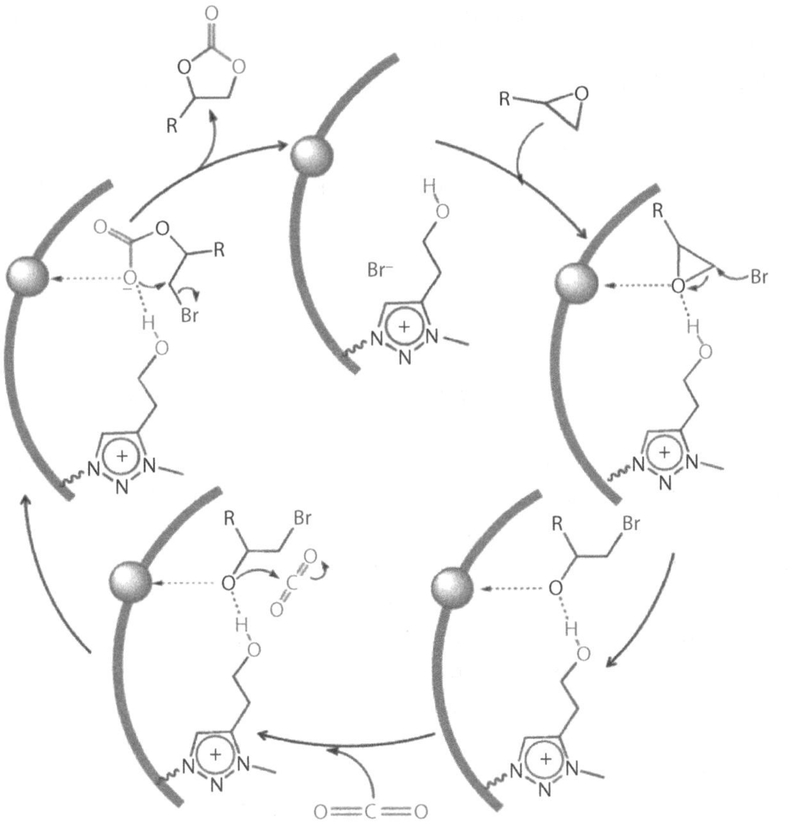

Figure 7.6 The role of free hydroxy function in activation of epoxy ring in CO_2 conversion reaction into organic carbonate. In this scheme, triazolium ring is bonded to the main chain of the framework and Br^- anions are counter ions in the pores to neutralize the framework charge. Hydroxy function is effective in activation of epoxide and Br^- anions attack to the epoxy ring to open it and make it ready for addition of CO_2 molecules [31].

owing to their highly charge separated nature. For example, the H atoms of hydroxyl functions of IR-MOF-8-OH and IR-MOF-14-OH are exchanged with Li^+ ions to improve the framework-H_2 interactions [32]. In another work, hydroxyl-functionalized (Zn_2(TCPB)(DPG)) framework, TCPB = 1,2,4,5-tetrakis(4-carboxyphenyl)-benzene and DPG = meso-1,2-bis(4-pyridyl)-1,2-ethanediol, shows improved affinity to CO_2 molecules after Li^+–H^+ cation exchange through enhanced solid–gas interactions [33]. In another work, Cu^+–H^+ exchange was conducted to provide proper catalytic sites in CO_2 hydrogenation reaction [34].

We know that nitrogen atom roles as electron-rich or Lewis basic site but in case of enoxide functional group, more electronegative O atom is bonded to N atom. As results, N atom is partially positive and O atom is partially negative with active non-bonding electron-pairs. These active non-bonding electron-pairs are effective in improvement of framework-CO_2 affinity through enoxide(O)·(C)CO_2 donor–acceptor interactions [14] On the other hand, the charge separated nature of N(δ^+)·O(δ^-) is beneficial for interaction with both anionic and cationic species [35–37].

7.1.2 Function–Structure Properties

Hard-soft chemistry is a very promising strategy to control structure and functionality of MOFs (Figure 7.7). It is well-known that MOF-5 is synthesized based on connection of H_2BDC ligand (1,4-benzenedicarboxylic acid) with Zn(II) ions. Functionalization of H_2BDC ligand with -OH and -SH results in 2,5-dihydroxy-1,4-benzenedicarboxylic acid (H_2BDC-$(OH)_2$) and 2,5-dimercapto-1,4-benzenedicarboxylic acid (H_2BDC-$(SH)_2$). Combination of H_2BDC-$(OH)_2$ with Co(II) ions or other metal ions like Mg(II) results in synthesis of MOF-74 (Figure 7.8). This is due to the fact hard hydroxy functions which are adjacent to carboxy groups, are coordinated to metal ions as well as carboxy groups. So, combination of metal ions with H_2BDC-$(OH)_2$ and H_2BDC ligands gives rise in MOFs with different topology. Combination of H_2BDC-$(SH)_2$ and Zr(IV) ions resulted in UiO-66$(SH)_2$ framework which is isostructure with UiO-66. In the structure of UiO-66$(SH)_2$, -SH group are free because -SH are soft and Zr(IV) centers are hard. So, despite the fact that -SH groups are adjacent to carboxy functions of the ligand, they do not coordinate to metal ions and remain free. To achieve hydroxy functionalized MOF-5, hydroxy functions should be protected. After synthesis of the framework with protected

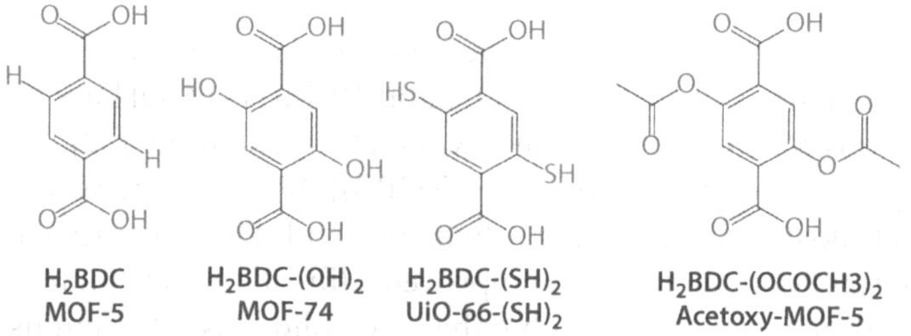

Figure 7.7 H_2BDC based ligand for construction of MOFs based on hard-soft chemistry post-synthesis methods.

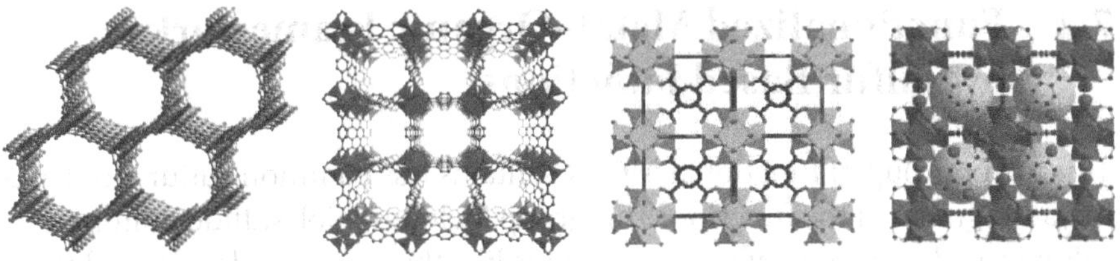

Figure 7.8 Structure of MOF-5, MOF-74, UiO-66 and UiO-66$(SH)_2$. In the UiO-66$(SH)_2$, the violet spheres are SH groups.

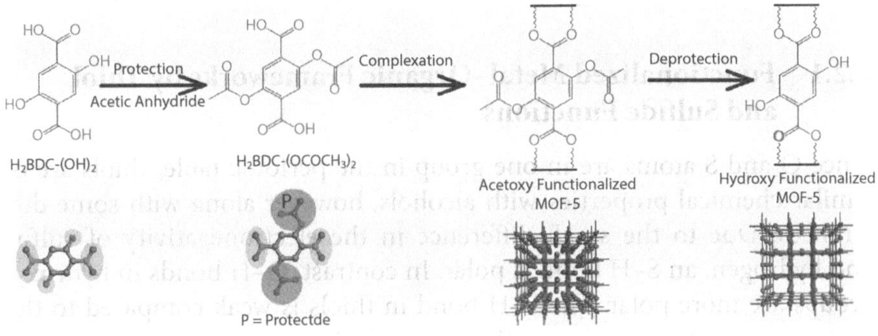

Figure 7.9 Protection of $H_2BDC-(OH)_2$ by acetoxy groups and deprotection of $H_2BDC-(OCOCH_3)_2$ ligand to achieve hydroxy functionalized MOF-5 [39].

acetoxy functions, they are subsequently deprotected. So the pore walls are decorated with hydroxy groups (Figure 7.9) [38, 39].

Due to the possible rotation in C-X-C motif, where X is O, S and NH, ether functionalized MOFs are flexible and able to structural transformation upon guest adsorption/removal or exposure to external stimuli. Possible rotations and vibrations around etheric C-O-C bond, made ether functionalized MOFs responsive against temperature or pressure [40, 41]. Foster and coworkers reported that incorporating alkyl ether functional groups between the layers is of benefit for exfoliation and interacting with solvent molecules to enhance dispersion [42]. In another work by Fischer and coworkers, a MOF functionalized with alkoxy function in side chain is developed which possesses extreme thermomechanical properties like temperature responsive breathing and thermal expansion because of increased thermal motions (vibrational and rotational movements) of alkoxy functions in side chains as a results of temperature lifting [43].

7.2 Functionalized Metal–Organic Frameworks by Sulfur Based Functions

This major group of functional MOFs contains less common sulfur atom as a major fragment of the function. This group includes thiol, sulfide, sulfonate or sulfonic acid and some of other functions like thiourea, thiadazole, and thiocathecole. The advantages of this group of functionalized MOFs is that they contain a soft and electron rich sulfur atom which is of benefit for designing electron rich and polar frameworks with soft guest interactive sites. However, despite such advantages, S-containing FMOFs are limited because (I) compared to O atom, S atom is heavier leading to lower surface area ($cm^2 \cdot g^{-1}$) and (II) the (–S–) fragment is partially oxidizable in higher temperatures.

7.2.1 Functionalized Metal–Organic Frameworks by Thiol and Sulfide Functions

Since O and S atoms are in one group in the periodic table, thiols are of similar chemical properties with alcohols, however along with some differences. Due to the small difference in the electronegativity of sulfur and hydrogen, an S–H bond is polar. In contrast, O–H bonds in hydroxyl groups are more polar. The S–H bond in thiols is weak compared to the O–H bond in alcohols. So, relative to the alcohols, thiols are more acidic.

With metal ions, thiolates behave as ligands to form transition metal thiolate complexes. The term mercaptan is derived from the Latin mercurium captans (capturing mercury) because the thiolate group bonds so strongly with mercury compounds. According to hard/soft acid/base (HSAB) theory, sulfur is a relatively soft (polarizable) atom. This explains the tendency of thiols to bind to soft elements/ions such as mercury, lead, cadmium and iodine. The stability of metal thiolates parallels that of the corresponding sulfide minerals.

Owing to hard-soft-acid-base (HSAB) theory, thiol and sulfide functions are electron-rich so that they can interact selectively with soft and heavy metal ions especially Hg(II). In this regard, soft-soft acid–base interactions or donor–acceptor coordination interactions are effective. Also, since thiols are acidic, they can adsorb metal ions through proton exchange mechanism. Moreover, since these functions are electron rich and polarizable they can interact with electron deficient species and other kind of soft and polarizable guests, respectively. In these conditions, physical electrostatic interactions are important as well as donor–acceptor interactions.

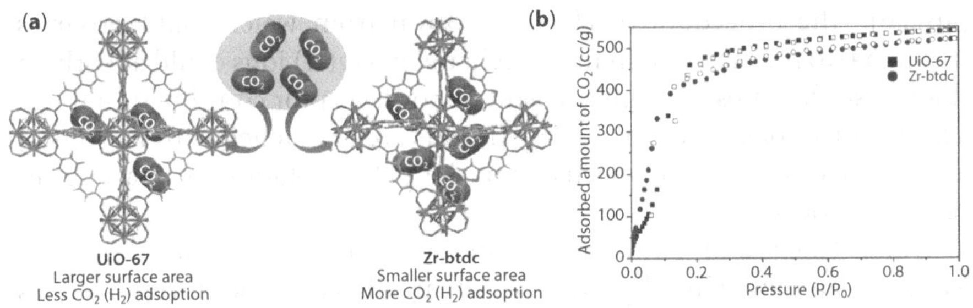

Figure 7.10 Application of Zr-btdc in gas adsorption. (a) Structural representation of UiO-67 and Zr-btdc. (b) CO_2 adsorption and desorption isotherms of Zr-btdc and UiO-67 at 195 K [44].

As a polarizable and electron-rich adsorbents, thiophene decorated MOFs applied in gas adsorption and the results show that presence of thiophene sites is effective in improvement of framework-CO_2 or H_2 interactions [44, 45]. Minyoung Yoon and coworkers synthesized a thiophene decorated MOF (Zr-btdc), isostructure with UiO-67 based on 2,2'-bithiophene-5,5'-dicarboxylic acid ligand (Figure 7.10) [44]. The results of N_2 adsorption show that BET surface area of Zr-btdc is 10% smaller than UiO-67. CO_2 adsorption measurements at 67 K reveal that in a pressure range between 0 and 0.1, Zr-btdc has higher adsorption capacity while total adsorbed CO_2 amount for UiO-67 (540 cc·g^{-1}) is larger than that of Zr-btdc (520 cc·g^{-1}). H_2 adsorption isotherms of UiO-67 and Zr-btdc at 77 K show that in spite of the smaller surface area, Zr-btdc shows a 1.5 times larger H_2 sorption capacity (2.30 wt%) than UiO-67 (1.61 wt%). Combination of the electrostatic potential surface map of both ligands applied in the structure of UiO-67 and Zr-btdc reveals that there is slightly negative potential on the surface of btdc ligand because of electron-rich electronegative S atoms while there is partially positive charge on the surface of applied ligand in the structure of UiO-67. Consistently, calculation of isosteric heats of adsorption shows Zr-btdc framework has higher affinity to both CO_2 and H_2 molecules (5.3 vs 7.2 kJ·mol^{-1} for H_2 and 17.1 vs 18.4 kJ·mol^{-1} for CO_2). Clearly, incorporation of electronegative atoms in a ligand can increase the interaction between gas molecules, resulting in an increase in the gas sorption capacity of MOFs.

Considering the soft electron-rich nature of sulfur atom, functional MOFs based on thiol and sulfide functions applied in removal and sensing of heavy metal ions, especially Hg(II). Removal of mercury ions is the most important and applied application for thiol/sulfide decorated MOFs

[46–56]. The procedure of Hg(II) removal from water could progress through two mechanism. In first mechanism, Hg(II) ions could adsorb on thiol or sulfide sites through soft-soft donor–acceptor coordination interactions. In second mechanism, Hg(II) ions are exchanged with H^+ ions of acidic (-SH) site. In this mechanism, the pH of solution decreases after removal process.

Qihui Chen and coworkers synthesized a novel sulfur decorated MOF based on SCN^- ligand, named as FJI-H12 and formulated as $[Co_3(Timt)_4(SCN)_6(H_2O)_{14}(EtOH)]_n$ (Timt is 2,4,6-tri(1-imidazolyl)-1,3,5-triazine), and applied it in efficient removal of Hg(II) ions (Figure 7.11) [47]. Structural analysis shows that the SCN^- ligand is coordinated to metal ions from N and S atoms are free. The results of metal ion adsorption experiments show that 200 mg of FJI-H12 can remove Hg(II) ions from a 50 ml solution of 20.45 ppb at room temperature after 12 h. Final concentration of Hg(II) in the solution was less than 0.02 ppb based on ICP instrument. They attributed such efficient removal process to the strong interaction between Hg(II) ions and S atom of SCN^- ligand. Based on FT-IR analysis, the typical stretch mode of SCN^- groups shows a large shift from 2,072 cm^{-1} in FJI-H12 to 2,125 cm^{-1} in FJI-H12-Hg, indicating the strong binding interactions between Hg(II) and SCN^- groups.

Hard-Soft acid-based theory is applied for construction of thiol functionalized UiO-66 framework with formula Zr-DMBD based on 2,5-dimercapto-1,4-benzenedicarboxylic acid (H_2DMBD) organic linker and Zr(IV) metal ions. This MOF applied for removal of Hg(II) metal ions (Figure 7.12) [54]. Adsorption analysis show that 10 mg Zr-DMBD can remove Hg(II) ions from aqueous solution with an efficiency of over 99% with initial concentration of 10 ppm. Characterization experiments of Hg(II)@Zr-DMBD sample based on Raman and FT-IR analysis reveal that

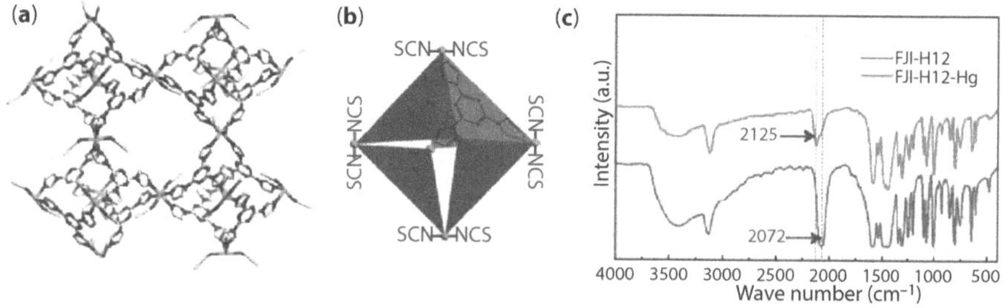

Figure 7.11 Application of FJI-H12 in Hg(II) removal. (a) Structure of FJI-H12. (b) Schematic representation of one cage with free-standing NCS^- groups. (c) FT-IR spectra of the material before and adsorption process [47].

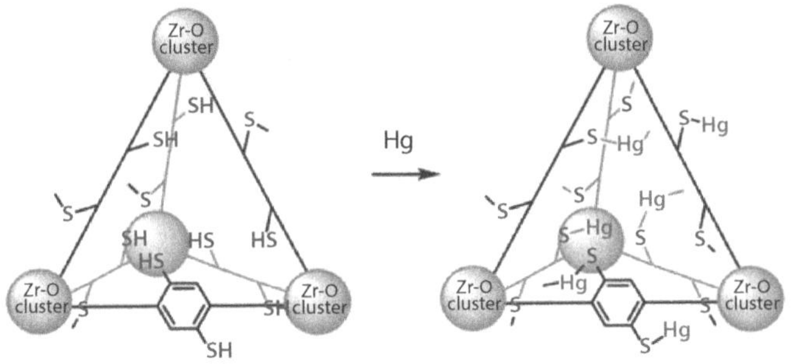

Figure 7.12 Adsorption of Hg ions on thiol sites of Zr-DMBD [54].

characteristic S-H stretching at 2,560 cm^{-3} for Zr-DBMD became absent in the Hg^{2+}@Zr-DMBD and a strong band at 355 cm^{-1} appeared which is consistent with the Hg-S stretching bond. These changes clearly show that Hg(II) ions are replaced with protons and Hg-S bond is formed. Moreover, Zr-DBMD can adsorb mercury vapors when immersed in a sand bath and heated up to 140 °C for 24 h.

Moreover than Hg(II), thiol and sulfide decorated MOFs applied in detection of other heavy metal ions [57–59]. Thioether-alkene functionalized ASMOF-5 had been synthesized based on 2,6-dithioalloxyterephthalic acid and applied for detection of metal ions (Figure 7.13) [58]. The results show that this MOF is able to detect Pd(II) ion through selective visual color change depending on the concentration of Pd(II) ions along with noticeable 1D quenching in photoluminescence spectra of the material. They repeated these experiments with other kind of ligands, non-functional ligand (MOF-5 framework) and thioether based ligand without alkene groups (SESMOF-5 framework), and observe no color change. They mentioned that presence of both thiol and alkene groups are essential because the sulfur-conjugated aromatic core features intense photoluminescence and alkene unit is able to interact with noble metal species like Pd(II) due to its distinct π–donor and π–acceptor characters.

Thiol or sulfide decorated MOFs not only applied in detection and removal of metal ions, but owing to their redox activity (particularly in case of -SH), polarizable nature and ability to interact through donor-acceptor or charge transfer interactions, they applied in removal and process of other chemicals like oxoanions (Cr(IV), As(V), Br(IV)) and Iodine [60–66]. Jinlou Gu and coworkers applied alkyl thiol-containing Zr-based MOFs (Zr-MSA and Zr-DMSA) for efficient removal of toxic Cr(VI) from aqueous media using mercaptosuccinic acid (MSA) and meso-dimercaptosuccinic acid (DMSA) as ligands (Figure 7.14) [67]. Owing to strong donor–acceptor interactions between Cr(IV) ions and (-SH) groups,

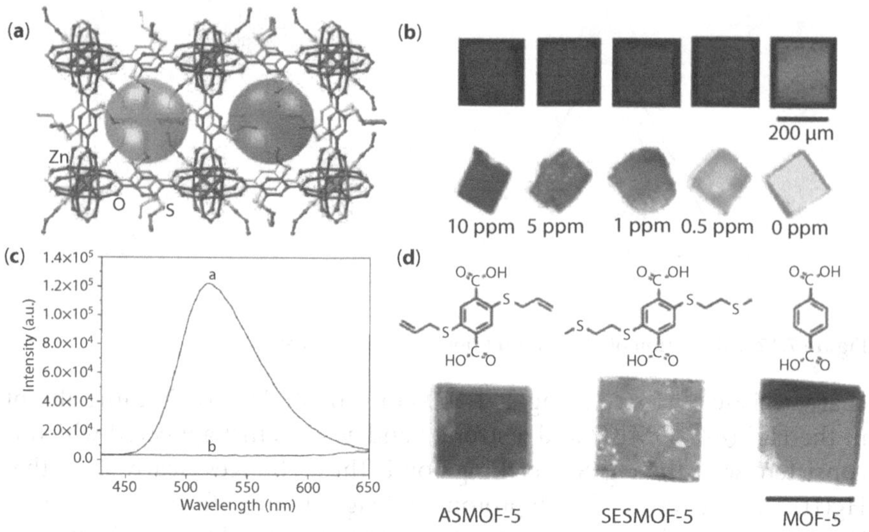

Figure 7.13 Application of ASMOF-5 in Pd(II) detection. (a) Crystal structure of ASMOF-5. (b) Photographs of the ASMOF-5 crystals under 365 nm UV radiation (top row) and under natural light (bottom row) before and after immersion at 80 °C in CH_2Cl_2 solutions of $PdCl_2(CH_3CN)_2$ of various concentrations (as labeled). Durations of immersion: top row, 1 h; bottom row, 4 h. (c) Room-temperature solid-state emission spectra (a) as-made ASMOF-5 (b) a $PdCl_2$ loaded sample of ASMOF-5 (excitation wavelength λ_{ex} = 360 nm). (d) Structures of the molecules (top) and photographs(bottom row) of the corresponding crystals of ASMOF-5, SESMOF-5 and MOF-5 (under natural light) after immersion (for 5 min at 80 °C, and then 3 h at rt) in saturated CH_3CN solutions of $Pd(CH_3CN)_2Cl_2$ (1.0 ml each, about 0.25% w/w). The scale bar is 200 μm [58].

Cr(VI)-thiolate complex are formed through cation exchange process which is followed by a redox reaction between thiol function and Cr(IV) species. As a result of this redox reaction, thiol is transformed to sulfonic acid and Cr(IV) is converted to Cr(III). The Cr(III) species could be removed from the solution through sulfonate(O)·Cr(III) interactions. As results of these host–guest interactions, Zr-MSA and Zr-DMSA could adsorb 202.0 and 138.7 mg·g^{-1} Cr(IV), respectively.

Owing to their binding strength and exchangeable acidic H atom, thiol functions applied as anchoring site for immobilization of metal ions for catalytic and photocatalytic applications [68, 69]. In this type of applications, -SH groups are not directly attended in the catalytic reaction, but they are stabilizer or the catalytically of photocatalytically active metal ions in the pores of the MOFs.

Up-to-date, the $-NH_2$ has been evidenced as the most efficient functional group to reduce the band gap of a given MOF as compared to other

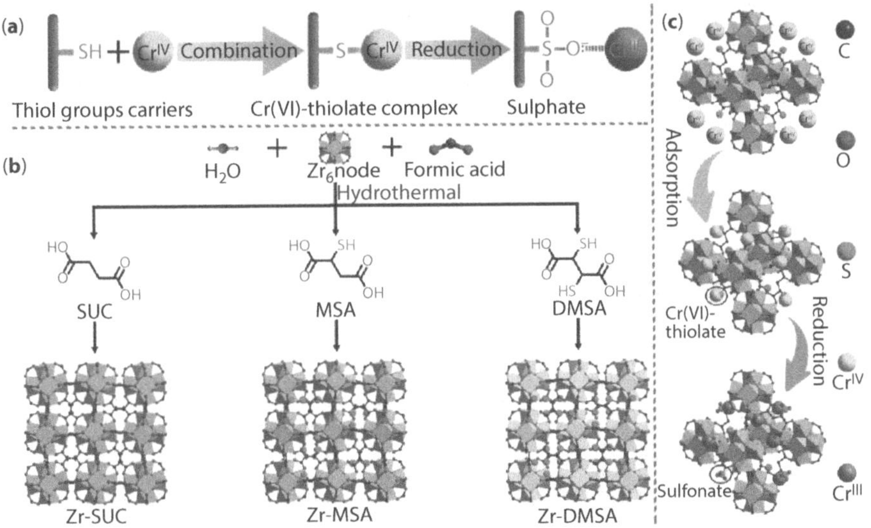

Figure 7.14 (a) Schematic illustration of the degradation of Cr(VI) by thiol-containing compounds. (b) The aqueous-phase synthesis of thiol-free and thiol-containing UiO-style MOFs. (c) Adsorption and degradation of Cr(VI) by thiol-containing MOFs [67].

groups such as –Br, -OH, –SH and $–NO_2$. Chong-Qing Wan and coworkers synthesized isostructural UiO-66-$(SO_3H)_2$, UiO-66-$(SH)_2$ and UiO-66-$(SCH_3)_2$ frameworks as n-type semiconductors to evaluate the role of S-based functions in photophysical properties of MOFs (Figure 7.15) [70]. They show that electronic properties of the functional groups play key roles in determination of the band gap (E_g) and photoactive properties of the MOFs. Combination of UV–Vis adsorption spectra of these functional MOFs and their parent UiO-66 framework shows that main adsorption band of the UiO-66 in UV region (edge of 320 nm), while the band edges of UiO-66-$(SO_3H)_2$, UiO-66-$(SH)_2$ and UiO-66-$(SCH_3)_2$ are around 425, 440 and 470 nm, respectively. This red shift for functional MOFs, 150 nm for UiO-66-$(SCH_3)_2$, 120 nm for UiO-66-$(SH)_2$ and 105 nm for UiO-66-$(SO_3H)_2$ is in according to the electron-donating capacity of the functions ((-SCH_3) > (-SH) > (-SO_3H)). These results are consistent with their yellow color changed from that white of UiO-66. In other word, functional groups with a larger electron-donating capacity to the aromatic linker may be beneficial, which results in a higher level of the highest occupied molecular orbital that thus markedly narrows the band gap (E_g). Measured band gap values are 2.80, 3.02 and 3.22 eV for UiO-66-$(SCH_3)_2$, UiO-66-$(SH)_2$ and UiO-66-$(SO_3H)_2$, respectively, being much lower than that 4.0 eV of the

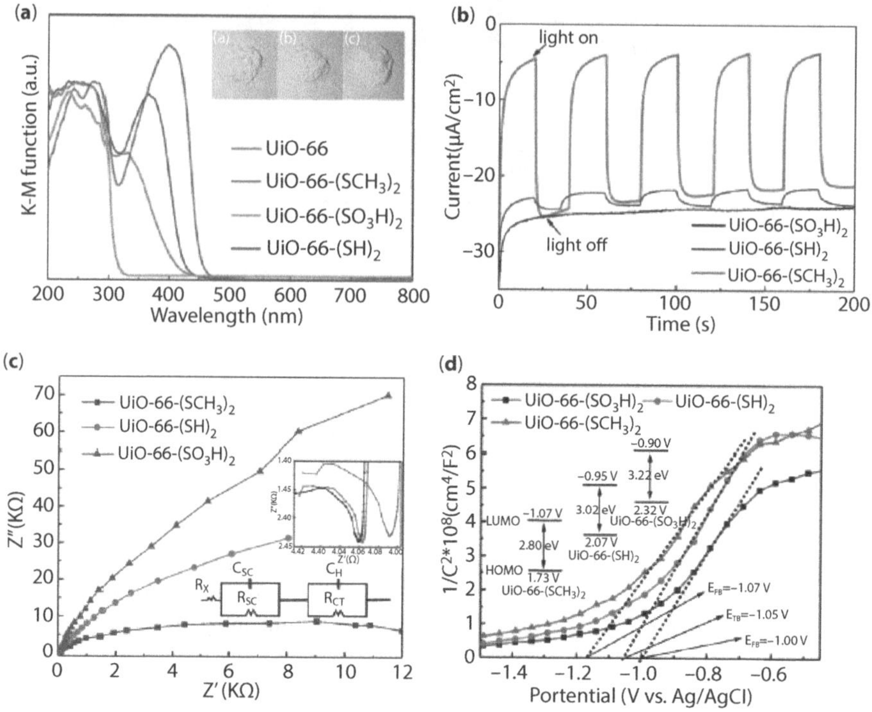

Figure 7.15 Photophysical properties of UiO-66-$(SO_3H)_2$, UiO-66-$(SH)_2$ and UiO-66-$(SCH_3)_2$. (a) UV–vis absorption spectra of functional UiO-66 frameworks with S-based functions. (b) Current/time plots of the materials measured at an applied bias potential of 0.19 V vs. RHE with and without visible-light irradiation (>400 nm) pulse of 20 s. (c) Nyquist and (d) Mott–Schottky plots of UiO-66-X2 in 0.1M H_2SO4 [70].

parent UiO-66. Consistently, the current-time curves for UiO-66-$(SCH_3)_2$, UiO-66-$(SH)_2$ and UiO-66-$(SO_3H)_2$ follow the same order. Such distinct photocurrents of functional MOFs reflect their different interface charge separation efficiencies. Electrochemical impedance spectroscopy shows that the capacitance arc measured for UiO-66-$(SCH_3)_2$ is much smaller than that of UiO-66-$(SH)_2$ and UiO-66-$(SO_3H)_2$, suggesting the highest electron mobility and a lowest electron-hole recombination rate for UiO-66-$(SCH_3)_2$. Calculated flat-band potential based on Mott-Schottky equation are −1.00, −1.05 and −1.17 V vs. Ag/AgCl for UiO-66-$(SO_3H)_2$, UiO-66-$(SH)_2$ and UiO-66-$(SCH_3)_2$, respectively. These results clearly show that -(SCH_3) groups with high electron-donating capacity are really effective in improvement of photophysical properties of MOFs through reduction is band gap and improved photo-induced electron generations.

7.2.2 Functionalized Metal–Organic Frameworks by Sulfonate-Sulfonic Acid Function

Sulfonate $\left(-SO_3^-\right)$/sulfonic acid ($-SO_3H$) is another S-based function which is extensively applied in the structure of MOFs as coordinating and guest interactive site. The kind of application and host–guest chemistry of this function highly depended on the fact that this function is protonated or not. So, sulfonate $\left(-SO_3^-\right)$ and sulfonic acid ($-SO_3H$) have their particular application in the structure of functional MOFs. In most cases of construction of porous sulfonate-based MOFs, one or more oxygens in each sulfonate function remain uncoordinated when forming metal–sulfonate nodes. As a result, the pore surface of these MOFs is enriched with free O atom which is beneficial for development of highly polar and hydrophilic pores. Such polar porosity could improve proton conductivity of sulfonate decorated MOFs as well as increased CO_2 affinity compared with their carboxylate-based counterparts.

Since the sulfonic acid ($-SO_3H$) is a strong Brønsted acid, sulfonic acid functionalized MOFs are applied as catalyst in acid catalyzed reactions [71–76] and since the Brønsted acidity of MOFs is highly correlated with their proton conductivity, sulfonic FMOFs are used as platforms for proton conductivity [75, 77–86]. Also, ($-SO_3H$) group as H-bond donor can interact with guest molecules containing electron-rich sites [87].

Chang Seop Hong and coworkers synthesized a sulfonic acid functionalized UiO-66 framework (UiO-66$(SO_3H)_2$) with the aim of proton conductivity (Figure 7.16) [77]. Owing to presence of sulfonic acid groups and hydrophilic nature of the pores of UiO-66$(SO_3H)_2$, this material shows very high affinity to water molecules which is beneficial to generate well-established hydrogen-bonded water networks along the proton transport pathways. Since the proton-donation ability of a functional group to the hydrogen-bonded water network is governed by the pK_a value, it anticipate that UiO-66$(SO_3H)_2$ has higher conductivity rather UiO-66$(SH)_2$, which in itself shows higher proton conductivity rather UiO-66. To determine the proton conductivity (σ) of each sample, alternating-current (ac) impedance spectrum of a pelletized sample was measured from 25 to 80 °C at 90%RH. The bulk conductivity (3.5×10^{-7} S·cm^{-1}) of UiO-66 at 25 °C and 90% RH increased to 4.3×10^{-6} S·cm^{-1} at 80 °C and 90% RH for UiO-66$(SH)_2$ by introduction of -SH groups. The conductivity of UiO-66$(SO_3H)_2$ increased by more than four orders of magnitude to 1.4×10^{-2} S·cm^{-1} at 25 °C and 90% RH when compared to that of UiO-66. At 80 °C, the conductivity was much more pronounced, with a value of 8.4×10^{-2} S·cm^{-1}. This remarkable conductivity results from the existence of strong

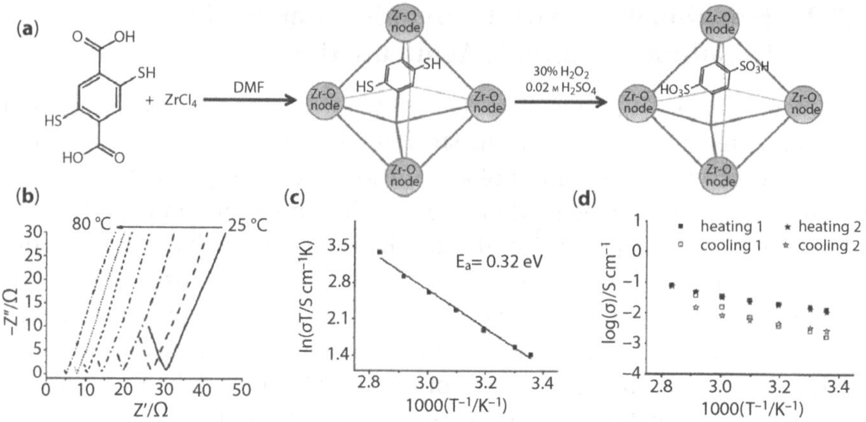

Figure 7.16 Application of UiO-66(SO_3H)$_2$ in proton conductivity. (a) Synthesis of UiO-66(SH)$_2$ and post-synthetic oxidative modification of UiO-66(SH)$_2$ to UiO-66(SO_3H)$_2$. Impedance spectra, (b) Arrhenius plot (c), and log-scaled proton conductivities (d) for the heating-cooling cycles of UiO-66(SO_3H)$_2$ at 90% RH [77].

Brønsted acid sites (-SO_3H groups) on the organic linkers. These acidic groups facilitate preferential adsorption of water molecules into the confined spaces, which enables organization of the hydrophilic domains and establishes favorable proton-transport pathways, similar to those observed in Nafion.

In another work, Samar K. Das and coworkers proved that moreover than introduction of sulfonic acid groups into the pores, the length of the side chain has remarkable effect on the proton conductivity of the material (Figure 7.17) [88]. In this regard, they post-synthetically modified Ui-66-NH_2 to PSM-1 and PSM-2 frameworks functionalized with pendant alkyl chain holding the -SO_3H group is of different length. PSM-1 displays the highest MOF-based proton conductivity (1.64×10^{-1} S·cm^{-1}) at 80 °C, which is comparable to commercially available Nafion, while PSM-2 shows significantly lower conductivity (4.6×10^{-3} S·cm^{-1}). The variation in length of the dangling alkyl chains can result in the change in several parameters like pK_a of each sulfonic acid group in the pores of PSM-1 and PSM-2. The theoretical pK_a values of the dangling -SO_3H group of PSM 1 and PSM-2 are 3.47 and 4.91. As a result of such higher pK_a, PSM-1 has higher conductivity and lower activation energy than PSM-2.

Proton conductivity of PSM-1 and PSM-2 highly depended to humidity. The water-assisted proton conductivity of PSM-1 can be compared to that of the UiO-66-(SO_3H)$_2$. Meanwhile, PSM-1 has only one-fourth concentration of Brønsted acid groups compared to UiO-66(SO_3H)$_2$ but almost twice

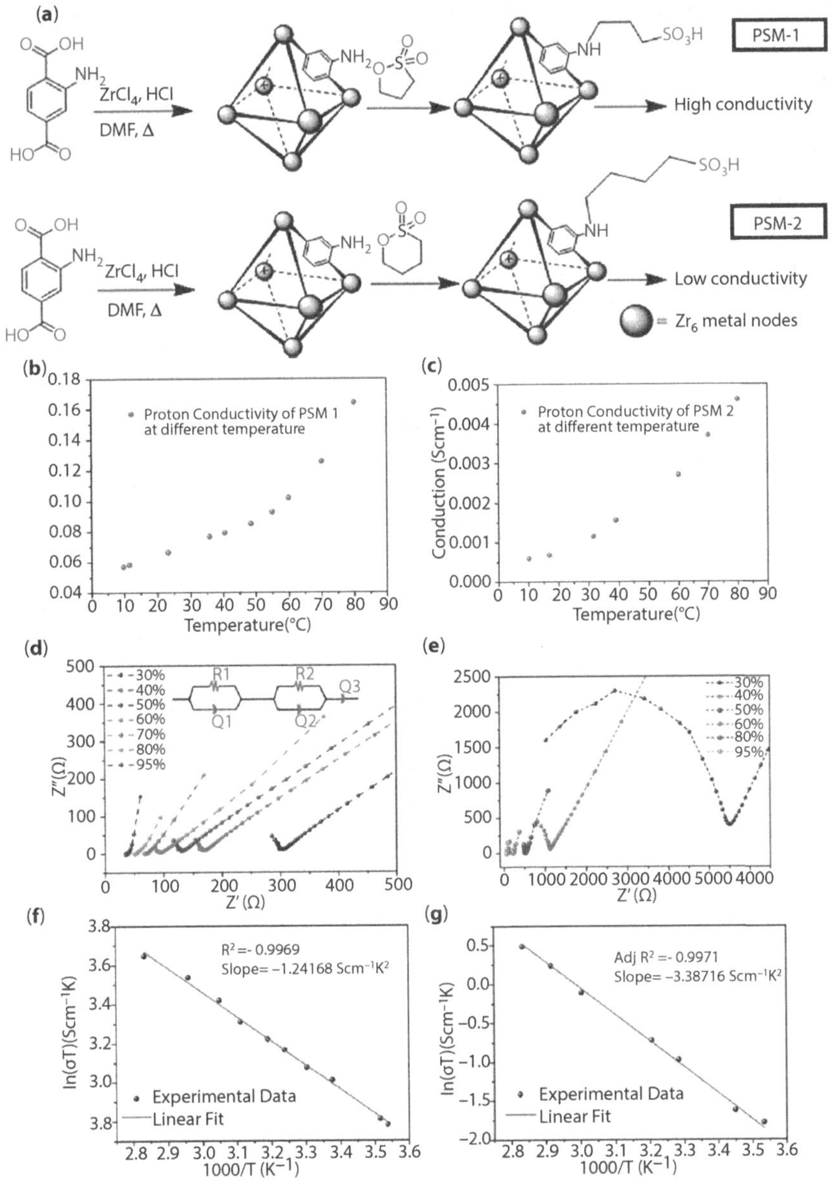

Figure 7.17 Application of PSM-1 and PSM-2 in proton conductivity. (a) Synthesis of UiO-66-NH_2 and its post-synthetic modification to PSM-1 (Top) and PSM-2 (Bottom). Proton conductivity recorded during heating cycle of (b) PSM-1 and (c) PSM-2. (d) Nyquist plot of impedance spectra of PSM-1 under different humidification conditions at 30 °C. Inset: equivalent circuit of impedance spectra of PSM-1 and PSM-2. (e) Nyquist plot of impedance spectra of PSM-2 under different humidification conditions at 30 °C. Arrhenius plots of temperature dependence of proton conduction of PSM-1 (f) and PSM-2 (g) [88].

efficient in comparison to UiO-66(SO_3H)$_2$. A comprehensive assessment of the two suggests that the extent of availability of labile protons plays the most influential role in the case of PSM-1 and not the concentration of the Brønsted acid groups as observed in the case of UiO-66- (SO_3H)$_2$.

Sulfonate $\left(-SO_3^-\right)$ shows different behaviors in gas adsorption, sensing and removal of analytes. As Lewis basic site, Sulfonate $\left(-SO_3^-\right)$ can detect metal ions through donor-acceptor coordination interactions. Also, electrostatic attraction between MOFs containing highly polar and active ($-SO_3^-$) groups is expected to facilitate the adsorption of cationic guests via electrostatic attraction. In addition, $\left(-SO_3^-\right)$ is ideally prefect to interact with H-bond donor guests. Considering these advantageous of Sulfonate $\left(-SO_3^-\right)$ group, sulfonate decorated MOFs applied in C_2H_2 [89], CO_2 [90–94] and H_2 adsorption [91, 95], removal of cationic dyes [96], detection and separation of metal ions [82, 97, 98] and small molecules like acetylacetone [66] and picric acid [99] which are able to interact with sulfonate as H-bond donor analytes.

Baiyan Li and coworkers synthesized MIL-101-SO_3Ag through post-synthesis cation (Ag(I)) exchange within the MIL-101-SO_3H and applied it for separation of unsaturated C2 hydrocarbons from saturated hydrocarbons (Figure 7.18) [89]. Under 1 atm pressure, the ethylene uptake amounts increase from 42 $cm^3{\cdot}g^{-1}$ at 296 K and 37 $cm^3{\cdot}g^{-1}$ at 318 K for MIL-101-Cr-SO_3H to 73 $cm^3{\cdot}g^{-1}$ at 296 K and 63 $cm^3{\cdot}g^{-1}$ at 318 K for MIL-101-Cr-SO_3Ag. Although the ethylene uptake had increased remarkably but the uptake amounts of ethane at 296 K and 318 K for two samples are comparable. These observations clearly show that –SO_3Ag are effective in improvement of ethylene capacity while would not increase the ethane uptake capacity. Calculation of isosteric heats of adsorption at zero loading

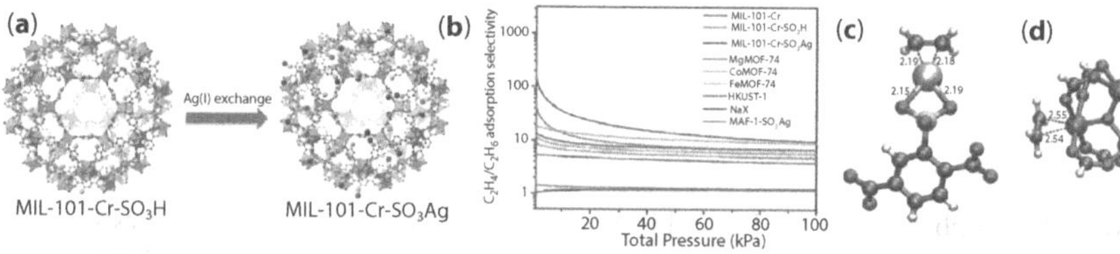

Figure 7.18 Application of MIL-101-Cr-SO_3Ag in selective uptake of ethylene. (a) Schematic illustration of Ag(I) exchange in MIL-101-Cr-SO_3H to afford MIL-101-Cr-SO_3Ag. (b) Comparison C_2H_4–C_2H_6 adsorption selectivities for MIL-101-Cr-SO_3Ag with MIL-101-Cr-SO_3H and other porous materials at 318 K. The optimized position of an ethylene molecule within (c) ($C_6H_3(CO_2)_2$)SO_3Ag and (d) $[Cr_3O(O_2CH)_6]^+$ clusters [89].

clearly shows that this for ethylene in MIL-101-Cr-SO_3Ag is 63 kJ·mol^{-1}, remarkably higher than that of MIL-101-Cr-SO_3H (10 kJ·mol^{-1}). *In situ* IR spectroscopic studies and computational calculations reveal that –$SO_3Ag(Ag^+)$·(π-bond)ethylene π-complexation contributes dominantly to the high ethylene–ethane adsorption selectivity as well as interaction between coordinatively unsaturated Cr(III) centers and ethylene π-bond.

In another work, Bing Li and coworkers synthesized MIL-101-SO_3H and applied it for removal of cationic dyes including methylene blue (MB) and malachite green (MG) (Figure 7.19) [96]. The results show that the adsorption capacity of these MOFs is highly dependent to the pH value. pH is a very important value in the dye adsorption efficiency by MOFs because the pH of solutions affects not only the structure of the dye molecules, but also the chemical properties of adsorbent, such as surface charge and the dissociation of functional groups of the adsorbent. Experimental measurements show that up on increase in the pH value of the solution, the adsorption capacity of the material is increased and in low pHs, the adsorption capacity is reduced. When the pH of dye solution is low, the presence of excess H^+ ions can restrain the ionization of the sulfonic acid, so the fully protonated sulfonic acid, -SO_3H, is predominance. The adsorption capacity of the dyes is low in the absence of electrostatic attraction between adsorbent and adsorbate. As the pH of the dye solution becomes higher, the sulfonic acid is gradually turned into $-SO_3^-$ and the negative charges on the surface of MIL-101-SO_3H increase, which is favorable for the adsorption of positively charged MB and MG due to electrostatic attraction.

As an antenna functional group within the pores of MOFs, sulfonic acid function is highly effective for improvement in photophysical property

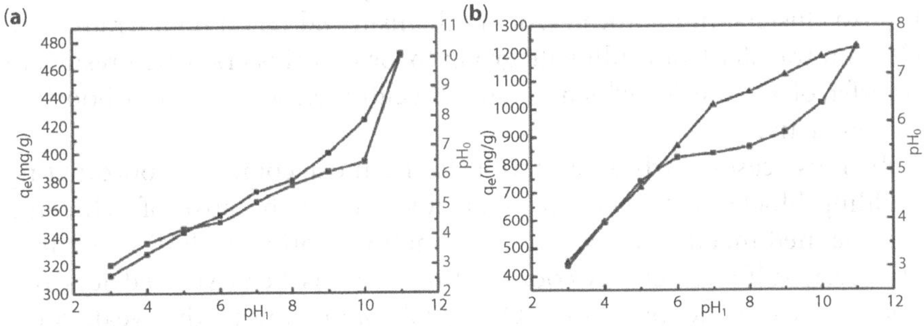

Figure 7.19 Application of MIL-101-SO_3H in removal of cationic dyes. Effect of pH on the adsorption capacities of MB (a) and MG (b), and the equilibrium pH_e after adsorption. (c_0 = 200 and 500 mg·L^{-1} for MB and MG, respectively; m = 20.0 mg; V = 50 ml) [96].

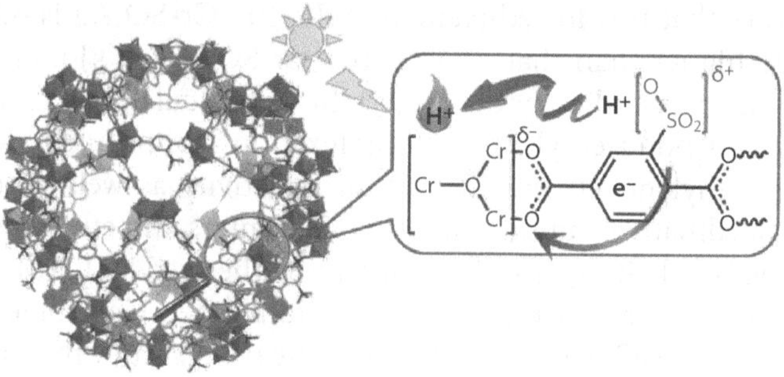

Figure 7.20 Light-enhanced acid-catalytic reaction over MIL-101-SO_3H, in which electron is transferred from $-SO_3^-$ groups to Cr-oxo clusters facilitating the release of protons from the MOF [100].

of the MOFs such as light-enhanced photocatalytic activity. Jiang and coworkers reported that esterification reaction between benzyl alcohol and acetic acid can be catalyzed by MIL-101-SO_3H as a photocatalyst platform (Figure 7.20) [100]. MIL-101 does not show any specific improved catalytic activity up on light irradiation in this reaction while investigations show that MIL-101-SO_3H represents noticeable catalytic activity as well as improved catalytic activity up on light irradiation. Without light irradiation and presence of MIL-101-SO_3H as catalyst, the achieved reaction conversion is 83.8% with 100% selectivity in 322 min while up on light irradiation and presence of MIL-101-SO_3H, the reaction conversion reaches to 97% just in 160 min. This observation, improved catalytic activity for MIL-101-SO_3H up on light irradiation but not for MIL-101, shows that the presence of $-SO_3H$ function is the reason of the light responsivity of MIL-101-SO_3H. Simulation calculations show that the HOMO of the MIL-101-SO_3H is dominated by $-SO_3H$ group and LUMO is dominated by Cr-oxo clusters. In photocatalytic mechanism and up on light irradiation, electron transfer from sulfonate group to Cr-oxo clusters which results in transfer of H^+ species which results in acidity enrichment and boosting catalyst activity.

In most cases, carboxy function is the most common coordinating building blocks in the synthesis of novel MOFs because of achieving well-defined metal–carboxylate coordination motifs in MOF structures. However, sulfonic acid as coordinating site has been studied less frequently in the chemistry of MOFs, probably owing to the weak coordination tendency of the sulfonate oxygens toward metal centers. As coordinating sites, sulfonate-based ligands have certain advantages in construction of CPs including [101].

Due to the weak metal–sulfonate interaction, sulfonate based frameworks are not sufficiently robust to tolerate permanent porosity. Also, the weak interaction is useful for development of soft sulfonate based coordination polymers with dynamic structural flexibility. Sulfonate group can be considered as a trioxy anionic unit with local C3v symmetry. So, in comparison with planner bidentate carboxylate groups, sulfonate group provides potentially higher framework dimensionality, higher possibilities in coordination modes and lower degree of structural predictability. This may lead to lower crystallinity and poorly regular inorganic assembly. Also, the pore surface of sulfonate based coordination polymers are more polar that carboxylate based coordination polymers because normally two oxygen atoms of carboxylate are coordinated to metal ions providing non-polar porosity while for sulfonate normally the pores contain pendant S-O groups making the pores polar. In comparison of carboxylate and sulfonate it is possible to mention that sulfonate-based frameworks are less predictable and crystalline with lower porosity yet higher flexibility, thermal stability, and polarity.

One strategy in combination of advantageous of carboxylate and sulfonate groups is to develop coordination polymers and MOFs based on sulfonate–carboxylate ligands. In this type of multifunctional ligands, strong crystalline coordination strength of carboxylate groups and its high crystallinity is combined with higher dimensionally and flexible nature of sulfonate groups for development of new material with desirable properties.

7.2.3 Functionalized Metal–Organic Frameworks by Other S-Based Functions

There are some of other S-based functions like thiourea, thiadazole and thiocathecole applied in the structure of MOFs. Thiourea function applied as strong H-bond donor organocatalyst site in Morita–Baylis–Hillman reaction (Figure 7.21) [102]. In the catalytic mechanism thiourea catalytic site activated benzaldehyde through thiourea(NH)·(O=C)benzaldehyde hydrogen bonding and in next step two EtOH molecules can add to the benzaldehyde to synthesis the products. Thiadazole applied as electron donor site to interact with electron deficient guests. So, they applied in detection of soft and heavy metal ions and CO_2 separation [27, 103].

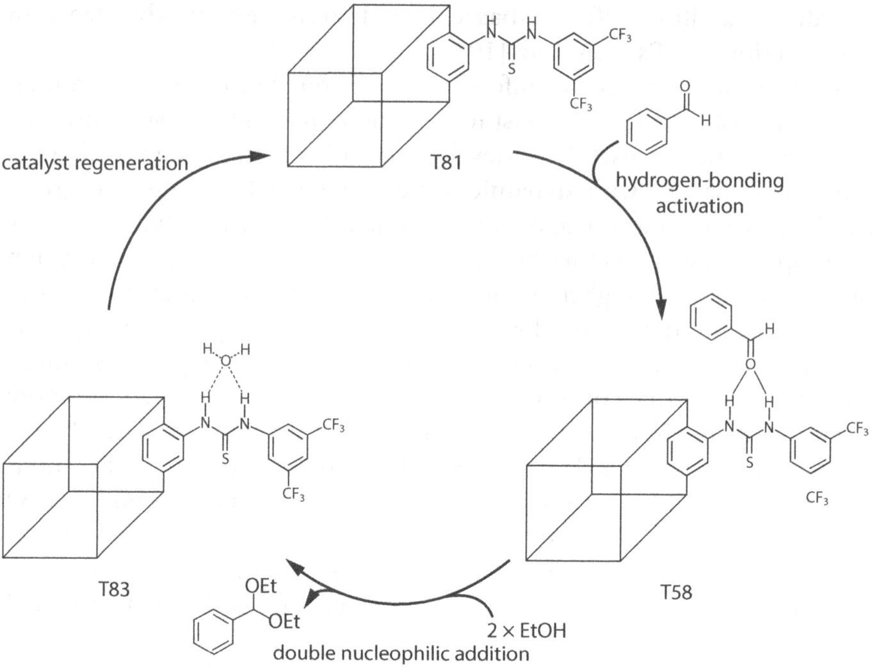

Figure 7.21 Proposed acetalization reaction mechanism and determination of first-order reaction kinetics of the catalyst [102].

References

1. Zhong, R., Liu, J., Huang, X., Yu, X., Sun, C., Chen, G., Zou, R., Experimental and theoretical investigation of a stable zinc-based metal–organic framework for CO2 removal from syngas. *CrystEngComm*, 17, 8221–8225, 2015.
2. Ling, Y., Yang, F., Deng, M., Chen, Z., Liu, X., Weng, L., Zhou, Y., Novel Iso-Reticular Zn(ii) Metal–Organic Frameworks constructed by Trinuclear-Triangular and Paddle-Wheel Units: Synthesis, Structure and Gas Adsorption. *Dalton Trans.*, 41, 4007–4011, 2012.
3. Chen, Z., Xiang, S., Arman, H.D., Li, P., Zhao, D., Chen, B., Significantly Enhanced CO2/CH4 Separation Selectivity within a 3D Prototype Metal–Organic Framework Functionalized with OH Groups on Pore Surfaces at Room Temperature. *Eur. J. Inorg. Chem.*, 2011, 2227–2231, 2011.
4. Wu, D., Gassensmith, J.J., Gouvêa, D., Ushakov, S., Stoddart, J.F., Navrotsky, A., Direct Calorimetric Measurement of Enthalpy of Adsorption of Carbon Dioxide on CD-MOF-2, a Green Metal–Organic Framework. *J. Am. Chem. Soc.*, 135, 6790–6793, 2013.

5. García-Pérez, E., Serra-Crespo, P., Hamad, S., Kapteijn, F., Gascon, J., Molecular simulation of gas adsorption and diffusion in a breathing MOF using a rigid force field. *Phys. Chem. Chem. Phys.*, 16, 16060–16066, 2014.
6. Yang, J., Yan, X., Xue, T., Liu, Y., Enhanced CO2 adsorption on Al-MIL-53 by introducing hydroxyl groups into the framework. *RSC Adv.*, 6, 55266–55271, 2016.
7. Lu, Z., Xing, Y., Du, L., He, H., Zhang, J., Hang, C., Isostructural functionalization by -OH and -NH2: Different contributions to CO2 adsorption. *RSC Adv.*, 7, 47219–47224, 2017.
8. Lan, Y.Q., Jiang, H.L., Li, S.L., Xu, Q., Mesoporous Metal–Organic Frameworks with Size-tunable Cages: Selective CO2 Uptake, Encapsulation of Ln3+ Cations for Luminescence, and Column-Chromatographic Dye Separation. *Adv. Mater.*, 23, 5015–5020, 2011.
9. Bai, L., Tu, B., Qi, Y., Gao, Q., Liu, D., Liu, Z., Zhao, L., Li, Q., Zhao, Y., Enhanced performance in gas adsorption and Li ion batteries by docking Li+ in a crown ether-based metal–organic framework. *Chem. Commun.*, 52, 3003–3006, 2016.
10. Henke, S., Schmid, R., Grunwaldt, J.-D., Fischer, R.A., Flexibility and Sorption Selectivity in Rigid Metal–Organic Frameworks: The Impact of Ether-Functionalised Linkers. *Chem. Eur. J.*, 16, 14296–14306, 2010.
11. Henke, S., Florian Wieland, D.C., Meilikhov, M., Paulus, M., Sternemann, C., Yusenko, K., Fischer, R.A., Multiple phase-transitions upon selective CO2 adsorption in an alkyl ether functionalized metal–organic framework—An *in situ* X-ray diffraction study. *CrystEngComm*, 13, 6399–6404, 2011.
12. Henke, S. and Fischer, R.A., Gated Channels in a Honeycomb-like Zinc-Dicarboxylate-Bipyridine Framework with Flexible Alkyl Ether Side Chains. *J. Am. Chem. Soc.*, 133, 2064–2067, 2011.
13. Thallapally, P.K., Tian, J., Radha Kishan, M., Fernandez, C.A., Dalgarno, S.J., McGrail, P.B., Warren, J.E., Atwood, J.L., Flexible (Breathing) Interpenetrated Metal–Organic Frameworks for CO2 Separation Applications. *J. Am. Chem. Soc.*, 130, 16842–16843, 2008.
14. Xiong, Y., Fan, Y.-Z., Yang, R., Chen, S., Pan, M., Jiang, J.-J., Su, C.-Y., Amide and N-oxide functionalization of T-shaped ligands for isoreticular MOFs with giant enhancements in CO2 separation. *Chem. Commun.*, 50, 14631–14634, 2014.
15. Munn, A.S., Amabilino, S., Stevens, T.W., Daniels, L.M., Clarkson, G.J., Millange, F., Lennox, M.J., Düren, T., Bourelly, S., Llewellyn, P.L., Walton, R.I., Metal–Organic Frameworks from Divalent Metals and 1,4-Benzenedicarboxylate with Bidentate Pyridine-N-oxide Co-ligands. *Cryst. Growth Des.*, 15, 891–899, 2015.
16. Song, C., Ling, Y., Jin, L., Zhang, M., Chen, D.-L., He, Y., CO2 adsorption of three isostructural metal–organic frameworks depending on the incorporated highly polarized heterocyclic moieties. *Dalton Trans.*, 45, 190–197, 2016.

17. Ma, L.N., Liu, Y., Li, Y.Z., Hu, Q.X., Hou, L., Wang, Y.Y., Three Lanthanide Metal–Organic Frameworks Based on an Ether-Decorated Polycarboxylic Acid Linker: Luminescence Modulation, CO2 Capture and Conversion Properties. *Chem.—Asian J.*, 15, 191–197, 2020.
18. Salih, Z.I., Guo, Y.-J., Zheng, J.-J., Zhao, X., Effect of modified linkers of MOF-5 on enhancing interaction energies: A theoretical study. *Comput. Theor. Chem.*, 1058, 28–33, 2015.
19. Orcajo, G., Montes-Andrés, H., Villajos, J.A., Martos, C., Botas, J.A., Calleja, G., Li-Crown ether complex inclusion in MOF materials for enhanced H2 volumetric storage capacity at room temperature. *Int. J. Hydrogen Energy*, 44.35, 19285–19293, 2019.
20. Luo, F., Chen, J.L., Dang, L.L., Zhou, W.N., Lin, H.L., Li, J.Q., Liu, S.J., Luo, M.B., High-performance Hg2+ removal from ultra-low-concentration aqueous solution using both acylamide- and hydroxyl-functionalized metal-organic framework. *J. Mater. Chem. A*, 3, 9616–9620, 2015.
21. Xu, H., Dong, Y., Wu, Y., Ren, W., Zhao, T., Wang, S., Gao, J., An -OH group functionalized MOF for ratiometric Fe3+ sensing. *J. Solid State Chem.*, 258, 258, 441–446, 2018.
22. Pal, S. and Bharadwaj, P.K., A Luminescent Terbium MOF Containing Hydroxyl Groups Exhibits Selective Sensing of Nitroaromatic Compounds and Fe(III) Ions. *Cryst. Growth Des.*, 16, 5852–5858, 2016.
23. Zhao, Y., Ouyang, H., Feng, S., Luo, Y., Shi, Q., Zhu, C., Chang, Y.-C., Li, L., Du, D., Yang, H., Rapid and selective detection of Fe (III) by using a smartphone-based device as a portable detector and hydroxyl functionalized metal–organic frameworks as the fluorescence probe. *Anal. Chim. Acta*, 1077, 160–166, 2019.
24. Mahmoud, M.E., Moussa, Z., Prakasam, T., Li, L., Abiad, M.G., Patra, D., Hmadeh, M., Lanthanides based metal organic frameworks for luminescence sensing of toxic metal ions. *J. Solid State Chem.*, 281, 121031, 2020.
25. Yu, C., Shao, Z., Hou, H., A functionalized metal–organic framework decorated with O− groups showing excellent performance for lead(ii) removal from aqueous solution. *Chem. Sci.*, 8, 7611–7619, 2017.
26. Mi, X., Sheng, D., Yu, Y.E., Wang, Y., Zhao, L., Lu, J., Li, Y., Li, D., Dou, J., Duan, J., Tunable light emission and multiresponsive luminescent sensitivities in aqueous solutions of two series of lanthanide metal–organic frameworks based on structurally related ligands. *ACS Appl. Mater. Interfaces*, 11, 7914–7926, 2019.
27. Song, C., Ling, Y., Jin, L., Zhang, M., Chen, D.-L., He, Y., CO 2 adsorption of three isostructural metal–organic frameworks depending on the incorporated highly polarized heterocyclic moieties. *Dalton Trans.*, 45, 190–197, 2016.
28. Zhao, Y., Wu, H., Emge, T.J., Gong, Q., Nijem, N., Chabal, Y.J., Kong, L., Langreth, D.C., Liu, H., Zeng, H., Li, J., Enhancing Gas Adsorption and Separation Capacity through Ligand Functionalization of Microporous Metal–Organic Framework Structures. *Chem. Eur. J.*, 17, 5101–5109, 2011.

29. Yang, S., Sun, J., Ramirez-Cuesta, A.J., Callear, S.K., David, W.I.F., Anderson, D.P., Newby, R., Blake, A.J., Parker, J.E., Tang, C.C., Schröder, M., Selectivity and direct visualization of carbon dioxide and sulfur dioxide in a decorated porous host. *Nat. Chem.*, 4, 887, 2012.
30. Berijani, K. and Morsali, A., Dual activity of durable chiral hydroxyl-rich MOF for asymmetric catalytic reactions. *J. Catal.*, 378, 28–35, 2019.
31. Zhou, L.J., Sun, W., Yang, N.N., Li, P., Gong, T., Sun, W.J., Sui, Q., Gao, E.Q., A Facile and Versatile "Click" Approach Toward Multifunctional Ionic Metal–organic Frameworks for Efficient Conversion of CO2. *ChemSusChem*, 12, 2202–2210, 2019.
32. Klontzas, E., Mavrandonakis, A., Tylianakis, E., Froudakis, G.E., Improving Hydrogen Storage Capacity of MOF by Functionalization of the Organic Linker with Lithium Atoms. *Nano Lett.*, 8, 1572–1576, 2008.
33. Bae, Y.-S., Hauser, B.G., Farha, O.K., Hupp, J.T., Snurr, R.Q., Enhancement of CO2/CH4 selectivity in metal–organic frameworks containing lithium cations. *Micropor. Mesopor. Mat.*, 141, 231–235, 2011.
34. Maihom, T., Wannakao, S., Boekfa, B., Limtrakul, J., Production of Formic Acid *via* Hydrogenation of CO2 over a Copper-Alkoxide-Functionalized MOF: A Mechanistic Study. *J. Phys. Chem. C*, 117, 17650–17658, 2013.
35. Hu, L., Lin, X.-M., Lin, J., Zhang, R.-Q., Zhang, D.-L., Cai, Y.-P., Structural diversity of Mn(ii), Zn(ii) and Pb(ii) coordination polymers constructed from isomeric pyridylbenzoate N-oxide ligands: Structures and electrochemical properties. *CrystEngComm*, 18, 9307–9315, 2016.
36. Aboutorabi, L., Morsali, A., Tahmasebi, E., Büyükgüngor, O., Metal–Organic Framework Based on Isonicotinate N-Oxide for Fast and Highly Efficient Aqueous Phase Cr(VI) Adsorption. *Inorg. Chem.*, 55, 5507–5513, 2016.
37. Shokouhfar, N., Aboutorabi, L., Morsali, A., Improving the capability of UiO-66 for Cr (VI) adsorption from aqueous solutions by introducing isonicotinate N-oxide as the functional group. *Dalton Trans.*, 47, 14549–14555, 2018.
38. Yamada, T. and Kitagawa, H.J.S.C., Synthesis of a novel isoreticular metal–organic framework by protection and complexation of 2, 5-dihydroxyterephthalic acid. *Supramol. Chem.*, 23, 315–318, 2011.
39. Yamada, T. and Kitagawa, H., Protection and Deprotection Approach for the Introduction of Functional Groups into Metal–Organic Frameworks. *J. Am. Chem. Soc.*, 131, 6312–6313, 2009.
40. Schneemann, A., Rudolf, R., Baxter, S.J., Vervoorts, P., Hante, I., Khaletskaya, K., Henke, S., Kieslich, G., Fischer, R.A., Flexibility control in alkyl ether-functionalized pillared-layered MOFs by a Cu/Zn mixed metal approach. *Dalton Trans.*, 48, 6564–6570, 2019.
41. Henke, S., Schneemann, A., Ẅtscher, A., Fischer, R.A., Directing the breathing behavior of pillared-layered metal–organic frameworks *via a* systematic library of functionalized linkers bearing flexible substituents. *J. Am. Chem. Soc.*, 134, 9464–9474, 2012.

42. Foster, J.A., Henke, S., Schneemann, A., Fischer, R.A., Cheetham, A.K., Liquid exfoliation of alkyl-ether functionalised layered metal–organic frameworks to nanosheets. *Chem. Commun.*, 52, 10474–10477, 2016.
43. Henke, S., Schneemann, A., Fischer, R.A., Massive Anisotropic Thermal Expansion and Thermo-Responsive Breathing in Metal–Organic Frameworks Modulated by Linker Functionalization. *Adv. Funct. Mater.*, 23, 5990–5996, 2013.
44. Yoon, M. and Moon, D., New Zr (IV) based metal–organic framework comprising a sulfur-containing ligand: Enhancement of CO2 and H2 storage capacity. *Micropor. Mesopor. Mat.*, 215, 116–122, 2015.
45. Parshamoni, S., Sanda, S., Jena, H.S., Konar, S., A copper based pillared-bilayer metal organic framework: Its synthesis, sorption properties and catalytic performance. *Dalton Trans.*, 43, 7191–7199, 2014.
46. Liu, T., Che, J.-X., Hu, Y.-Z., Dong, X.-W., Liu, X.-Y., Che, C.-M., Alkenyl/Thiol-Derived Metal–Organic Frameworks (MOFs) by Means of Postsynthetic Modification for Effective Mercury Adsorption. *Chem. Eur. J.*, 20, 14090–14095, 2014.
47. Liang, L., Chen, Q., Jiang, F., Yuan, D., Qian, J., Lv, G., Xue, H., Liu, L., Jiang, H.-L., Hong, M., In situ large-scale construction of sulfur-functionalized metal–organic framework and its efficient removal of Hg(ii) from water. *J. Mater. Chem. A*, 4, 15370–15374, 2016.
48. Li, M.-Q., Wong, Y.-L., Lum, T.-S., Sze-Yin Leung, K., Lam, P.K.S., Xu, Z., Dense thiol arrays for metal–organic frameworks: Boiling water stability, Hg removal beyond 2 ppb and facile crosslinking. *J. Mater. Chem. A*, 6, 14566–14570, 2018.
49. Ke, F., Qiu, L.-G., Yuan, Y.-P., Peng, F.-M., Jiang, X., Xie, A.-J., Shen, Y.-H., Zhu, J.-F., Thiol-functionalization of metal–organic framework by a facile coordination-based postsynthetic strategy and enhanced removal of Hg2+ from water. *J. Hazard. Mater.*, 196, 36–43, 2011.
50. Huang, L., He, M., Chen, B., Hu, B., A mercapto functionalized magnetic Zr-MOF by solvent-assisted ligand exchange for Hg2+ removal from water. *J. Mater. Chem. A*, 4, 5159–5166, 2016.
51. Sohrabi, M.R., Preconcentration of mercury(II) using a thiol-functionalized metal–organic framework nanocomposite as a sorbent. *Microchim. Acta*, 181, 435–444, 2014.
52. Ding, L., Luo, X., Shao, P., Yang, J., Sun, D., Thiol-functionalized Zr-based metal–organic framework for capture of Hg(II) through a proton exchange reaction. *ACS Sustainable Chem. Eng.*, 6.7, 8494–8502, 2018.
53. Halder, S., Mondal, J., Ortega-Castro, J., Frontera, A., Roy, P., A Ni-based MOF for selective detection and removal of Hg2+ in aqueous medium: A facile strategy. *Dalton Trans.*, 46, 1943–1950, 2017.
54. Yee, K.-K., Reimer, N., Liu, J., Cheng, S.-Y., Yiu, S.-M., Weber, J., Stock, N., Xu, Z., Effective mercury sorption by thiol-laced metal–organic frameworks: In strong acid and the vapor phase. *J. Am. Chem. Soc.*, 135, 7795–7798, 2013.

55. Jiang, S.-Y., He, W.-W., Li, S.-L., Su, Z.-M., Lan, Y.-Q., Introduction of Molecular Building Blocks to Improve the Stability of Metal–Organic Frameworks for Efficient Mercury Removal. *Inorg. Chem.*, 57, 6118–6123, 2018.
56. Li, G.-P., Zhang, K., Zhang, P.-F., Liu, W.-N., Tong, W.-Q., Hou, L., Wang, Y.-Y., Thiol-Functionalized Pores *via* Post-Synthesis Modification in a Metal–Organic Framework with Selective Removal of Hg (II) in Water. *Inorg. Chem.*, 58, 3409–3415, 2019.
57. Zhang, J., Xiong, Z., Li, C., Wu, C., Exploring a thiol-functionalized MOF for elimination of lead and cadmium from aqueous solution. *J. Mol. Liq.*, 221, 43–50, 2016.
58. He, J., Zha, M., Cui, J., Zeller, M., Hunter, A.D., Yiu, S.-M., Lee, S.-T., Xu, Z., Convenient detection of Pd (II) by a metal–organic framework with sulfur and olefin functions. *J. Am. Chem. Soc.*, 135, 7807–7810, 2013.
59. Xie, Y., Ye, G., Peng, S., Jiang, S., Wang, Y., Hu, X., Postsynthetic functionalization of water stable zirconium metal–organic frameworks for high performance copper removal. *Analyst*, 144, 4552–4558, 2019.
60. Munn, A.S., Millange, F., Frigoli, M., Guillou, N., Falaise, C., Stevenson, V., Volkringer, C., Loiseau, T., Cibin, G., Walton, R.I., Iodine sequestration by thiol-modified MIL-53(Al). *CrystEngComm*, 18, 8108–8114, 2016.
61. Yang, X.-Y., Yuan, S., Qin, J.-S., Lollar, C., Alsalme, A., Zhou, H.-C., A flexible thioether-based MOF as a crystalline sponge for structural characterization of liquid organic molecules. *Mater. Chem. Front.*, 1, 1764–1767, 2017.
62. Xiao, X., Deng, Y., Xue, J., Gao, Y., Adsorption of chromium by functionalized metal–organic frameworks from aqueous solution. *Environ. Technol.*, 1–13, 2019.
63. Munn, A.S., Millange, F., Frigoli, M., Guillou, N., Falaise, C., Stevenson, V., Volkringer, C., Loiseau, T., Cibin, G., Walton, R.I., Iodine sequestration by thiol-modified MIL-53 (Al). *CrystEngComm*, 18, 8108–8114, 2016.
64. Han, P. and Xia, Y., Thiol-functionalized metal–organic framework for highly efficient removal of bromate from water. *J. Environ. Chem. Eng.*, 6, 3384–3391, 2018.
65. Audu, C.O., Nguyen, H.G.T., Chang, C.-Y., Katz, M.J., Mao, L., Farha, O.K., Hupp, J.T., Nguyen, S.T., The dual capture of As V and As III by UiO-66 and analogues. *Chem. Sci.*, 7, 6492–6498, 2016.
66. Chakraborty, G., Das, P., Mandal, S.K., Polar Sulfone-Functionalized Oxygen-Rich Metal–Organic Frameworks for Highly Selective CO2 Capture and Sensitive Detection of Acetylacetone at ppb Level. *ACS Appl. Mater. Interfaces*, 12, 11724–11736, 2020.
67. Yang, P., Shu, Y., Zhuang, Q., Li, Y., Gu, J., Metal–Organic Frameworks Bearing Dense Alkyl Thiol for the Efficient Degradation and Concomitant Removal of Toxic Cr(VI). *Langmuir*, 35, 16226–16233, 2019.
68. Liu, D.C., Ouyang, T., Xiao, R., Liu, W.J., Zhong, D.C., Xu, Z., Lu, T.B., Anchoring CoII Ions into a Thiol-Laced Metal–Organic Framework for

Efficient Visible-Light-Driven Conversion of CO2 into CO. *ChemSusChem*, 12, 2166–2170, 2019.

69. Fei, H. and Cohen, S.M., Metalation of a Thiocatechol-Functionalized Zr(IV)-Based Metal–Organic Framework for Selective C–H Functionalization. *J. Am. Chem. Soc.*, 137, 2191–2194, 2015.
70. Chen, T.-F., Han, S.-Y., Wang, Z.-P., Gao, H., Wang, L.-Y., Deng, Y.-H., Wan, C.-Q., Tian, Y., Wang, Q., Wang, G., Modified UiO-66 frameworks with methylthio, thiol and sulfonic acid function groups: The structure and visible-light-driven photocatalytic property study. *Appl. Catal. B: Environ.*, 259, 118047, 2019.
71. Wang, J.-H., Tang, G.-M., Wang, Y.-T., Cui, Y.-Z., Wang, J.-J., Ng, S.W., A series of phenyl sulfonate metal coordination polymers as catalysts for one-pot Biginelli reactions under solvent-free conditions. *Dalton Trans.*, 44, 17829–17840, 2015.
72. Zhang, G., Yang, H., Fei, H., Unusual Missing Linkers in an Organosulfonate-Based Primitive–Cubic (pcu)-Type Metal–Organic Framework for CO2 Capture and Conversion under Ambient Conditions. *ACS Catal.*, 8, 2519–2525, 2018.
73. Zhang, G., Wei, G., Liu, Z., Oliver, S.R.J., Fei, H., A Robust Sulfonate-Based Metal–Organic Framework with Permanent Porosity for Efficient CO2 Capture and Conversion. *Chem. Mater.*, 28, 6276–6281, 2016.
74. Zang, Y., Shi, J., Zhang, F., Zhong, Y., Zhu, W., Sulfonic acid-functionalized MIL-101 as a highly recyclable catalyst for esterification. *Catal. Sci. Technol.*, 3, 2044–2049, 2013.
75. Goesten, M.G., Juan-Alcañiz, J., Ramos-Fernandez, E.V., Sai Sankar Gupta, K.B., Stavitski, E., van Bekkum, H., Gascon, J., Kapteijn, F., Sulfation of metal–organic frameworks: Opportunities for acid catalysis and proton conductivity. *J. Catal.*, 281, 177–187, 2011.
76. Liu, F., Ma, X., Li, H., Wang, Y., Guo, M., Yaxin, H., Lu, W., Zhou, S., Yu, M., Dilute sulfonic acid post functionalized metal organic framework as a heterogeneous acid catalyst for esterification to produce biodiesel. *Fuel*, 266, 117149, 2020.
77. Phang, W.J., Jo, H., Lee, W.R., Song, J.H., Yoo, K., Kim, B., Hong, C.S., Superprotonic Conductivity of a UiO-66 Framework Functionalized with Sulfonic Acid Groups by Facile Postsynthetic Oxidation. *Angew. Chem. Int. Ed.*, 54, 5142–5146, 2015.
78. Wong, N., Hurd, J.A., Vaidhyanathan, R., Shimizu, G.K.H., A proton-conducting cesium sulfonate metal organic framework. *Can. J. Chem.*, 93, 988–991, 2015.
79. Zhang, G. and Fei, H., Missing metal-linker connectivities in a 3-D robust sulfonate-based metal–organic framework for enhanced proton conductivity. *Chem. Commun.*, 53, 4156–4159, 2017.
80. Dong, X.-Y., Wang, R., Li, J.-B., Zang, S.-Q., Hou, H.-W., Mak, T.C., A tetranuclear Cu 4 (μ 3-OH) 2-based metal–organic framework (MOF) with

sulfonate–carboxylate ligands for proton conduction. *Chem. Commun.*, 49, 10590–10592, 2013.
81. Meng, X., Song, S.-Y., Song, X.-Z., Zhu, M., Zhao, S.-N., Wu, L.-L., Zhang, H.-J., A tetranuclear copper cluster-based MOF with sulfonate–carboxylate ligands exhibiting high proton conduction properties. *Chem. Commun.*, 51, 8150–8152, 2015.
82. Dong, X.-Y., Wang, R., Wang, J.-Z., Zang, S.-Q., Mak, T.C.W., Highly selective Fe3+ sensing and proton conduction in a water-stable sulfonate-carboxylate Tb–organic-framework. *J. Mater. Chem. A*, 3, 641–647, 2015.
83. Li, Z., He, G., Zhao, Y., Cao, Y., Wu, H., Li, Y., Jiang, Z., Enhanced proton conductivity of proton exchange membranes by incorporating sulfonated metal–organic frameworks. *J. Power Sources*, 262, 372–379, 2014.
84. Shimizu, G.K., Taylor, J.M., Kim, S., Proton conduction with metal–organic frameworks. *Science*, 341, 354–355, 2013.
85. Shalini, S., Dhavale, V.M., Eldho, K.M., Kurungot, S., Ajithkumar, T.G., Vaidhyanathan, R., 1000-fold enhancement in proton conductivity of a MOF using post-synthetically anchored proton transporters. *Sci. Rep.*, 6, 32489, 2016.
86. Sikdar, N., Dutta, D., Haldar, R., Ray, T., Hazra, A., Bhattacharyya, A.J., Maji, T.K., Coordination-Driven Fluorescent J-Aggregates in a Perylenetetracarboxylate-Based MOF: Permanent Porosity and Proton Conductivity. *J. Phys. Chem. C*, 120, 13622–13629, 2016.
87. Ahmed, I., Hasan, Z., Khan, N.A., Jhung, S.H., Adsorptive denitrogenation of model fuels with porous metal–organic frameworks (MOFs): Effect of acidity and basicity of MOFs. *Appl. Catal. B: Environ.*, 129, 123–129, 2013.
88. Mukhopadhyay, S., Debgupta, J., Singh, C., Sarkar, R., Basu, O., Das, S.K., Designing UiO-66-Based Superprotonic Conductor with the Highest Metal–Organic Framework Based Proton Conductivity. *ACS Appl. Mater. Interfaces*, 11, 13423–13432, 2019.
89. Zhang, Y., Li, B., Krishna, R., Wu, Z., Ma, D., Shi, Z., Pham, T., Forrest, K., Space, B., Ma, S., Highly selective adsorption of ethylene over ethane in a MOF featuring the combination of open metal site and π-complexation. *Chem. Commun.*, 51, 2714–2717, 2015.
90. Yang, Q., Wiersum, A.D., Llewellyn, P.L., Guillerm, V., Serre, C., Maurin, G., Functionalizing porous zirconium terephthalate UiO-66(Zr) for natural gas upgrading: A computational exploration. *Chem. Commun.*, 47, 9603–9605, 2011.
91. Neofotistou, E.D., Malliakas, C.N., Trikalitis, P., Unprecedented Sulfone-Functionalized Metal–Organic Frameworks and Gas-Sorption Properties. *Chem. Eur. J.*, 15, 4523–4527, 2009.
92. Wang, B., Huang, H., Lv, X.-L., Xie, Y., Li, M., Li, J.-R., Tuning CO2 Selective Adsorption over N2 and CH4 in UiO-67 Analogues through Ligand Functionalization. *Inorg. Chem.*, 53, 9254–9259, 2014.

93. Lu, W., Yuan, D., Sculley, J., Zhao, D., Krishna, R., Zhou, H.-C., Sulfonate-Grafted Porous Polymer Networks for Preferential CO2 Adsorption at Low Pressure. *J. Am. Chem. Soc.*, 133, 18126–18129, 2011.
94. Zhang, N., Peng, D., Wu, H., Ren, Y., Yang, L., Wu, X., Wu, Y., Qu, Z., Jiang, Z., Cao, X., Significantly enhanced CO2 capture properties by synergy of zinc ion and sulfonate in Pebax-pitch hybrid membranes. *J. Membr. Sci.*, 549, 670–679, 2018.
95. Mavrandonakis, A., Klontzas, E., Tylianakis, E., Froudakis, G.E., Enhancement of Hydrogen Adsorption in Metal–Organic Frameworks by the Incorporation of the Sulfonate Group and Li Cations. A Multiscale Computational Study. *J. Am. Chem. Soc.*, 131, 13410–13414, 2009.
96. Luo, X.-P., Fu, S.-Y., Du, Y.-M., Guo, J.-Z., Li, B., Adsorption of methylene blue and malachite green from aqueous solution by sulfonic acid group modified MIL-101. *Micropor. Mesopor. Mat.*, 237, 268–274, 2017.
97. Kang, C., Peng, Y., Tang, Y., Huang, H., Zhong, C., Sulfate-Rich Metal–Organic Framework for High Efficiency and Selective Removal of Barium from Nuclear Wastewater. *Ind. Eng. Chem. Res.*, 56, 13866–13873, 2017.
98. Wang, H.-H., Zhou, L.-J., Wang, Y.-L., Liu, Q.-Y., Terbium-biphenyl-3,3′-disulfonyl-4,4′-dicarboxylate framework with sulfonate sites for luminescent sensing of Cr3+ ion. *Inorg. Chem. Commun.*, 73, 94–97, 2016.
99. Singh, N., Singh, U.P., Butcher, R.J., Luminescent sulfonate coordination polymers: Synthesis, structural analysis and selective sensing of nitroaromatic compounds. *CrystEngComm*, 19, 7009–7020, 2017.
100. Xu, C., Sun, K., Zhou, Y.-X., Ma, X., Jiang, H.-L., Light-enhanced acid catalysis over a metal–organic framework. *Chem. Commun.*, 54, 2498–2501, 2018.
101. Ali Akbar Razavi, S. and Morsali, A., Linker functionalized metal–organic frameworks. *Coord. Chem. Rev.*, 399, 213023, 2019.
102. Luan, Y., Zheng, N., Qi, Y., Tang, J., Wang, G., Merging metal–organic framework catalysis with organocatalysis: A thiourea functionalized heterogeneous catalyst at the nanoscale. *Catal. Sci. Technol.*, 4, 925–929, 2014.
103. Wei, N., Zhang, Y.-R., Han, Z.-B., Thiadiazole-functional porous metal–organic framework as luminescent probe for Cd2+. *CrystEngComm*, 15, 8883–8886, 2013.

8

Urea and Amide Decorated Metal-Organic Frameworks

Abstract
In this chapter, we discussed about metal-organic frameworks decorated with urea, amide, oxalamide and squaramide functions. Host-guest chemistry of these functions revolves around presence of carbonyl and amine motifs in their structure. Also, structure-function properties of these functions clearly differs form on another. This chapter deeply discusses about findings about functionalized metal organic frameworks with these functions.

Keywords: Amide, urea, oxalamide, squaramide, functional metal-organic frameworks, carbon dioxide capture, heterogeneous organocatalysis

A considerable number of the functions introduced inside the structure of MOFs are carbonyl-containing functions including urea, amide, oxalamide, squaramide, ketone, imide and carboxy. Due to the presence of carbonyl group, all of these functions have partially mutual host–guest chemistry. However, the difference in their host–guest chemistry is because of the various groups connected to the carbonyl function. Direct connections between carbonyl and other groups including amine(s) and hydroxy provide different properties for this group of FMOFs. In this regard, we classified carbonyl based functions in two group; those that C=O group is not connected to -NH and those C=O group is connected to –NH function. In this chapter we discuss about functions like urea and amide in which C=O group is connected to –NH function(s). Other carbonyl based functions are discussed in next chapter.

Ali Morsali and Sayed Ali Akbar Razavi. Functional Metal-Organic Frameworks: Structure, Properties and Applications, (165–200) © 2021 Scrivener Publishing LLC

8.1 Functionalized Metal–Organic Frameworks by Amide Function

8.1.1 General Chemical Properties of Amide Function

Amide (-CO-NH-) is a chemically low-reactive and structurally relatively rigid function which possesses carbonyl (-CO-) and (-NH-) as two different guest-interactive sites with different chemistry. Because of the greater electronegativity of oxygen, the carbonyl group (C=O) has a stronger dipole rather than N–C dipole. So, because of high dipole moment and oxygen lone pairs of carbonyl site, amide function chemistry is consist of electrical field interaction and enhanced lone pair polarization with other dipole/quadruple molecules. Moreover, electron rich oxygen atom of carbonyl motif of amide group is electron rich/donor, so can as Lewis basic site. As results, basic environment is provided by the amide.

The amide group possesses two types of hydrogen bonding sites: the (-NH-) group acts as a hydrogen bond donor and the (-CO-) group acts as hydrogen bond acceptor. The homogeneous amide containing hosts tend to form hydrogen bonds among themselves and interact negligibly with guest molecules. By avoiding amide–amide interactions, amide functional group would constitute attractive interaction sites for different kind of host–guest interactions.

8.1.2 Function–Application Properties

Owing to different chemical properties because of presence of both carbonyl (C=O) and amine (NH) groups, like strong dipole moment and high polarizing ability, electron donor and Lewis basicity and ability to interact as both H-bond donor and acceptor functional group, amide functionalized MOFs are applied in different types of applications. Amide decorated MOFs applied as heterogeneous Lewis basic catalytic site in base assisted catalytic reactions owing to Lewis basicity of carbonyl site [1–5]. This Lewis basicity help amide decorated MOFs apply as sensor or adsorbent for metal ions [6–8]. Moreover, since carbonyl site is a H-bond acceptor site, amide decorated MOFs applied for separation and detection of guest molecules containing H-bond donor sites especially alcohol [9] and picric acid [3, 10]. Owing to the polarity, Lewis basicity and H-bond donor/acceptor properties amide functionalized MOFs applied extensively in field of gas adsorption.

Based on reviewing of published articles, it is found that amide decorated MOFs applied as polar hosts for improvement of effective interaction between MOF skeleton and gas molecules to improve carbon dioxide separation capacity and selectivity, to increase the storage capacity of energy gases like hydrogen and methane and safe handling of explosive acetylene.

CO_2 capture and separation is the filed that amide FMOFs are mostly applied [11–31]. Amide functional groups provide polar surface for adsorbent MOF which is beneficial for selective capture of CO_2 in binary mixture with nonpolar gases N_2 or CH_4 gases. Generally based on reviewing the published paper about application of amide decorated MOFs in CO_2 separation we can claim that two types of host–guest interactions are involved in improved CO_2 separation by amide functionalized MOFs. Although CO_2 is nonpolar, but it's charge-separated and quadrupolar nature is critical to interact with amide function. In first possible interaction between amide FMOFs and CO_2 molecules, the carbonyl-O atom of amide acts as Lewis basic or electron rich site to interact with partially positive C-atom of carbon dioxide in amide(C=O)·C(CO_2) donor–acceptor interaction. In the second possible interaction between amide FMOFs and CO_2 molecules, the NH site of amide function acts as hydrogen bond donor site and CO_2 act as hydrogen bond acceptor site through its partially negative O atoms in (amide)NH·O(CO_2) hydrogen bonding.

Martin Schroder and coworkers synthesized a series of isoreticular MFM-MOFs to investigate the roles of porosity and functionality in CO_2 separation. The structure of applied ligand and each MOF is illustrated in Figure 8.1 [11]. These isoreticular MOFs are classified in two group; MFM-MOFs with short

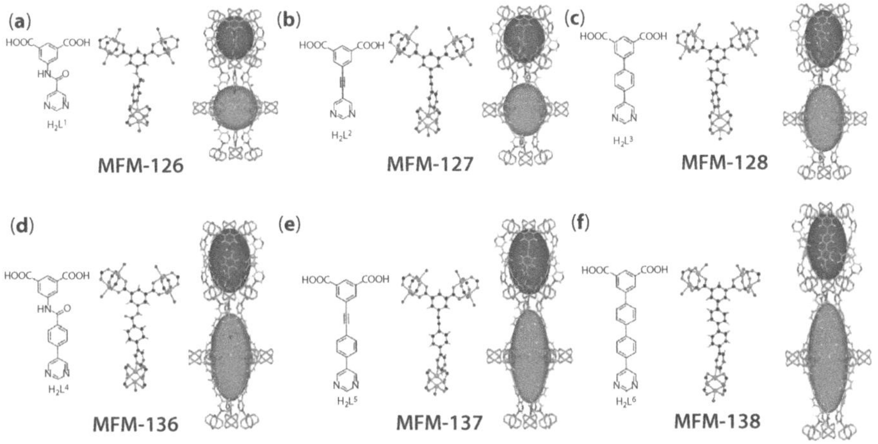

Figure 8.1 Coordination environment and crystal structure of isoreticular MFM-frameworks in CO2 adsorption. (a) Structural representation and ligand structure for MFM-126. (b) Structural representation and ligand structure for MFM-127. (c) Structural representation and ligand structure for MFM-128. (d) Structural representation and ligand structure for MFM-136. (e) Structural representation and ligand structure for MFM-137. (f) Structural representation and ligand structure for MFM-138. Coordination environment and crystal structure of isoreticular MFM-frameworks in CO_2 adsorption [11].

ligands (MFM-126, MFM-127 and MFM-128) and MFM-MOFs with long ligands (MFM-136, MFM-137 and MFM-138). MFM-128 and MFM-138 are developed based on phenyl containing ligands. In MFM-127 and MFM-137 on phenyl ring is replaced with alkyne bond. MFM-126 and MFM-136 are decorated with amide function. The reported BET surface areas for these MOFs are 1,004 $m^2 \cdot g^{-1}$ for MFM-126, 1,557 $m^2 \cdot g^{-1}$ for MFM-127, 1,491 $m^2 \cdot g^{-1}$ for MFM-128, 1,634 $m^2 \cdot g^{-1}$ for MFM-136, 1,749 $m^2 \cdot g^{-1}$ for MFM-137 and 1,590 $m^2 \cdot g^{-1}$ for MFM-138. In both series with short and long ligands, alkyne decorated MOFs (MFM-127 and MFM-137) represent higher surface area because alkyne group is space-efficient and can increase the hypothetical maximum surface area of MOFs. Gas Adsorptions at high pressure (20 bar) show that MOFs with higher surface area represent higher CO_2 capacity (Figure 8.2). This observation shows that the pore functionality had little effect on high pressure gas adsorption, where porosity is the dominant factor. However, in low pressure region (0–1 bar) the role of functional groups and stronger

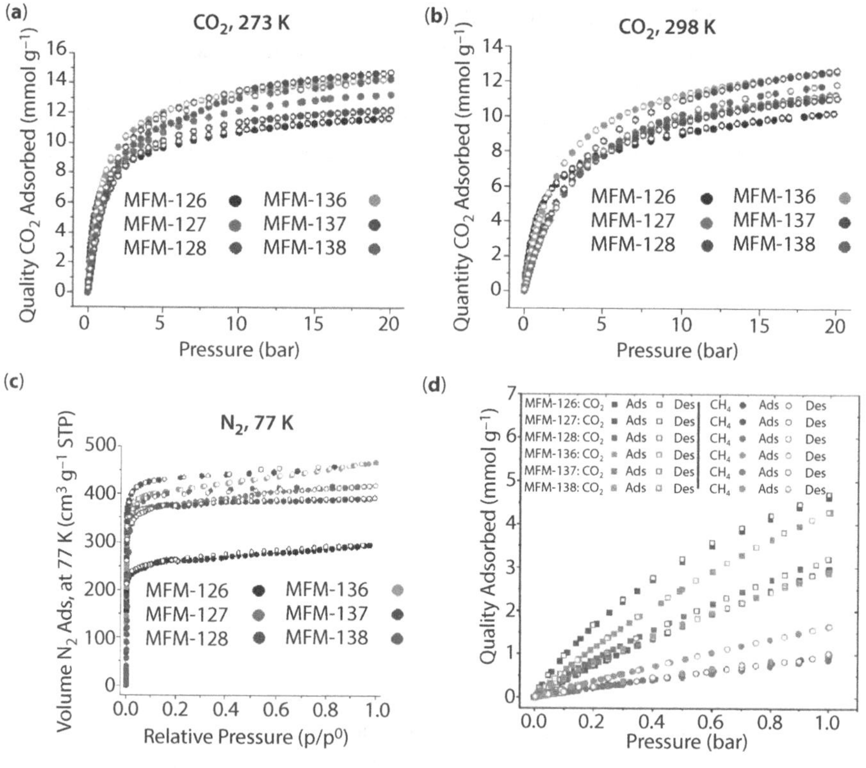

Figure 8.2 Application of isoreticular MFM-frameworks in gas adsorption. (a) High pressure CO_2 adsorption at 273 K. (b) High pressure CO_2 adsorption at 298 K. (c) Nitrogen adsorption at 77 K. (d) Low pressure CO_2 adsorption at 298 K [11].

interactions between gas molecules and the host framework is more important. Amide decorated MFM-126 with lowest surface area has the highest CO_2 uptake at 1 bar and 298 K (4.63 $mmol{\cdot}g^{-1}$ for MFM-126, 4.28 $mmol{\cdot}g^{-1}$ for MFM-136, 3.20 $mmol{\cdot}g^{-1}$ for MFM-128, 2.97 $mmol{\cdot}g^{-1}$ for MFM-127, 2.92 $mmol{\cdot}g^{-1}$ for MFM-137 and 2.89 $mmol{\cdot}g^{-1}$ for MFM-138) suggesting that amide decorated frameworks possesses higher affinity toward CO_2 molecules. Among amide decorated MFM-126 and MFM-136 frameworks, despite the lower surface area, MFM-126 has higher CO_2 uptake that MFM-136 clarifying the importance of gas-framework interactions at low pressures. MFM-126 shows the higher isosteric heat of CO_2 adsorption (30.7 $kJ{\cdot}mol^{-1}$ for MFM-126, 30.0 $kJ{\cdot}mol^{-1}$ for MFM-138, 25.8 $kJ{\cdot}mol^{-1}$ for MFM-127, 20.4 $kJ{\cdot}mol^{-1}$ for MFM-128, 20.1 $kJ{\cdot}mol^{-1}$ for MFM-136 and 19.2 $kJ{\cdot}mol^{-1}$ for MFM-137) and CO_2/N_2 15:85 selectivity at 298 K (39.6 for MFM-126, 23.2 for MFM-136, 18.9 for MFM-128, 15.7 for MFM-137, 15.5 for MFM-138 and 7.65 for MFM-127) or CO_2/CH_4 50:50 selectivity at 298 K (11.7 for MFM-126, 4.53 for MFM-128, 4.08 for MFM-137, 3.87 for MFM-138, 3.35 for MFM-136 and 3.33 for MFM-127). *In situ* inelastic neutron scattering and synchrotron FT-IR microspectroscopy were employed to elucidate dynamic interactions of adsorbed CO_2 molecules within amide decorated MFM-126. CO_2 molecules interact with amide functions through (amide)NH·O(CO_2) hydrogen bonding.

Junfeng Bai and coworkers synthesized an amide decorated MOF ($[Sc_3(\mu_3\text{-}O)(L)_{1.5}(H_2O)_3Cl]$ named as NJU-Bai49 where H_4L = 5-(3,5-dicarboxybenzamido)isophthalic acid) along with coordinatively unsaturated metal sites and applied it for highly selective CO_2 adsorption (Figure 8.3) [12]. The maximum N_2 uptake amount of NJU-Bai49 is 300 $cm^3{\cdot}g^{-1}$ with Brunauer–Emmett–Teller (BET) surface area of 1,189 $m^2{\cdot}g^{-1}$ and estimated pore volume of 0.46 $cm^3{\cdot}g^{-1}$. NJU-Bai49 could adsorb 43.3 $cm^3{\cdot}g^{-1}$ CO_2 (8.4 wt%) and 23.1 $cm^3{\cdot}g^{-1}$ CO_2 (4.5 wt%) at 273 and 298 K under 0.15 bar, which is larger than many amide decorated MOFs like NJU-Bai21 (4.4 wt %), LIFM-11(Cu) (4.3 wt %), NJU-Bai0 (4.2 wt %), HNUST-1 (3.7 wt %), NJUBai22 (3.1 wt %), and isoreticular SNNU-61 (4.0 wt %) and only smaller than that of HHU-2 (6.3 wt %). CO_2 uptake capacity of NJU-Bai49 under 1 bar and 273 K and 298 K reaches 137.5 and 88.2 $cm^3{\cdot}g^{-1}$ CO_2, corresponding to 27.0 and 17.3 wt %. The CO_2 adsorption enthalpy for NJU-Bai49 is calculated to be 33.4 $kJ{\cdot}mol^{-1}$ at zero loading, which shows the strong interaction between the CO_2 molecules and NJU-Bai49 framework. However, with more CO_2 molecule loading, the enthalpy values gradually reduce over the entire loading range, which indicates the heterogeneity of CO_2 binding sites within the framework. Theoretical studies based on grand canonical Monte Carlo (GCMC) simulations show that amide function could interact with CO_2 molecules differently. In crystal structure of NJU-Bai49, one cage as primary

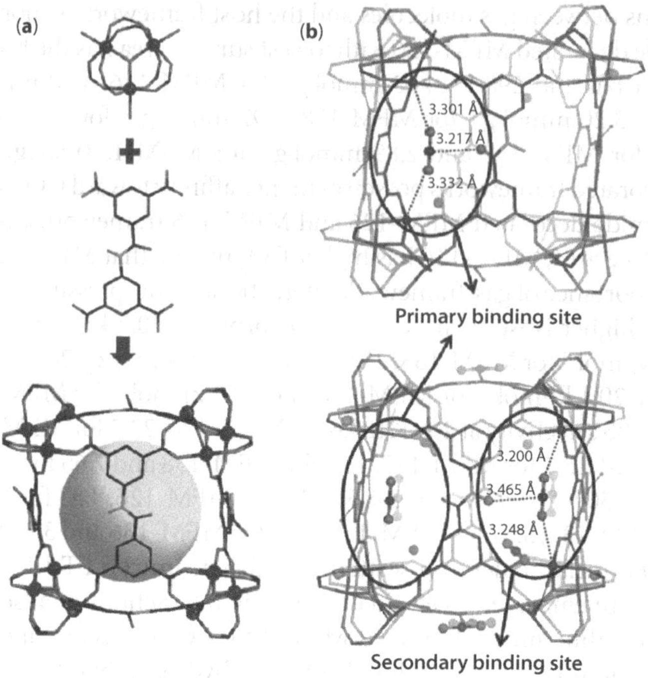

Figure 8.3 Application of NJU-Bai49 in selective CO_2 adsorption. (a) $[Sc_3(\mu_3\text{-}O)(COO)_6]$ cluster and deprotonated ligand L^{4-} in the X-ray crystal structure of NJUBai49. (b) Preferential CO_2 molecule binding sites simulated with GCMC at 0.15 bar (Primary bunding site) and 1 bar (secondary binding site). Color code: C, black; O, red; N, turquoise; Sc, blue; Cl, green; H, gray. Partial H atoms are omitted for clarity [12].

binding site is consist of the C=O bond from amid function and exposed metal sites and the next cage as secondary binding site composed of the N-H bond from amid function and exposed metal sites. At low CO_2 loading in pressures lower than 0.15 bar, CO_2 molecules occupy primary binding sites. In primary binding site, the distance between exposed open Sc(III) sites and O-atom of CO_2 are larger that amide (C=O) site and C-atom of CO_2 whereas the distance between NH site of amide group in secondary binding site is larger than exposed open Sc(III) sites and O-atom of CO_2. This work clearly shows that both amide decorated pores and exposed unsaturated metal sites play critical roles in enhancement of CO_2 capacity at low pressures.

Many amide functionalized MOFs have been applied in CO_2 adsorption and separation. All of them clearly show that introduction of amide function inside the MOF structures can be as effective as taking advantage of Lewis basic sites like amine ($-NH_2$) and azine (=N-N=) and polar functions like sulfune ($-SO_3$) and hydroxyl (-OH). Interestingly, another work reported that amide function did not show positive effects on CO_2 adsorption and there was no direct binding between the adsorbed CO_2 and CH_4 molecules and the pendant amide group in the pore [32]. Clearly, it can be realize that not only the presence of functional group, but also accessibility of functions in the cavities and effective orientation of functional groups as guest-interactive sites inside the pores are of special importance for improvement of guest-framework host–guest chemistry. In other word, not only the guest-host interactions of applied functional group, but combination of geometry, pore size and functionality are critical factor to improve gas sorption properties of functional MOFs.

Although amide decorated MOFs show their best performance in CO_2 separation, however they represent improving effects in storage and adsorption of acetylene [33, 34], methane [35, 36] and hydrogen [37]. Junfeng Bai and coworkers tried to compare the role of acetylene and amide functions in the high-performance acetylene storage. In this regard they introduced alkyne and amide functions into the structure of MOF-505 with NbO topology. Amide decorated framework is denoted as NJU-Bai-17 (formulated as $[Cu_2(DBAI)(H_2O)_2]\cdot 4DMF\cdot 4H_2O$ where H_4DBAI is (5-(3,5-dicarboxybenzamido)isophthalic acid) and alkyne decorated framework is denoted as compound 1 (Figure 8.4) [33]. BET surface area ($m^2\cdot g^{-1}$) of NJU-Bai-17 and compound 1 are 2,423 $m^2\cdot g^{-}1$ and 2,604 $m^2\cdot g^{-1}$, respectively. MOF-505 takes up 148 and 177 $cm^3\cdot g^{-1}$ C_2H_2 at 296 and 273 K. By expanding with the functional C≡C group, the C_2H_2 uptake capacity of compound 1 increases up to 160 and 252 $cm^3\cdot g^{-1}$ at 296 and 273 K. Amide decorated NJU-Bai-17 shows higher C_2H_2 uptake of 222.4 and 295 $cm^3\cdot g^{-1}$ under 1 bar at 296 and 273 K. These results clearly show that introduction of functional groups especially amide is very beneficial for improvement of C_2H_2 capacity. The results of theoretical and

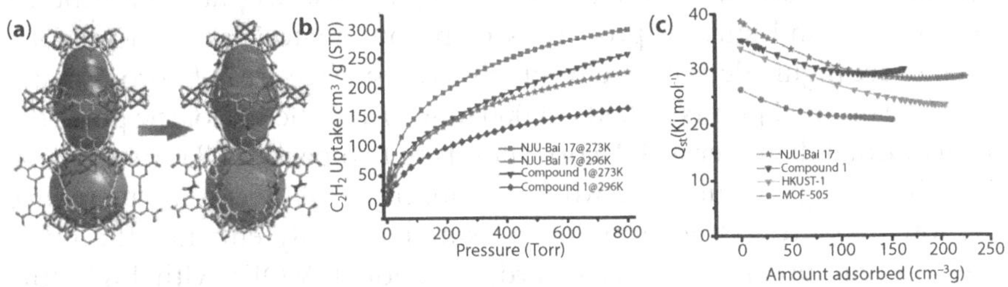

Figure 8.4 Application of NJU-Bai-17 in acetylene storage. (a) The X-ray crystal structures of compound 1 (left) and NJU-Bai 17 (right). (b) C_2H_2 adsorption isotherms for compound 1 and NJU-Bai 17 at 273 K and 296 K. (c) The comparison of Q_{st} for C_2H_2 adsorption of NJU-Bai 17, compound 1, MOF-505 and HKUST-1 [33].

experimental studies show that the acetylene uptake by the pore space is improved over 70% by replacing C≡C with amide group in NJU-Bai 17. This observation is occurred because of higher affinity between amide decorated framework of NJU-Bai-17 and C_2H_2 molecules through polarization on the C_2H_2 molecule by the amide groups due to the quadrupole–electrical field interactions between C_2H_2 and amide.

In another work by Schroder and coworkers, they designed a high surface area MOF, named as MFM-188 with formula $[Cu_4\mathbf{L}(H_2O)_4]\cdot 12H_2O$ ($H_8\mathbf{L}$ is 5,5',5'',5'''-([1,1'-biphenyl]-3,3',5,5'-tetracarbonyl)tetrakis (azanediyl) tetraisophthalic acid) and BET surface area of 2,568 $m^2\cdot g^{-1}$. MFM-188 is decorated with amide functional groups. Application of MFM-188 in CO_2 and C_2H_2 adsorption up to 1 bar shows that in both cases the isotherms showed fully reversible adsorption with a CO_2 uptake of 120 $cm^3\cdot g^{-1}$ (23.7 wt% or 86.7 v/v) recorded at 298 K and a C_2H_2 uptake of 232 $cm^3\cdot g^{-1}$ (27.0 wt% or 166.7 v/v) at 295 K. The affinity between MFM-188 framework and carbon dioxide and acetylene molecules evaluated using isosteric heats of adsorption at low pressure which are 21.0 and 32.5 $kJ\cdot mol^{-1}$ for CO_2 and C_2H_2, respectively. Experimental characterizations show that NH site of amide function interact with C_2D_2 and CO_2 molecules through $C_2D_2(\pi)\cdot$(NH)amide and $CO_2(O)\cdot$(NH)amide hydrogen bonds.

As mentioned earlier, amide function is a good choice to polarize the pore surface of MOFs. In this regard amide decorated MOFs display good affinity toward quadrupolar CO_2 and C_2H_2 molecules. However, this surface polarity of amide functional MOFs could improve the CH_4 working capacity of amide decorated MOFs in different way.

CH_4 volumetric working capacity is the most important factor to evaluate the performance of the CH_4 storage materials in real conditions. CH_4 volumetric working capacity is defined as the difference of CH_4 uptakes between high-pressure and low-pressure. Low pressure uptake is dependent on CH_4–framework interaction and high pressure uptake is attributed to pore size and surface properties of porous materials. High-pressure uptakes are highly determined by gravimetric surface area of porous materials (the optimal points is 2,500–3,000 $m^2\cdot g^{-1}$) and the size of the pore (the ideal optimal diameter is 8 Å corresponds to two packed CH_4 molecules). To achieve high volumetric working capacity, it is necessary to reduce the low-pressure gas adsorption and simultaneously enhance the high-pressure gas uptake. In this regard, functional MOFs with high surface area must be synthesized with functions which increase the methane–framework interaction at high pressure and reduce the methane-framework interaction at low pressure. Since amide function is hydrophilic and polar, it is theoretically ideal function to reduce the interaction with nonpolar CH_4 molecules at low pressure. With this idea, Mingxing Zhang and coworkers introduce an amide groups inside the pores of NOTT-101 to generate amide decorated NJU-Bai-45 (Figure 8.5) [35]. Non-functionalized NOTT-101 and NJU-Bai-45 are isostructure. Introduction of amide function in the main chain of the applied ligand, results in longer ligand for NJU-Bai-45 with higher pore volume (1.203 vs 1.080 $cm^3 \cdot g^{-1}$) and surface area (3,073 vs 2,805 $m^2 \cdot g^{-1}$) rather NOTT-101. After introduction of amide function into the structure of NJU-Bai-45, the gravimetric surface area reaches the optimal level. Also, by introduction of amide group into the pores of NJU-Bai-45 the polarity of the pores surface is increased which is suitable for reduction in methane-framework affinity at low pressures. The results of gas adsorption measurements indicate that NJU-Bai-45 has improved total CH_4 uptake at 65 bar and 298 K rather NOTT-101 (395 $cm^3 \cdot g^{-1}$ (251 $cm^3 \cdot cm^{-3}$) vs 344 $cm^3 \cdot g^{-1}$ (237 $cm^3 \cdot cm^{-3}$)). Moreover, the initial Q_{st} value of NJU-Bai-45 (15.0 $kJ \cdot mol^{-1}$) for CH_4 adsorption is slightly smaller than NOTT-101 (15.5 $kJ \cdot mol^{-1}$). So, as a result of such higher CH_4 storage capacity at high pressures and lower methane-framework affinity at low pressures, the working capacity of NJU-Bai-45 is increased rather NOTT-101 (298 $cm^3 \cdot g^{-1}$(190 $cm^3 \cdot cm^{-3}$) vs 251 $cm^3 \cdot g^{-1}$ (164 $cm^3 \cdot cm^{-3}$)).

Space and coworkers reported that presence of polar amide groups inside the structure of Cu-TPBTM (with formula $[Cu_3(tbptm^{6-})(H_2O)_8]$ where H_6tpbtm is N,N',N"-tris(isophthalyl)-1,3,5-benzenetricarboxamide) indirectly increases the low-pressure physisorption of hydrogen in comparison with non-functional PCN-61 framework (with formula $[Cu(H_2O)]_3(btei) \cdot 5DMF \cdot 4H_2O$ where H_6btei is 5,5',5"-benzene-1,3,5-triyl-tris(1-ethynyl-2-isophthalate)) (Figure 8.6) [37] PCN-61 and Cu-TPBTM contain open Cu(II) sites and hydrogen molecules bind onto the interior of exposed open Cu(II) sites at low loading. Presence of the negatively

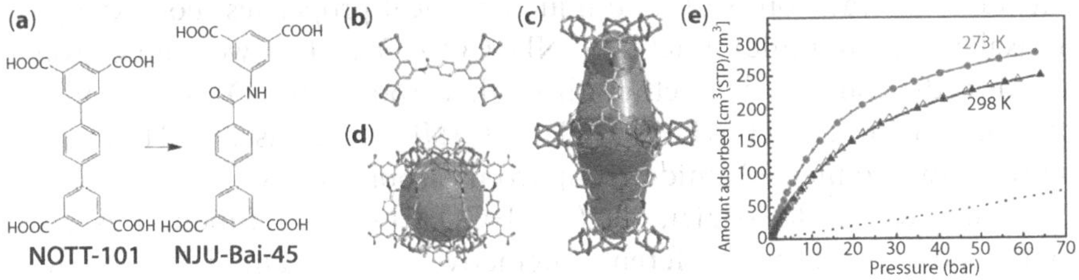

Figure 8.5 Application of NJU-Bai-45 in methane storage. (a) Applied ligands in the structure of NOTT-101 and NJU-Bai-45. (b) Ligand and four connected Cu(II) paddlewheel. (c) Shuttle-shaped cage with the size of around 11 × 24 Å. (d) Globe-shaped cage with a diameter of 14 Å. (e) Total CH_4 adsorption isotherms (between 5.8 and 65 bar) for NJU-Bai-45. Black line represents the pure CH_4 stored in the gas tank [35].

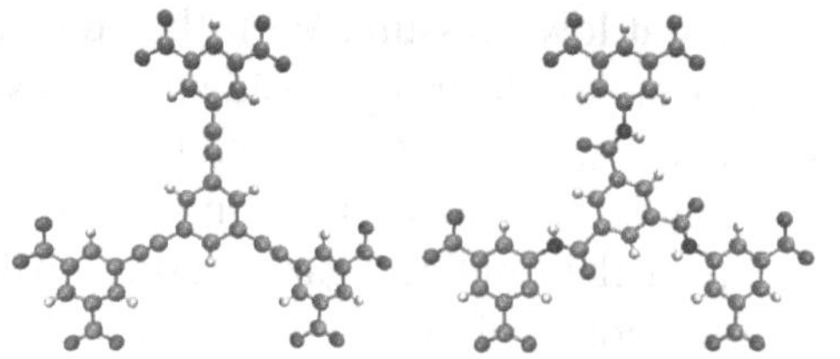

Figure 8.6 Representation of the applied ligand in the structure of PCN-61 and Cu-TPBTM for hydrogen storage [37].

charged oxygen atom of the amide group in the structure of tpbtm^{6-} hexacarboxylate ligand causes the interior Cu^{2+} ions to exhibit a higher positive charge through an inductive effect. As a result, the H_2–Cu^{2+} interactions are strengthened in Cu–TPBTM framework rather PCN-61.

As we saw, host–guest interactions of amide functions are versatile so that amide decorated MOFs can interact with different types of analytes. However, gas adsorption, especially CO_2 separation, is the most practical application for amide decorated MOFs. This is due to the fact that amide decorated MOFs are stable after activation. Also, owing to the polarity of the amide function, pore surface of these MOFs are polar and so effective in interaction with quadrupolar CO_2 molecules. In addition, since the amide function could easily locate between the phenyl rings of the ligand in the main chain of the framework, they increase the length of the ligand and so the pore volume and surface area of the MOFs and they did not occupy the accessible pore volume inside the pores of MOF. So, amide is an ideal polar function to increase the polarity and porosity of the material for high capacity CO_2 separation.

There are some functional groups based on amide function. In these functional groups, two amide functions are connected together in different ways and each on is of its unique chemical properties. For example, in oxalamide (oxamide) function (-NH-CO-CO-NH-), two amide group are directly connected together through carbonyl (C=O) sites and in malonamide function (-NH-CO-CH_2-CO-NH-) there is a (-CH_2-) motif between two connected amide group from carbonyl (C=O) site.

Similar to amide, oxalamide (-NH-CO-CO-NH-) function is also highly polar with two different interactive (-CO-) and (-NH-) sites. Therefore, similar to amide, host–guest chemistry of oxalamide functionalized MOFs are based on hydrogen bonding, donor–acceptor, polarization, and dipole–quadrupole interactions. Considering these chemical properties oxalamide decorated MOFs applied in gas adsorption [36, 38, 39] especially CO_2, [38, 40, 41] CO_2 conversion and metal ion removal.

Previously, we applied oxalamide decorated TMU-56 (with formula $[Zn_2(ox)(oba)_2]$.2DMF where ox is (N1, N2-di(pyridine-4-yl)oxalamide and H_2oba is 4,4′-oxybisbenzoic acid) in high capacity Pb(II) removal (Figure 8.7) [42]. Adsorption analyses show that TMU-56 is able to adsorb 1,130 $mg{\cdot}g^{-1}$ Pb(II) just in 20s which is much higher adsorption capacity and shorter removal time than other Pb(II) adsorbents like MOF-5 (659 $mg{\cdot}g^{-1}$) and MnO_2-MOF (917 $mg{\cdot}g^{-1}$). The results of FT-IR analysis clarify that oxalamide function are able to coordinate to Pb(II) metal ions. High affinity of oxophil Pb(II) cations to be chelated to oxalamide function is the reason for such efficient removal process. However, other mechanisms like Pb(II)–Zn(II) exchange are responsible for such high adsorption capacity.

Schröder and coworkers applied an oxalamide functionalized MOF with formula $[Cu_2(H_2O)_2(bdpo)]$ (H_4bdpo is N,N′-bis(3,5-dicarboxyphenyl)-oxalamide) which is denoted as NOTT-125 for selective CO_2 separation (Figure 8.8) [40]. Reported BET surface area ($m^2{\cdot}g^{-1}$) and pore volume ($cm^3{\cdot}g^{-1}$) for NOTT-125 are 2471 and 1.1, respectively. NOTT-101 could adsorb 40.0 wt% CO_2 at 273 K and 18.19 wt% CO_2 at 298 K at low pressure (1 bar) with adsorption enthalpy of 25.35 $kJ{\cdot}mol^{-1}$ at zero coverage. Simulations show that there are some strong interactions between CO_2 and the oxalamide functional group which are not possible with a single amide. In these interactions, CO_2 molecule interacts with both amide motifs of oxalamide function through simultaneous hydrogen bond CO_2(O)·(NH)amide(1) and dipole–quadrupole CO_2(C)·(O)amide(2) interactions. However in case of methane adsorption, the binding identified between NOTT-125 and CH_4 is dependent on the presence of only one of the amide groups of the linker.

Wang and coworkers reported that oxalamide function can act as photosensitizer chromophore in the structure of MOFs. They synthesized three oxalamide functionalized MOFs with formula $[Ln_2(bdpo)_{1.5}(DMA)_3(H_2O)]{\cdot}5H_2O$ based on N,N′-bis(3,5-dicarboxyphenyl)-oxalamide (H_4bdpo) ligand and different lanthanide metal center (Ln = Eu, Gd and Tb) (Figure 8.9) [43].

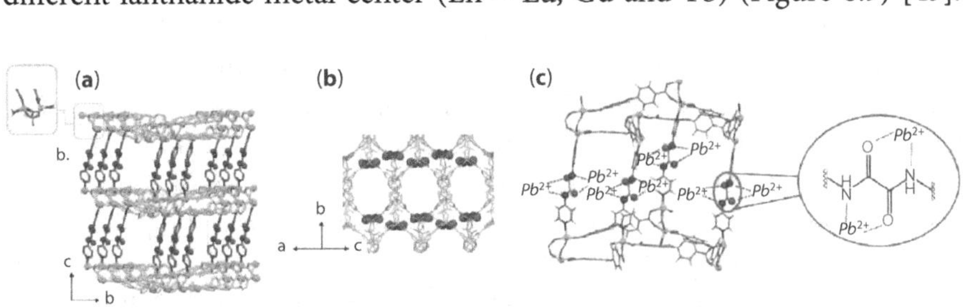

Figure 8.7 Application of TMU-56 in Pb(II) removal. (a) Crystal structure of TMU-56. (b) Oxalamide decorated pores of TMU-56. (c) Chelation mechanism of oxalamide functions to Pb(II) ions [42].

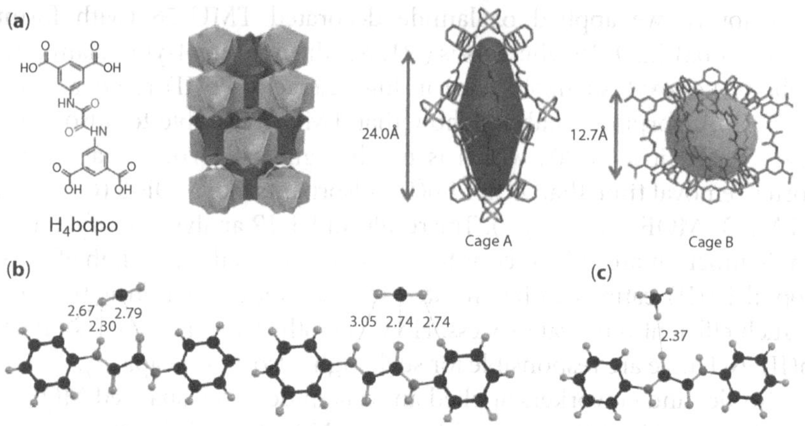

Figure 8.8 Structural representation of NOTT-125 and possible interactions between oxalamide and gas molecules and related distances. (a) Structure of H_4bdpo ligand (left) and fof topology of NOTT-125 (right). (b) Oxalamide-carbon dioxide interactions which are not possible with single amide. (c) Oxalamide–methane interactions [40].

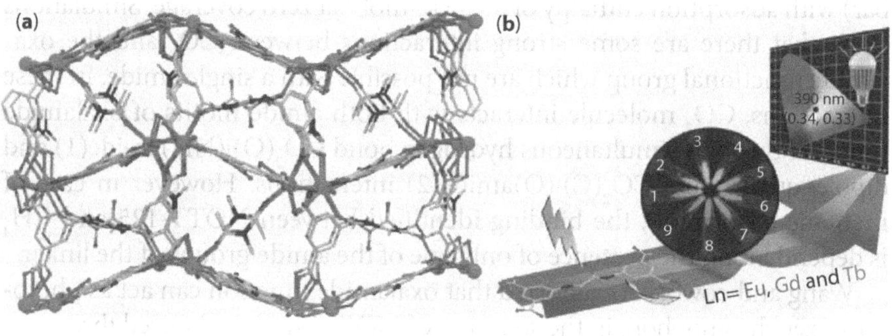

Figure 8.9 (a) 3D framework of $[Ln_2(bdpo)_{1.5}(DMA)_3(H_2O)]{\cdot}5H_2O$ (b) H_4bdpo to Ln metal ions energy transfer and unusually fluent change of luminescent color [43].

Depending on the nature of lanthanide metal center, the MOF emit different luminescent color. Such luminescent behavior is due the efficient energy transfer from bdpo^{4-} linkers to Ln(III) ions. Since oxalamide function is planar and π-conjugated, it helps to effective π-electron delocalization in the conjugated framework as strong absorbing chromophore. Simulations verified that H_4bdpo linkers are an efficient sensitizer for absorption and transference of energy to lanthanide ions.

Malonamide (-NH-CO-CH_2-CO-NH-) is another amide based functions. The carbonyl (C=O) sites in this function are headed in same direction. So, they are available for chelation to metal ions. In this regard, we synthesized and applied a malonitrile decorated MOF for high capacity

Hg(II) removal [44, 45]. This hypothesis is confirmed by FT-IR analyses. Also, the two hydrogen atoms of central methylene are acidic owing to presence of two polar (C=O) sites. Considering this chemical property of malonamide function, we synthesized and applied another malonamide decorated MOFs for catalyzed epoxide ring opening reaction [46]. Malonitrile function could activate epoxides through hydrogen bonding between hydrogen atoms of methylene atoms and oxygen atom of epoxides.

Our efforts to direct solvothermal synthesis of a malonitrile MOF using N1,N3-di(pyridine-4-yl)malonamide (L) was not successful. So, we introduced this ligand into the structure of urea decorated TMU-32 ([Zn(oba)(L2)]·2DMF·H_2O where H_2oba: 4,4'-oxydibenzoic acid; L2 = 1,3-di(pyridin-4-yl)urea) through solvent-assisted linker exchange process (Figure 8.10) [47]. The optimum exchange percentage is 65%. So, best structure after exchange

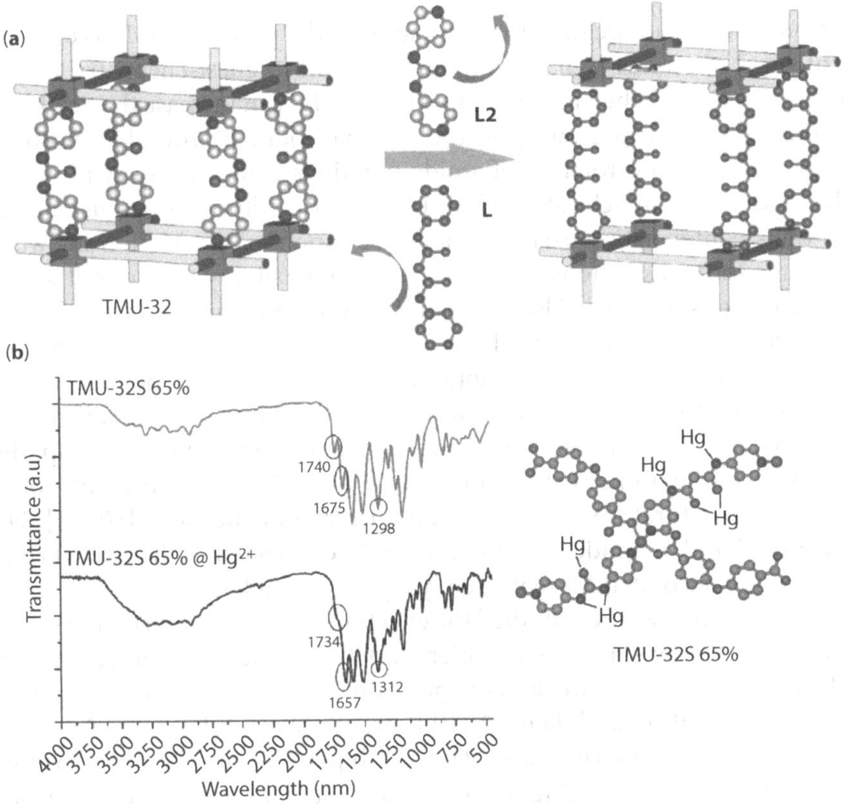

Figure 8.10 Application of TMU-32S-65% in Hg(II) removal. (a) L2-L ligand exchange process. (b) Changes in FT-IR spectra of TMU-32S-65% before and after Hg(II) adsorption along with possible coordination modes [47].

process is TMU-32S-65% in which 65% of urea L2 ligands are exchanged with malonitrile L ligands. The results of Hg(II) removal show that TMU-32S-65% could adsorb 1,428 mg·g^{-1} Hg(II) from aqueous solution. FT-IR analyses confirm the strong interactions between carbonyl groups of urea and malonitrile with Hg(II) ions. Coordination of carbonyl groups to Hg(II) ions can limit the C=O vibration modes and subsequently decrease their vibrational frequency, resulting in a significant red shift from 1,675 to 1,657 cm^{-1} for malonamide and 1,740 to 1,734 cm^{-1} for the carbonyl group of the urea functions. Also, the peak at 1,294 cm^{-1}, which was ascribed to the characteristic C–N vibration mode, shifted to 1,312 cm^{-1} after loading Hg(II). Hence, excellent selectivity towards Hg(II) was displayed due to the synergistic effects of both hydrophilic urea and malonamide organic functional groups.

8.1.3 Function–Structure Properties

Introduction of amide functional groups into the pores of MOFs is of some structural effects as well as enrichment in host–guest-chemistry of the MOFs. First of all, the introduction of the amides can increase the flexibility of the entire structure owing to the possible rotation around C-N bond. Such flexibility and rotation of amide functions could rise to synthesis of chiral MOFs from achiral building blocks [48]. Although because of the cis-trans conformation exchange of amide function, the MOF framework tolerates some distortions and requires some flexibility, the overall structure pattern is retained. Also, since amide functions can act as a hydrogen bond donor and acceptor site, the amide groups can lead to the formation of hydrogen bonds and generate supramolecular coordination polymers [49].

Some of other factors could affect on the flexibility of the amide decorated MOFs. For example, if they are connected to the methyl group, the flexibility of the framework is enhanced in a way that the framework could undergo structural transformation up on certain conditions [50, 51]. On the other hand, if amide function is directly connected to the phenyl ring, final amide decorated MOF is usually rigid and stable.

Forgan and coworkers studied the effects of amide function on the structure of MOFs by synthesis of different MOFs including non-functional [Zn_2(**L1**)$_2$(**P1**)] framework with *pcu* topology and amide functionalized [Zn(**L1**)(**P2**)] and [Zn(**L2**)(**P1**)] frameworks, where H_2**L2**= 4-(4-carboxybezamido)benzoic acid, **P1** = 4,4'-bipyridine and **P2**= N-(pyridin-4-yl)-isonicotinamide (Figure 8.11) [52]. Structural characterizations reveal that amide function introduces some distortions such as changes in orientation of the SBUs with respect to one another but, the overall structure of amide decorated framework is similar to the non-functionalized framework.

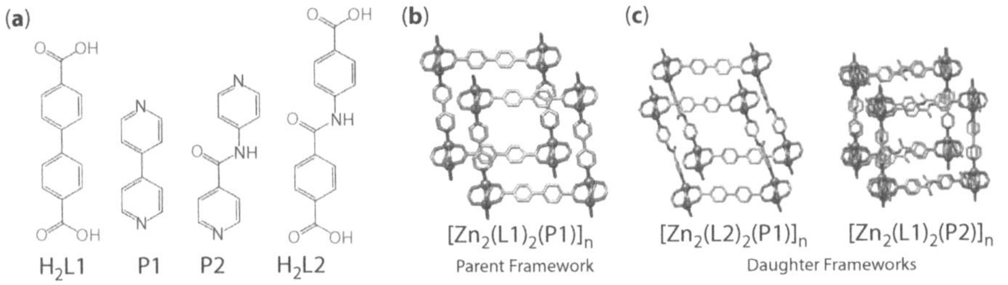

Figure 8.11 Effect of introducing amide function in the main-chain of the MOF structure. (a) Representation of H_2**L1**, H_2**L2**, **P1** and **P2** ligands, (b) $[Zn_2(L_1)_2(P_1)]$ framework as parent MOF with *pcu* topology. (c) $[Zn_2$(**L1**)$_2$(**P2**)] showing disposition of the SBUs induced by the amide pillar and $[Zn_2$(**L2**)$_2$(**P1**)] daughter frameworks which have a more regular arrangement of SBUs compared to $[Zn_2$(**L1**)$_2$(**P1**)] [52].

One of the most important factors that affects on the structure of amide decorated MOFs is the conformation of the amide function. Cis conformation and trans conformation of the amide function can lead to develop a MOF with different structure. On way to change the conformation of amide function is replacement of H atom of NH groups with alkyl groups. Myoung Soo Lah and coworkers show that combination of amide ligand (H_2**L2**, N,N',N"-methyl-4,4',4"-[1,3,5-benzenetriyltris(carbonylimino)]trisbenzoic acid) with cis-conformation with 3d metal ions can develop a metal–organic cluster with formula $[Ni_{14}(\mu^3\text{-}OH)_8$ (**L2**)$_6$(formate)$_2$(DMF)$_{10}$(H_2O)$_2$ (Figure 8.12) [53]; but, that combination of amide ligand with trans-conformation (H_3**L1**, 4,4',4"-[1,3,5-benzenetriyltris (carbonylimino)]trisbenzoic acid) with 3d metal ions can develop a three dimensional metal–organic framework ($[Cu_3(L1)_2(H_2O)_3]\cdot 14DMF\cdot 16H_2O$). This work indicates that the flexibility and conformation of amide function are of critical effects of the structure and dimensionality of resultant material.

Based on reviewing the literature, it observed that for amide functionalized MOFs the structure of the parent framework is retained if amide function is introduced between or instead of phenyl ring of organic linkers. It is necessary to mention that in this condition, amide functions are in trans conformation and (-CO-NH-) from. This observation indicates that amide decorated MOFs with special topology and structure can be developed using isoreticular chemistry. Also, their polarity and stability up on activation is the reasons why amide decorated MOFs applied extensively for gas adsorption.

Here, we gathered some amide decorated MOFs with different topologies and show that introduction of amide functions has not significant roles on the structure of the parent non-functionalized MOF. In MOFs with NbO topology, NOTT-100 and NOTT-101 are the parent non-functionalized frameworks. Introduction of amide function(s) between

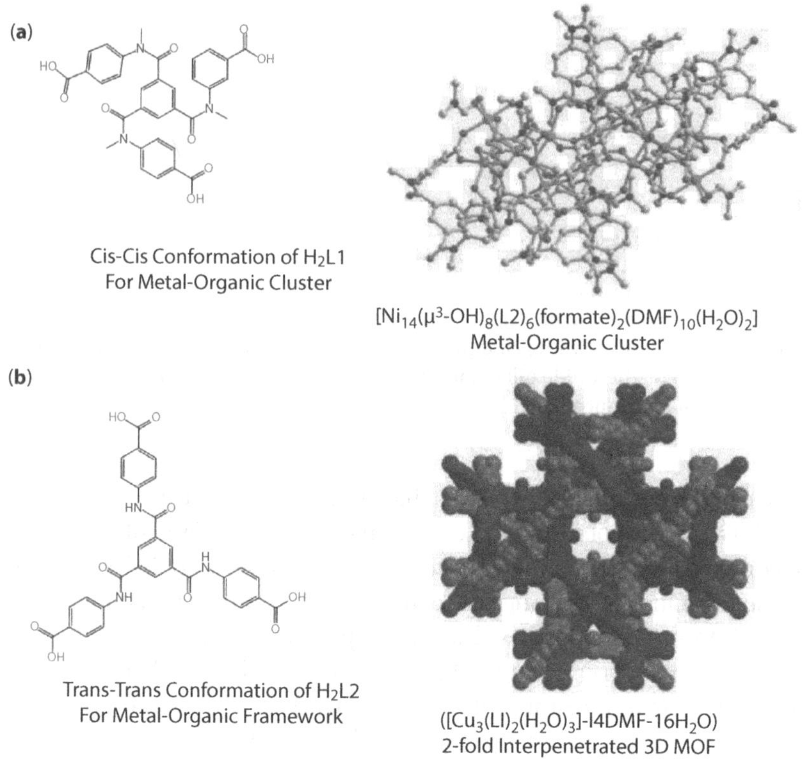

Figure 8.12 Effects of conformation of amide-function on the constructed compound. (a) metal–organic cluster based on amide ligands with cis–cis conformation. (b) Two-fold interpenetrated metal-organic framework based on amide ligands with trans-trans conformation, [53].

phenyl results in synthesis of NJU-Bai-17 and HNUST-1. Structural characterization show that the structure and topology of amide decorated NJU-Bai-17 and HNUST-1 frameworks is identical with non-functionalized NOTT-100 and NOTT-101 frameworks as well as similar geometry for functional and non-functional ligands (Figure 8.13) [25, 33, 54]. Similar observations attained in case of MOFs with rht topology (Figure 8.14) [28].

8.2 Functionalized Metal–Organic Frameworks by Urea Function

8.2.1 General Chemical Properties of Urea Function

Urea function (H_2N-CO-NH_2) is structurally consist of hydrogen bond donor (NH) sites and hydrogen bond acceptor (C=O) sites. It is proved

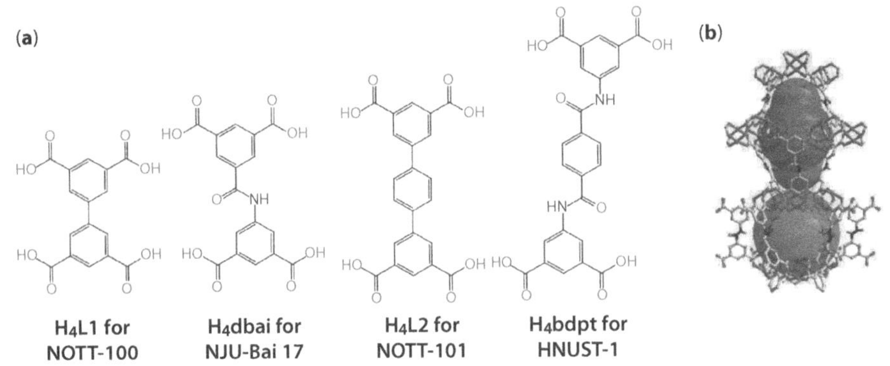

Figure 8.13 (a) Non-functional and amide-functionalized ligands for construction of *Nbo*-type MOFs. (b) Representation of crystal structure with NbO topology of NJU-Bai-17. NOTT-100 with formula $[Cu_2(\mathbf{L1})(H_2O)_2]$ where H_4**L1** is Biphenyl-3,3',5,5'-tetracarboxylic acid, NJU-Bai 17 with formula $[Cu_2(dbai)(H_2O)_2]\cdot 4DMF\cdot 4H_2O$ where H_4dbai is 5-(3,5-dicarboxybenzamido)isophthalic acid, NOTT-101 with formula $[Cu_2(\mathbf{L2})(H_2O)_2]$ where H_4**L2** is Terphenyl-3,3",5,5"-tetracarboxylic acid, and HNUST-1 with formula $[Cu_2(bdpt)(H_2O)_2]$ where H_4bdpt is (bis(3,5-dicarboxyphenyl) terephthalamide.

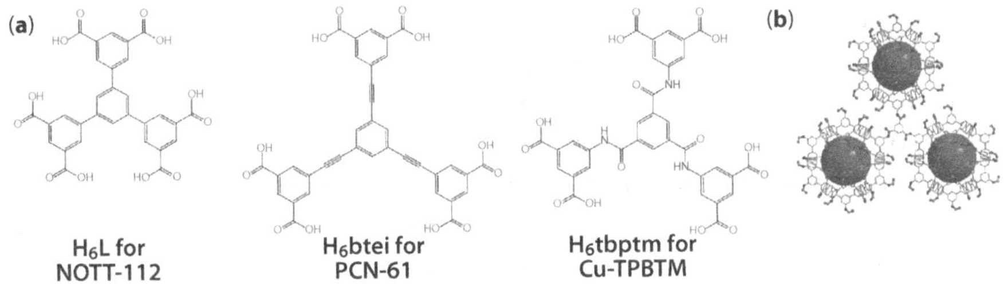

Figure 8.14 (a) Non-functional and amide-functionalized ligands for construction of *rht*-type MOFs. (b) Representation of crystal structure with rht topology of Cu-TPBTM with formula $[Cu_3(tbptm^{6-})(H_2O)_8]$ where H_6tpbtm is N,N',N"-tris(isophthalyl)-1,3,5-benzenetricarboxamide). NOTT-112 (with formula $[Cu_3(\mathbf{L})(H_2O)_3)]\cdot 8DMSO\cdot 15DMF\cdot 3H_2O$ where H6**L** is 1,3,5-tris(1,3,5-dicarboxy[1,1'-biphenyl]-4-yl)benzene), PCN-61 with formula $[Cu(H_2O)]_3(btei)\cdot 5DMF\cdot 4H_2O$ where H_6btei is 5,5',5"-benzene-1,3,5-triyltris(1-ethynyl-2-isophthalate).

that urea can interact with oxoanions through (urea)NH·O(oxoanion) hydrogen bond as well as other negatively charged species (Figure 8.15a). Moreover, owing to presence of two NH sites and C=O groups, urea functions benefits from enriched non-covalent host–guest chemistry giving rise in extensive application of urea in supramolecular chemistry.

Since organocatalysts interact with anionic moieties, the ability of urea to interact with negatively charged species and oxoanions, make it a

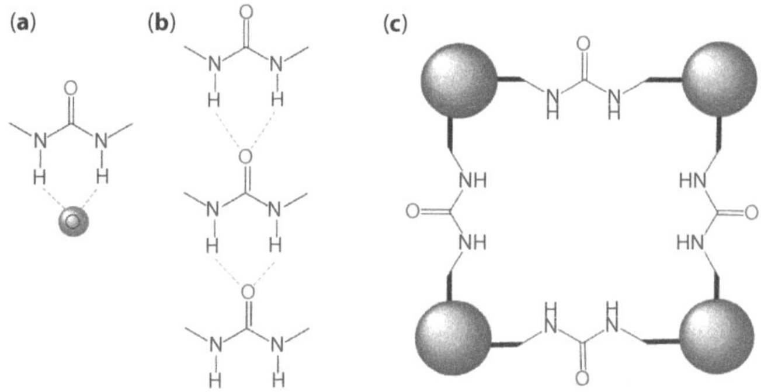

Figure 8.15 Host–guest interactions of urea. (a) Hydrogen bonding interaction between urea and electron-rich species. (b) Aggregation and self-quenching of urea groups. (c) Separated urea groups inside the pore-walls of FMOFs.

desirable function for development of organocatalysts. Among organocatalysts, hydrogen bond organocatalysts are of special importance because they are promising biomimetic alternative to Lewis acid catalysis. So, urea function applied as a two-point hydrogen bond donor organocatalytic site in synthetic organic chemistry.

Despite such advantages of urea in non-covalent interactions, homogeneous urea organocatalysts suffer from self-association. As a result of self-association of urea function through urea(-NH_2)·(-OC-)urea hydrogen bonds, the solubility and reactivity of homogeneous urea will decrease. An effective strategy to solve this problem is immobilization of urea functions in the pores of MOFs. In this condition, urea function does not interact with each other and they remain free for hydrogen bond organocatalytic purposes (Figure 8.15c).

8.2.2 Function–Application Properties

Urea function (-N-CO-NH-) is an ideal guest-interactive site inside the pores of MOFs. As mentioned this function can interact with both hydrogen bond donor and acceptor guests. Also, through carbonyl site (C=O) it can interact with electron deficient and cationic species and through (NH) sites it can interact with anionic and electron rich species. Owing to the rich host–guest chemistry of urea, this function applied in removal of metal ions, different types of catalytic reactions and detection of anions and electron-rich small organic molecules.

We have recently showed that urea decorated MOFs could represent high adsorption capacity to Pb(II) and Hg(II) metal ions. In this regard, we synthesized urea decorated TMU-31 ([Zn(L1)(L2)]·DMF, where L1 = 4,4'-(carbonylbis(azanediyl))dibenzoic acid and L2 = 3-di(pyridin-4-yl)urea) and TMU-32 ([Zn(oba)(L2)]·2DMF·H_2O, where H_2oba = 4,4'-oxidibenzoic acid and L2 = 1,3-di(pyridin-4-yl)urea) frameworks [55]. These MOFs represent high adsorption capacity to Hg(II) (476 mg·g^{-1} for TMU-31 and 416 mg·g^{-1} for TMU-32) and Pb(II) (909 mg·g^{-1} for TMU-31 and 909 mg·g^{-1} for TMU-32) ions. FT-IR analyses show that these high capacities are related to the strong coordination interaction between C=O and NH group of urea and Pb(II) or Hg(II) metal ions. In other work we proved that incorporation of magnetic Fe_3O_4 nanoparticles into pores of TMU-32 is a delicate strategy to improve the removal capacity of Fe_3O_4@TMU-32 (45 % and 54 % toward Pb(II) and Hg(II)) rather pristine TMU-32 framework [56].

As strong H-bond donor/acceptor host, urea decorated MOFs applied for detection and removal of different types of analytes [57–61]. It is well-kwon that one urea by its NH groups could interact with C=O site of another urea through multiple hydrogen bonding. We applied this term in different ways. First, we applied urea decorated MOFs as H-bond donor to detect carbonyl and nitro compounds through urea(NH)·(C=O, N-O^-) hydrogen bonding [62–64]. Also, we applied urea decorated MOFs to detect phenolic compounds through urea(NH)·(O)phenolic hydroxy and urea(C=O)·(H)phenolic hydroxy hydrogen bonding [65].

Janiak and coworkers reported that the urea-functionalized MOF, $[Zn_2(\mathbf{L1})_2(bipy)]$ where $H_2\mathbf{L1}$ = [1,3]diazepine-3,9-dicarboxylic acid, exhibits an uptake of 10.9 mmol·g^{-1} (41 wt%) of SO_2 at 293 K and 1 bar (Figure 8.16) [66]. Compounds $[Zn_2(\mathbf{L1})_2(bipy)]$ and $[Zn_2(\mathbf{L1})_2(bpe)]$ adsorb 14.3 mmol·g^{-1} (20 wt%) and 17.8 mmol·g^{-1} (23 wt%) NH_3, respectively. These high uptake values are traced to the urea functionality and its hydrogen bonding interactions with the adsorbents.

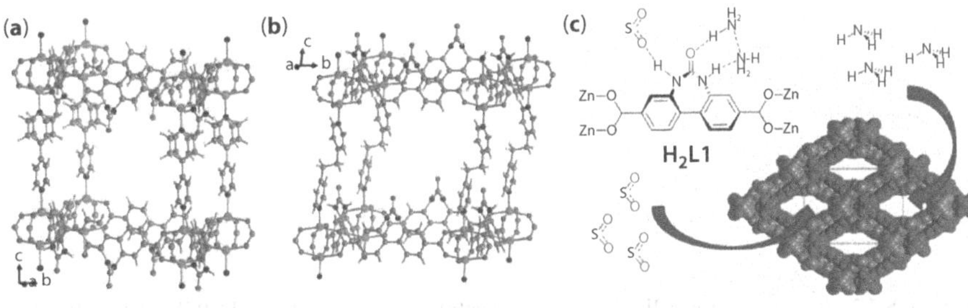

Figure 8.16 Structural representation of $[Zn_2(\mathbf{L1})_2(bipy)]$ (a) and $[Zn_2(\mathbf{L1})_2(bpe)]$ (b). Hydrogen bonding between urea group of H_2**FL1** and SO_2 and NH_3 guests (c) [66].

Urea function extensively applied in the field of anion recognition because urea has strong binding capability with negatively (partially) charged heteroatoms [67–72]. In construction of urea decorated MOFs for anion recognition some parameters must be considered such as accessibility and direction of urea in to the pores of MOFs, structure and topology of guest anion, the strength of hydrogen bond between urea and anions. These factors are very critical to sensitize a urea decorated MOF to specific anion. Urea function is sensitive to anions, especially oxoanions in a way that there are different reports about the effect of counter anion on the structure of urea decorated MOFs and the effects of the MOF structure and urea direction on the anion bonding properties.

A Ni– coordination network, $[Ni(EDPU)_2(H_2O)_2][(SO_4)(H_2O)_2](H_2O)_{3.5}(EtOH)_{0.4}$ which is assembeled form EDPU (EDPU = ethylenedi(m-pyridylurea)) as a bis-urea ligand, can selectively recognizes and separates Y-shaped and highly hydrophilic sulfate oxoanion through multiple hydrogen bonds from aqueous solution (Figure 8.17) [72]. This framework also can selectively and efficiently separate sulfate from aqueous solution containing F^-, Cl^-, Br^-, I^-, NO_3^- and ClO_4^-. The topology of the framework and accessibility of urea function inside the pores of MOF as well as anion topologies require the construction of directed hydrogen bonds for strategic design of selective anion recognizer urea decorated MOFs.

Urea function is a favorite organocatalytic site which draws lots of attention among chemists. Urea function could interact with electron rich sites of substrates through its hydrogen bond donor NH sites. This way of activation of electron rich substrates is similar to activation by metal ions as

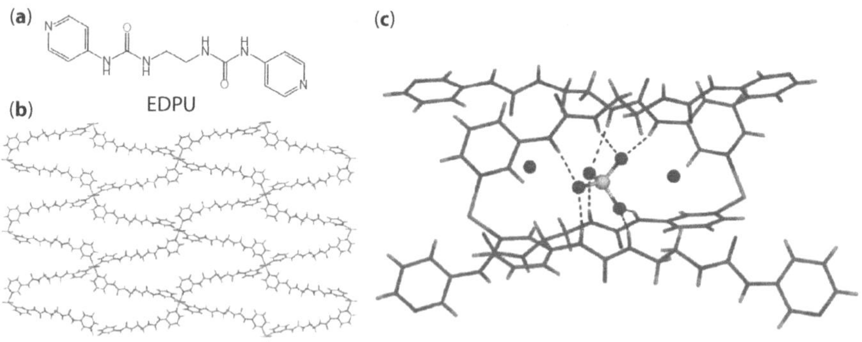

Figure 8.17 Structure of EDPU ligand (a), crystal structure of $[Ni(EDPU)_2(H_2O)_2][(SO_4)(H_2O)_2](H_2O)_{3.5}(EtOH)_{0.4}$ (b) and Interaction of sulfate *via* eight hydrogen bonds from four urea groups (c) [72].

catalytic Lewis acid sites. Anyway, owing to the sustainable development and green chemistry roles, scientists prefer to activate electron rich substrates through organocatalytic sites like urea. In this line, immobilization of urea function inside the pores of MOFs is beneficial for development of MOF-based heterogeneous organocatalysts. Urea decorated MOFs applied as H-bond donor catalyst in different type of reactions like epoxide ring opening [73–75], Friedel–Crafts [68, 76–81], Henry condensation [74, 82, 83] and cyanosilylation [74].

Hupp and coworkers applied urea decorated NU-601 (with formula $Zn_2(bipy)_2(H_4\mathbf{L})$ where $H_4\mathbf{L}$ is 5,5'-(carbonylbis(azanediyl))diisophthalic acid) as a hydrogen bond donor heterogeneous organocatalyst for Freidel–Craft reaction (alkylation between nitro-alkenes and pyrroles) (Figure 8.18) [81]. Findings show that NU-601 has a higher activity as heterogeneous catalyst compared to the diphenyl urea as homogeneous catalyst (90% conv. at 18 h vs 65% conv. at 18 h) and control reaction (21% conv. at 48 h). Since then, urea functionalized MOFs have been applied as hydrogen-bond organocatalysts in different reactions.

Chengfeng Zhu and coworkers synthesized a urea decorated MOF from a V-shaped urea-functionalized ligand and applied it as a hydrogen bond donor organocatalytic site in different reactions like epoxide ring opening, Henry condensation and cyanosilylation (Figure 8.19) [74]. The MOF shows high catalytic activity in all three reactions. To verify that whether the catalytic activity of the MOF is because of urea function, author applied the urea ligand as homogeneous organocatalyst and observed that the reaction conversion highly decreased. This observation clearly shows that immobilization of urea functions into the pores as heterogeneous hydrogen bond donor catalytic site is the main reason for catalytic activity of the material.

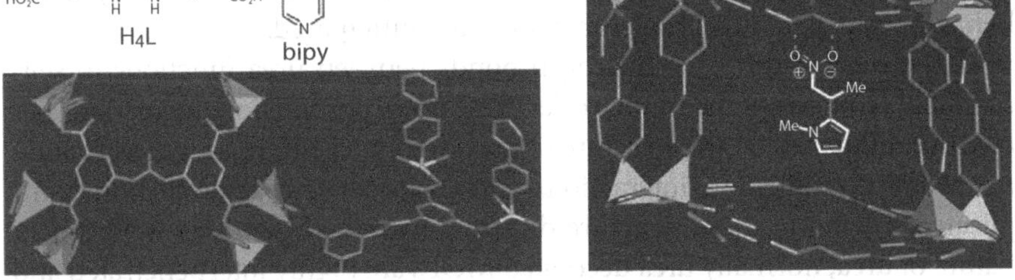

Figure 8.18 Structure, applied ligands in synthesis of NU-601 (a) Application of NU-601 as hydrogen bond organocatalyst in alkylation between nitro-alkenes and pyrroles (b) [81].

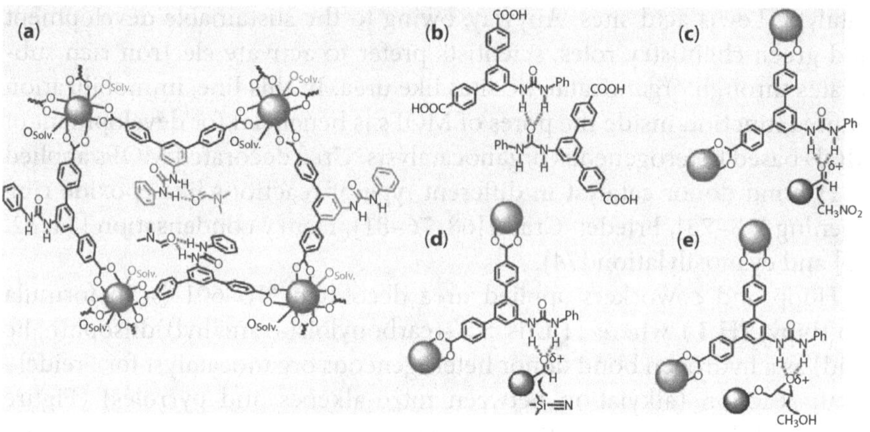

Figure 8.19 Application of urea decorated MOFs as hydrogen bond donor organocatalytic site in different reactions. (a) Structure of the material and urea decorated pores. (b) Self-association of urea-containing ligand in homogeneous catalysis. Substrate activation by the urea unit through hydrogen bonding for, Henry reaction (c), cyanosilylation reaction (d) and methanolysis reaction (e), when urea has been immobilized in MOF [74].

8.2.3 Structure–Function Properties

Due to the bent nature of urea function and strong structure-directing hydrogen bonds with other parts of the framework, introduction of urea function into the MOF pores has complicated effects on the structure, topology, stability, and flexibility of the MOFs.

There are some factors that affect on the structural properties of urea decorated MOFs and because of them structural properties of urea decorated MOFs are complicated. These factors are: (I) bent nature of urea function; owing to this characteristic, relative direction of coordinating sites changes and structure and topology of urea decorated MOFs changes in comparison with non-functionalized framework.

(II) possible rotation around C–N bonds of urea function; owing to these rotations urea decorated MOFs can be flexible and structural dynamic behavior and structural transformation are anticipated.

(III) structure directing hydrogen bonds between urea functions in self-assembly synthesis of the framework; These non-coordination interactions has effective influences on the structure of urea decorated MOFs as well as ligand–metal ion coordination interactions and make structural prediction complicated. Owing to these structure directing hydrogen bonds as well as bent nature of urea, normally urea decorated MOFs are highly interpenetrated and polycatenated. Although, the interpenetration can results in limited surface area and porosity but it can lead to synthesis of stable MOFs as well.

(IV) up on activation of crystal structure of urea decorated MOFs, they collapse or at least the crystallinity of the framework reduces noticeably; Urea function can interact with polar solvent molecules (like N,N'-dimethylformamide) which are trapped in the pore through hydrogen bonding as well as other urea functions of the framework. During activation process, these structure-directing/stabilizing hydrogen bonds can be interrupted giving rise in freedom of urea function for rotation around C–N bonds. So, in most cases, the structure of urea decorated MOFs will collapse. So, urea functionalized MOFs did not present good stability and rigidity during activation and gas adsorption measurements in most cases.

Forgan and coworkers synthesized non-functionalized parent [Zn_2(**L1**)$_2$(**P1**)] with *pcu* topology and urea decorated daughter [Zn(**L1**)(**P3**)] frameworks to investigate the roles of urea function of the structure and topology of the MOFs (**P1**= 4,4′-bipyridine, H_2**L1**=[1,1'-biphenyl]-4,4'-dicarboxylic acid and **P3**= N,N′-bis(4-pyridyl)urea) (Figure 8.20) [52]. Introduction of urea functions change the topology from *pcu* for parent framework to

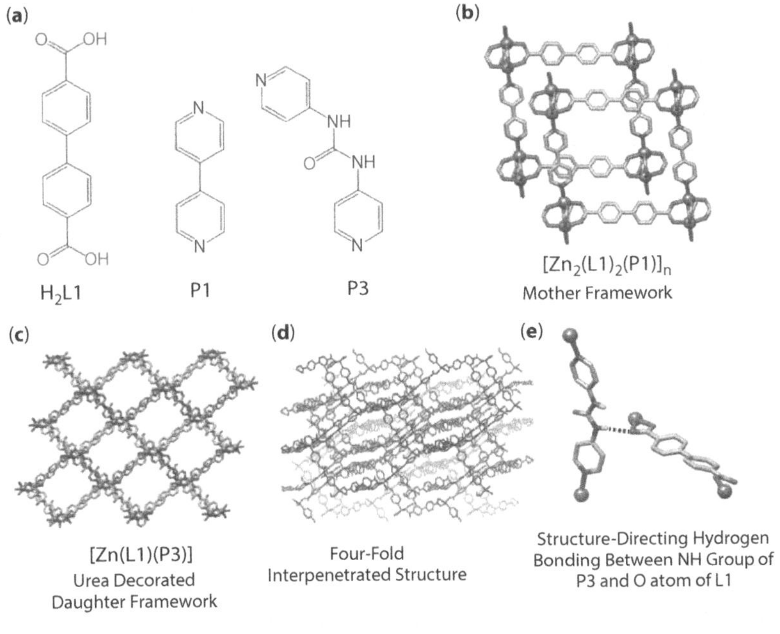

Figure 8.20 Effects of urea function in the main-chain on the structure of MOFs. (a) Structure of H_2**L1**, **P1** and **P3** ligands. (b) [Zn_2(**L1**)$_2$(**P1**)] framework as parent MOF with *pcu* topology. (c) [Zn_2(**L1**)$_2$(**P3**)] framework with dia topology. (d) Four-fold interpenetrated urea decorated daughter framework. (e) Structure-directing interaction between urea NH of each **P3** ligand and a carboxylic O of **L1** ligand [52].

dia for daughter framework. They mentioned that this structure change is because of bent nature of urea function and supramolecular hydrogen bonds between the urea NH of each **P3** ligand and a carboxylic O of **L1** ligand. These structure directing hydrogen bonds along with bent nature of urea make the interpenetrated *dia* topology more favorable. So, in contrast with amide functionl group, direct synthesis of urea functionalized MOFs with isoreticular synthesis strategy can be complicated because of bent nature of the urea and possible structure directing hydrogen bonds of urea containing linkers with each other or other parts of the struture.

Here based on reviewing the literature, we compared three isostructure MOFs with different number of urea function to examine the effects of urea function on the structural stability of MOFs (Figure 8.21) [68, 84, 85]. For this purpose, we compared NTU-105 as a functionalized framework with three triazole groups with Cu-UBTA functionalized with one urea and two triazole functions and Cu-TUH functionalized with three urea functions. The results of gas adsorption reveal that up on increasing the number of urea functions in the framework, the porosity of the material is decreased. Triazole functionalized NTU-101 is highly porous to N_2(861 cm^3/g, BET ($m^2 \cdot g^{-1}$) = 3,543) and CO_2 (187 $cm^3 \cdot g^{-1}$). Replacement of one urea group inside the H_6**L1** linker, leads to the construction of triazole-urea functionalized Cu-UBTA framework with reduced BET surface area (3,134 $m^2 \cdot g^{-1}$) and adsorption capacity (805 $cm^3 \cdot g^{-1}$ for N_2 and 165 $cm^3 \cdot g^{-1}$ for CO_2). In next step, replacement of three urea groups inside the H_6**L1** linker, leads to the construction of Cu-TUH framework. Cu-TUH show very negligible

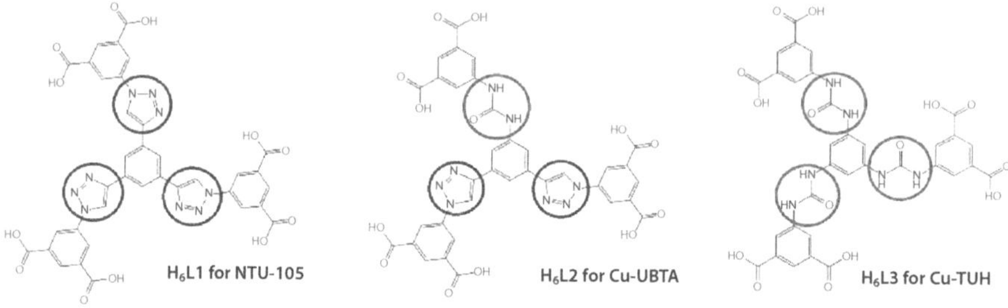

Figure 8.21 Representation of urea and triazole functionalized ligands for construction of *rht*- type MOFs. NTU-105 with formula Cu_3(**L1**)$(H_2O)_3$·11DMF·8H_2O, where H_6**L1** linker is 5,5',5"-(4,4',4"-(benzene-1,3,5-triyl)tris(1H-1,2,3-triazole-4,1-diyl))triisophthalic acid, Cu-UBTA with formula (Cu_3(**L2**)$(H_2O)_3$·10DMF·9H_2O) based on the unsymmetrical H_6**L2** hexacarboxylic acid linker (5,5'-(4,4'-(5-(3-(3,5-dicarboxyphenyl)ureido)-1,3-phenylene)bis(1H-1,2,3-triazol-4,1-diyl))diisophthalic acid) and Cu-TUH with formula Cu_3(**L3**)$(H_2O)_3$·5DMF·10H_2O) where H_6**L3** is 5,5',5"-(((benzene-1,3,5-triyltris(azanediyl)) tris(carbonyl))tris-(azanediyl))-triisophthalic acid.

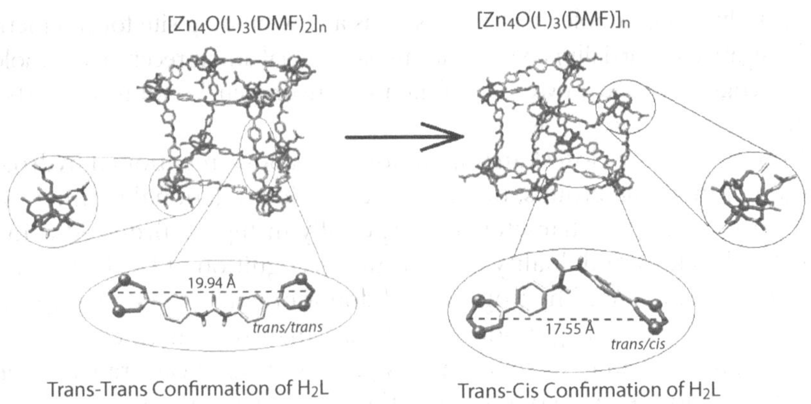

Figure 8.22 Structural change of $[Zn_4O(\mathbf{L})_3(DMF)_2]_n$ after activation by CS_2 [86].

surface area and adsorption capacity (BET ($m^2 \cdot g^{-1}$) = 68, 40 $cm^3 \cdot g^{-1}$ for N_2 and 37 $cm^3 \cdot g^{-1}$ for CO_2). It is decelerated that this observation is related to the higher flexibility of urea function rather triazole.

In another work, structural characterization reveals that up on exchange of DMF molecules within the pores of urea decorated MOF ($[Zn_4O(\mathbf{L})_3(DMF)_2]_n$ where $H_2\mathbf{L}$ is N,N′-bis(4-carboxyphenyl)urea), the structure is changed but is not collapsed (Figure 8.22) [86]. This structural change is possible due to the rotation of urea function from trans/trans to the trans/cis conformation which is accompanied with propensity of $[Zn_4O(RCO_2)_6]$ cluster toward solvation. So, overall structure is changed because of conformational change of urea functions, but is not collapsed. Also, it is observed that some of urea units, which show dramatic configurational rearrangement from the trans/trans to the trans/cis conformation, lose their capability as binding guests for organocatalysis.

In some cases like TMU-32 ([Zn(oba)(**L2**)]·2DMF·H_2O (H_2oba: 4,4′-oxidibenzoicacid; **L2** = 1,3-di(pyridin-4-yl)urea), the polycatenated framework and strong structure-directing hydrogen bonds are reasons for saved crystallinity of the structure after activation and microporosity measurements (BET ($m^2 \cdot g^{-1}$) = 432).

8.3 Functionalized Metal–Organic Frameworks by Squaramide Function

Squaramide, with the strong hydrogen bond donor ability from its two (-NH-) sites and remarkable hydrogen bond accepting character from its two (-CO-)

sites, apply in supramolecular chemistry as a new binding site for interaction with negatively (partially) charged atoms as artificial anion receptors in molecular recognition and powerful bifunctional hydrogen bonding catalysts in asymmetric organocatalysis.

Compared to its closest analogs among the above-mentioned hydrogen bonding functional groups, urea, the squaramide functionality differs significantly in various characteristics especially in higher Brønsted acidity, structural rigidity and duality in molecular recognition. It has been shown that squaramide-based hydrogen bond donor organocatalysts are approximately 6 orders of magnitude more acidic than their urea counterparts and the greater acidity of squaramides promotes significant rate acceleration relative to the urea counterpart. The aromatic squaramide has shown superiority over urea counterparts due to its stronger H-bond donor ability that was enhanced by its conformationally rigid square-shaped structure upon binding. While urea function reveals excellent anion-binding affinities, their ability to recognize cations is much more limited. However, the squaramide functionality shows dual ditopic binding with anion and cations (Figure 8.23). Squaramide function has been a prominent feature in symmetric or asymmetric organocatalysts and molecular recognition of analytes like Cu(II), NH_4^+ and different anions.

If these abilities of squaramide can be successfully transferred to crystalline solids with predetermined structures like MOFs, it may lead to a whole new generation of functional materials with targeted properties. As mentioned about urea, merging organocatalysis and MOF heterogeneous catalysis offers great opportunities to take advantage from both catalytic processes. But, despite much progress in this area for urea, the generation of a

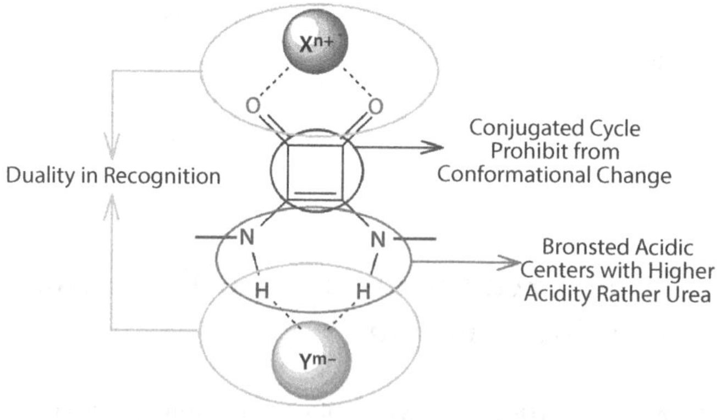

Figure 8.23 Significantly different characteristics between urea and squaramide.

squaramide structural motif as part of the MOF catalyst has attracted much less attention.

Successful transfer of such properties of squaramide into the structure of MOFs offers opportunities to simultaneously take advantage of both heterogeneous nature of MOFs and chemical characteristics of squaramide. However, despite considerable progress in the area of urea FMOFs, generation of a squaramide structural motif as part of the MOF catalyst has attracted much less attention.

Squaramide functionalized MOFs have been applied as active hydrogen bonding organocatalysts for the Friedel–Crafts reaction [87, 88] and Henry condensation [89]. Hupp and coworkers synthesized UiO-67 framework functionalized with squaramide function in side chain and applied it as a MOF-based heterogeneous organocatalyst in Friedel–Crafts reaction between indole and β-nitrostyrene (Figure 8.24) [88]. The results show that squaramide functionazlied UiO-67 reaches higher conversion rather similar urea functionazlied UiO-67 framework (78% vs. 38%) because higher stronger hydrogen bonds of squaramide.

Activity of organocatalytic hydrogen bond donor sites is proportional to their acidity. Since the acidity of squaramide is higher than its urea homologs, it can be understand that introduction of squaramide function into the pores of MOFs is a powerful strategy for designing highly efficient organocatalysts and molecular recognizer groups. Considering the similarity between squaramide, urea, amide and carbonyl functions, we compare the in structural and practical views in Figure 8.25.

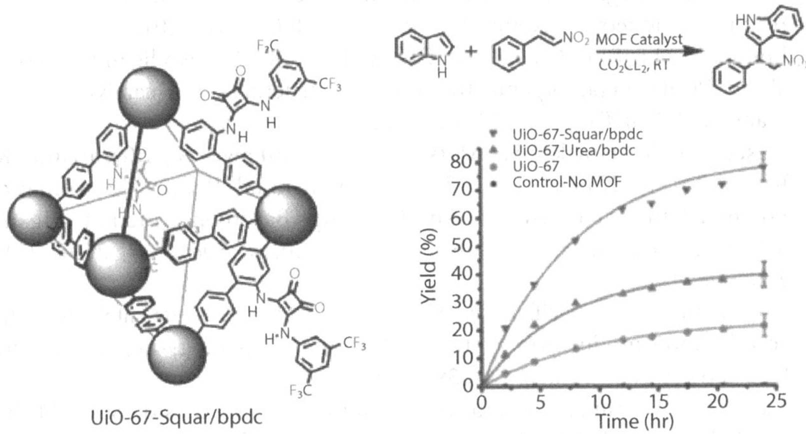

Figure 8.24 Application of UiO-67-Squar/bpdc in Friedel–Crafts reaction.

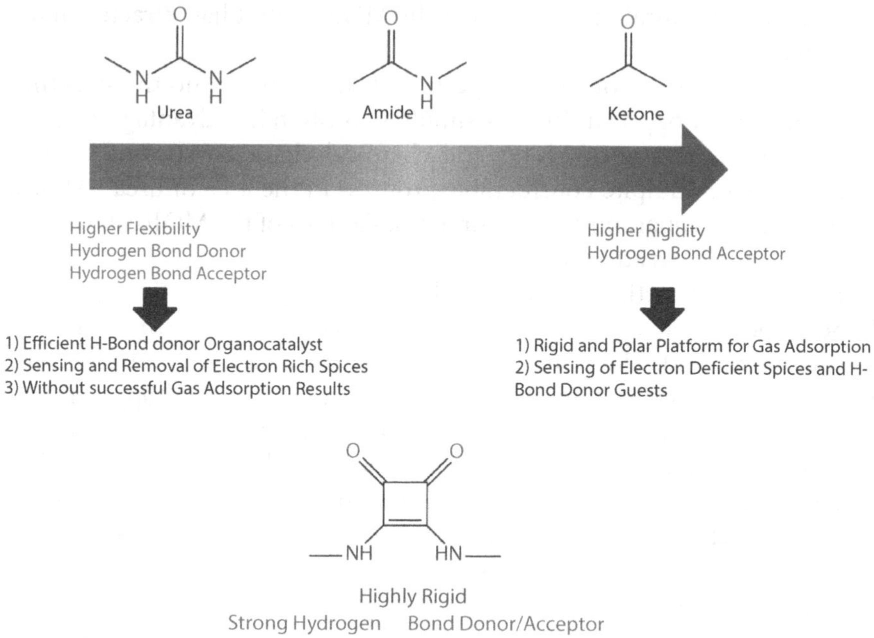

Figure 8.25 Comparison between carbonyl based functions.

References

1. Paul, A., Karmakar, A., da Silva, M.F.C.G., Pombeiro, A.J., Amide functionalized metal–organic frameworks for diastereoselective nitroaldol (Henry) reaction in aqueous medium. *RSC Adv.*, 5, 87400–87410, 2015.
2. Lin, X.-M., Li, T.-T., Chen, L.-F., Zhang, L., Su, C.-Y., Two ligand-functionalized Pb(ii) metal–organic frameworks: structures and catalytic performances. *Dalton Trans.*, 41, 10422–10429, 2012.
3. Hasegawa, S., Horike, S., Matsuda, R., Furukawa, S., Mochizuki, K., Kinoshita, Y., Kitagawa, S., Three-Dimensional Porous Coordination Polymer Functionalized with Amide Groups Based on Tridentate Ligand: Selective Sorption and Catalysis. *J. Am. Chem. Soc.*, 129, 2607–2614, 2007.
4. Ghorbanloo, M., Safarifard, V., Morsali, A., Heterogeneous catalysis with a coordination modulation synthesized MOF: morphology-dependent catalytic activity. *New J. Chem.*, 41, 3957–3965, 2017.
5. Karmakar, A., Paul, A., Rúbio, G.M., Soliman, M., Guedes da Silva, M.F.C., Pombeiro, A.J., Highly Efficient Bifunctional Amide functionalized Zn and Cd Metal Organic Frameworks for One-pot Cascade Deacetalization—Knoevenagel Reactions. *Front. Chem.*, 7, 699, 2019.

6. Esrafili, L., Safarifard, V., Tahmasebi, E., Esrafili, M.D., Morsali, A., Functional group effect of isoreticular metal–organic frameworks on heavy metal ion adsorption. *New J. Chem.*, 42, 8864–8873, 2018.
7. Khezerloo, E., Mousavi-khoshdel, S., Safarifard, V., Sensitive and selective detection of metal ions and small molecules in aqueous media using a hydrolytically stable amide-functionalized metal–organic framework. *Polyhedron*, 166, 166–174, 2019.
8. Yang, L.X., Feng, X.F., Yin, W.H., Tao, Y., Wu, H.Q., Li, J.Q., Ma, L.F., Luo, F., Metal–organic framework containing both azo and amide groups for effective U (VI) removal. *J. Solid State Chem.*, 265, 148–154, 2018.
9. Liu, H., Gao, G., Liu, J., Bao, F., Wei, Y., Wang, H., Amide-functionalized ionic indium–organic frameworks for efficient separation of organic dyes based on diverse adsorption interactions. *CrystEngComm*, 21, 2576–2584, 2019.
10. Buragohain, A., Yousufuddin, M., Sarma, M., Biswas, S., 3D Luminescent Amide-Functionalized Cadmium Tetrazolate Framework for Selective Detection of 2,4,6-Trinitrophenol. *Cryst. Growth Des.*, 16, 842–851, 2016.
11. Humby, J.D., Benson, O., Smith, G.L., Argent, S.P., da Silva, I., Cheng, Y., Rudić, S., Manuel, P., Frogley, M.D., Cinque, G., Host–guest selectivity in a series of isoreticular metal–organic frameworks: observation of acetylene-to-alkyne and carbon dioxide-to-amide interactions. *Chem. Sci.*, 10, 1098–1106, 2019.
12. Chen, C., Zhang, M., Zhang, W., Bai, J., Stable amide-functionalized metal–organic framework with highly selective CO2 adsorption. *Inorg. Chem.*, 58, 2729–2735, 2019.
13. Liu, H.-Y., Gao, G.-M., Bao, F.-L., Wei, Y.-H., Wang, H.-Y., Enhanced water stability and selective carbon dioxide adsorption of a soc-MOF with amide-functionalized linkers. *Polyhedron*, 160, 207–212, 2019.
14. Zheng, B., Yang, Z., Bai, J., Li, Y., Li, S., High and selective CO2 capture by two mesoporous acylamide functionalized rht-type metal–organic frameworks. *Chem. Commun.*, 48, 7025–7027, 2012.
15. Liu, X.-T., Jia, Y.-Y., Zhang, Y.-H., Ren, G.-J., Feng, R., Zhang, S.-Y., Zaworotko, M.J., Bu, X.-H., A new Co(ii) metal–organic framework with enhanced CO2 adsorption and separation performance. *Inorg. Chem. Front.*, 3, 1510–1515, 2016.
16. Keceli, E., Hemgesberg, M., Grünker, R., Bon, V., Wilhelm, C., Philippi, T., Schoch, R., Sun, Y., Bauer, M., Ernst, S., Kaskel, S., Thiel, W.R., A series of amide functionalized isoreticular metal–organic frameworks. *Micropor. Mesopor. Mat.*, 194, 115–125, 2014.
17. Duan, J., Yang, Z., Bai, J., Zheng, B., Li, Y., Li, S., Highly selective CO2 capture of an agw-type metal–organic framework with inserted amides: Experimental and theoretical studies. *Chem. Commun.*, 48, 3058–3060, 2012.
18. Park, J., Li, J.-R., Chen, Y.-P., Yu, J., Yakovenko, A.A., Wang, Z.U., Sun, L.-B., Balbuena, P.B., Zhou, H.-C., A versatile metal–organic framework for carbon

dioxide capture and cooperative catalysis. *Chem. Commun.*, 48, 9995–9997, 2012.

19. Deria, P., Li, S., Zhang, H., Snurr, R.Q., Hupp, J.T., Farha, O.K., A MOF platform for incorporation of complementary organic motifs for CO2 binding. *Chem. Commun.*, 51, 12478–12481, 2015.
20. Zheng, B., Wang, H., Wang, Z., Ozaki, N., Hang, C., Luo, X., Huang, L., Zeng, W., Yang, M., Duan, J., A highly porous rht-type acylamide-functionalized metal–organic framework exhibiting large CO2 uptake capabilities. *Chem. Commun.*, 52, 12988–12991, 2016.
21. Jin, W.-G., Chen, W., Xu, P.-H., Lin, X.-W., Huang, X.-C., Chen, G.-H., Lu, F., Chen, X.-M., An Exceptionally Water Stable Metal–Organic Framework with Amide-Functionalized Cages: Selective CO2/CH4 Uptake and Removal of Antibiotics and Dyes from Water. *Chem. Eur. J.*, 23, 13058–13066, 2017.
22. Lu, Z., Bai, J., Hang, C., Meng, F., Liu, W., Pan, Y., You, X., The Utilization of Amide Groups To Expand and Functionalize Metal–Organic Frameworks Simultaneously. *Chem. Eur. J.*, 22, 6277–6285, 2016.
23. Jia, Y.-Y., Liu, X.-T., Feng, R., Zhang, S.-Y., Zhang, P., He, Y.-B., Zhang, Y.-H., Bu, X.-H., Improving the Stability and Gas Adsorption Performance of Acylamide Group Functionalized Zinc Metal–Organic Frameworks through Coordination Group Optimization. *Cryst. Growth Des.*, 17, 2584–2588, 2017.
24. Safarifard, V., Rodríguez-Hermida, S., Guillerm, V., Imaz, I., Bigdeli, M., Azhdari Tehrani, A., Juanhuix, J., Morsali, A., Casco, M.E., Silvestre-Albero, J., Ramos-Fernandez, E.V., Maspoch, D., Influence of the Amide Groups in the CO2/N2 Selectivity of a Series of Isoreticular, Interpenetrated Metal–Organic Frameworks. *Cryst. Growth Des.*, 16, 6016–6023, 2016.
25. Zheng, B., Liu, H., Wang, Z., Yu, X., Yi, P., Bai, J., Porous NbO-type metal–organic framework with inserted acylamide groups exhibiting highly selective CO2 capture. *CrystEngComm*, 15, 3517–3520, 2013.
26. Lee, C.-H., Huang, H.-Y., Liu, Y.-H., Luo, T.-T., Lee, G.-H., Peng, S.-M., Jiang, J.-C., Chao, I., Lu, K.-L., Cooperative Effect of Unsheltered Amide Groups on CO2 Adsorption Inside Open-Ended Channels of a Zinc (II)–Organic Framework. *Inorg. Chem.*, 52, 3962–3968, 2013.
27. Xuan, Z.-H., Zhang, D.-S., Chang, Z., Hu, T.-L., Bu, X.-H., Targeted Structure Modulation of "Pillar-Layered" Metal–Organic Frameworks for CO2 Capture. *Inorg. Chem.*, 53, 8985–8990, 2014.
28. Zheng, B., Bai, J., Duan, J., Wojtas, L., Zaworotko, M.J., Enhanced CO2 Binding Affinity of a High-Uptake rht-Type Metal–Organic Framework Decorated with Acylamide Groups. *J. Am. Chem. Soc.*, 133, 748–751, 2011.
29. Chen, Z., Adil, K., Weseliński, Ł.J., Belmabkhout, Y., Eddaoudi, M., A supermolecular building layer approach for gas separation and storage applications: The eea and rtl MOF platforms for CO2 capture and hydrocarbon separation. *J. Mater. Chem. A*, 3, 6276–6281, 2015.
30. Lu, Z., Zhang, J., Duan, J., Du, L., Hang, C., Pore space partition *via* secondary metal ions entrapped by pyrimidine hooks: Influences on structural

flexibility and carbon dioxide capture. *J. Mater. Chem. A*, 5, 17287–17292, 2017.
31. Moreau, F., da Silva, I., Al Smail, N.H., Easun, T.L., Savage, M., Godfrey, H.G.W., Parker, S.F., Manuel, P., Yang, S., Schröder, M., Unravelling exceptional acetylene and carbon dioxide adsorption within a tetra-amide functionalized metal–organic framework. *Nat. Commun.*, 8, 14085, 2017.
32. Benson, O., da Silva, I., Argent, S.P., Cabot, R., Savage, M., Godfrey, H.G.W., Yan, Y., Parker, S.F., Manuel, P., Lennox, M.J., Mitra, T., Easun, T.L., Lewis, W., Blake, A.J., Besley, E., Yang, S., Schröder, M., Amides Do Not Always Work: Observation of Guest Binding in an Amide-Functionalized Porous Metal–Organic Framework. *J. Am. Chem. Soc.*, 138, 14828–14831, 2016.
33. Zhang, M., Li, B., Li, Y., Wang, Q., Zhang, W., Chen, B., Li, S., Pan, Y., You, X., Bai, J., Finely tuning MOFs towards high performance in C2H2 storage: Synthesis and properties of a new MOF-505 analogue with an inserted amide functional group. *Chem. Commun.*, 52, 7241–7244, 2016.
34. Xiong, S., He, Y., Krishna, R., Chen, B., Wang, Z., Metal–Organic Framework with Functional Amide Groups for Highly Selective Gas Separation. *Cryst. Growth Des.*, 13, 2670–2674, 2013.
35. Zhang, M., Chen, C., Shi, Z., Huang, K., Fu, W., Zhou, W., Inserting Amide into NOTT-101 to Sharply Enhance Volumetric and Gravimetric Methane Storage Working Capacity. *Inorg. Chem.*, 58, 13782–13787, 2019.
36. Zhang, M., Zhou, W., Pham, T., Forrest, K.A., Liu, W., He, Y., Wu, H., Yildirim, T., Chen, B., Space, B., Pan, Y., Zaworotko, M.J., Bai, J., Fine Tuning of MOF-505 Analogues To Reduce Low-Pressure Methane Uptake and Enhance Methane Working Capacity. *Angew. Chem.*, 129, 11584–11588, 2017.
37. Pham, T., Forrest, K.A., Nugent, P., Belmabkhout, Y., Luebke, R., Eddaoudi, M., Zaworotko, M.J., Space, B., Understanding Hydrogen Sorption in a Metal–Organic Framework with Open-Metal Sites and Amide Functional Groups. *J. Phys. Chem. C*, 117, 9340–9354, 2013.
38. Wang, Z., Zheng, B., Liu, H., Lin, X., Yu, X., Yi, P., Yun, R., High-Capacity Gas Storage by a Microporous Oxalamide-Functionalized NbO-Type Metal–Organic Framework. *Cryst. Growth Des.*, 13, 5001–5006, 2013.
39. Li, X.-Y., Li, Y.-Z., Yang, Y., Hou, L., Wang, Y.-Y., Zhu, Z., Efficient light hydrocarbon separation and CO2 capture and conversion in a stable MOF with oxalamide-decorated polar tubes. *Chem. Commun.*, 53, 12970–12973, 2017.
40. Alsmail, N.H., Suyetin, M., Yan, Y., Cabot, R., Krap, C.P., Lü, J., Easun, T.L., Bichoutskaia, E., Lewis, W., Blake, A.J., Schröder, M., Analysis of High and Selective Uptake of CO2 in an Oxamide-Containing {Cu2(OOCR)4}-Based Metal–Organic Framework. *Chem. Eur. J.*, 20, 7317–7324, 2014.
41. Fu, D., Xu, Y., Zhao, M., Chang, Z., Bu, X., Enhancement of gas-framework interaction in a metal–organic framework by cavity modification. *Sci. Bull.*, 61, 1255–1259, 2016.
42. Afshariazar, F., Morsali, A., Wang, J., Junk, P.C., Highest and Fastest Removal Rate of PbII Ions through Rational Functionalized Decoration

of a Metal–Organic Framework Cavity. *Chem.—Eur. J.*, 26, 1355–1362, 2020.
43. Li, X.-Y., Shi, W.-J., Wang, X.-Q., Ma, L.-N., Hou, L., Wang, Y.-Y., Luminescence Modulation, White Light Emission, and Energy Transfer in a Family of Lanthanide Metal–Organic Frameworks Based on a Planar π-Conjugated Ligand. *Cryst. Growth Des.*, 17, 4217–4224, 2017.
44. Esrafili, L., Gharib, M., Morsali, A., The targeted design of dual-functional metal–organic frameworks (DF-MOFs) as highly efficient adsorbents for Hg 2+ ions: Synthesis for purpose. *Dalton Trans.*, 48, 17831–17839, 2019.
45. Esrafili, L., Gharib, M., Morsali, A., Selective detection and removal of mercury ions by dual-functionalized metal–organic frameworks: Design-for-purpose. *New J. Chem.*, 43, 18079–18091, 2019.
46. Gharib, M., Esrafili, L., Morsali, A., Retailleau, P., Solvent-assisted ligand exchange (SALE) for the enhancement of epoxide ring-opening reaction catalysis based on three amide-functionalized metal–organic frameworks. *Dalton Trans.*, 48, 8803–8814, 2019.
47. Esrafili, L., Gharib, M., Morsali, A., The targeted design of dual-functional metal–organic frameworks (DF-MOFs) as highly efficient adsorbents for Hg2+ ions: Synthesis for purpose. *Dalton Trans.*, 48, 17831–17839, 2019.
48. Yun, R., Lu, Z., Pan, Y., You, X., Bai, J., Formation of a metal–organic framework with high surface area and gas uptake by breaking edges off truncated cuboctahedral cages. *Angew. Chem.*, 125, 11492–11495, 2013.
49. Xue, D.-X., Wang, Q., Bai, J., Amide-functionalized metal–organic frameworks: Syntheses, structures and improved gas storage and separation properties. *Coord. Chem. Rev.*, 378, 2–16, 2019.
50. Sun, R., Li, Y.-Z., Bai, J., Pan, Y., Synthesis, Structure, Water-Induced Reversible Crystal-to-Amorphous Transformation, and Luminescence Properties of Novel Cationic Spacer-Filled 3D Transition Metal Supramolecular Frameworks from N, N ‘, N “-Tris (carboxymethyl)-1, 3, 5-benzenetricarboxamide. *Cryst. Growth Des.*, 7, 890–894, 2007.
51. Duan, J., Zheng, B., Bai, J., Zhang, Q., Zuo, C., Metal-dependent dimensionality in coordination polymers of a semi-rigid dicarboxylate ligand with additional amide groups: Syntheses, structures and luminescent properties. *Inorg. Chim. Acta*, 363, 3172–3177, 2010.
52. Forgan, R.S., Marshall, R.J., Struckmann, M., Bleine, A.B., Long, D.-L., Bernini, M.C., Fairen-Jimenez, D., Structure-directing factors when introducing hydrogen bond functionality to metal–organic frameworks. *CrystEngComm*, 17, 299–306, 2015.
53. Rajput, L., Kim, D., Lah, M.S., Conformational control of ligands to create a finite metal–organic cluster and an extended metal–organic framework. *CrystEngComm*, 15, 259–264, 2013.
54. Lin, X., Jia, J., Zhao, X., Thomas, K.M., Blake, A.J., Walker, G.S., Champness, N.R., Hubberstey, P., Schröder, M., High H2 Adsorption by Coordination-Framework Materials. *Angew. Chem. Int. Ed.*, 45, 7358–7364, 2006.

55. Hakimifar, A. and Morsali, A., Urea-Based Metal–Organic Frameworks as High and Fast Adsorbent for Hg2+ and Pb2+ Removal from Water. *Inorg. Chem.*, 58.1, 180-187, 2018.
56. Abdollahi, N., Akbar Razavi, S.A., Morsali, A., Hu, M.-L., High capacity Hg(II) and Pb(II) removal using MOF-based nanocomposite: Cooperative effects of pore functionalization and surface-charge modulation. *J. Hazard. Mater.*, 387, 121667, 2020.
57. Liu, W., Huang, X., Chen, C., Xu, C., Ma, J., Yang, L., Wang, W., Dou, W., Liu, W., Function-Oriented: The Construction of Lanthanide MOF Luminescent Sensors Containing Dual-Function Urea Hydrogen-Bond Sites for Efficient Detection of Picric Acid. *Chem.—Eur. J.*, 25, 1090–1097, 2019.
58. Liu, W., Huang, X., Xu, C., Chen, C., Yang, L., Dou, W., Chen, W., Yang, H., Liu, W., A Multi-responsive Regenerable Europium–Organic Framework Luminescent Sensor for Fe3+, CrVI Anions, and Picric Acid. *Chem.—Eur. J.*, 22, 18769–18776, 2016.
59. Seo, P.W., Khan, N.A., Hasan, Z., Jhung, S.H., Adsorptive removal of artificial sweeteners from water using metal–organic frameworks functionalized with urea or melamine. *ACS Appl. Mater. Interfaces*, 8, 29799–29807, 2016.
60. Yang, X. and Xia, Y., Urea-modified metal-organic framework of type MIL-101(Cr) for the preconcentration of phosphorylated peptides. *Microchim. Acta*, 183, 2235–2240, 2016.
61. Seo, P.W., Khan, N.A., Jhung, S.H., Removal of nitroimidazole antibiotics from water by adsorption over metal–organic frameworks modified with urea or melamine. *Chem. Eng. J.*, 315, 92–100, 2017.
62. Azhdari Tehrani, A., Esrafili, L., Abedi, S., Morsali, A., Carlucci, L., Proserpio, D.M., Wang, J., Junk, P.C., Liu, T., Urea Metal–Organic Frameworks for Nitro-Substituted Compounds Sensing. *Inorg. Chem.*, 56, 1446–1454, 2017.
63. Hakimifar, A. and Morsali, A., High-sensitivity detection of nitroaromatic compounds (NACs) by the pillared-layer metal–organic framework synthesized *via* ultrasonic method. *Ultrason. Sonochem.*, 52, 62–68, 2019.
64. Tehrani, A.A., Abbasi, H., Esrafili, L., Morsali, A., Urea-containing metal-organic frameworks for carbonyl compounds sensing. *Sens. Actuators, B*, 256, 706–710, 2018.
65. Esrafili, L., Tehrani, A.A., Morsali, A., Ultrasonic assisted synthesis of two urea functionalized metal organic frameworks for phenol sensing: A comparative study. *Ultrason. Sonochem.*, 39, 307–312, 2017.
66. Glomb, S., Woschko, D., Makhloufi, G., Janiak, C., Metal–Organic Frameworks with Internal Urea-Functionalized Dicarboxylate Linkers for SO2 and NH3 Adsorption. *ACS Appl. Mater. Interfaces*, 9, 37419–37434, 2017.
67. Custelcean, R., Moyer, B.A., Bryantsev, V.S., Hay, B.P., Anion Coordination in Metal–Organic Frameworks Functionalized with Urea Hydrogen-Bonding Groups. *Cryst. Growth Des.*, 6, 555–563, 2006.
68. Wang, X.-J., Li, J., Li, Q.-Y., Li, P.-Z., Lu, H., Lao, Q., Ni, R., Shi, Y., Zhao, Y., A urea decorated (3, 24)-connected rht-type metal–organic framework

exhibiting high gas uptake capability and catalytic activity. *CrystEngComm*, 17, 4632–4636, 2015.

69. Krishna Kumar, D., Das, A., Dastidar, P., Exploring conformationally flexible hydrogen-bond-functionalized ligand and counter anions in metal–organic frameworks of Cu(ii). *New J. Chem.*, 30, 1267–1275, 2006.
70. Custelcean, R., Moyer, B.A., Hay, B.P., A coordinatively saturated sulfate encapsulated in a metal–organic framework functionalized with urea hydrogen-bonding groups. *Chem. Commun.*, 48, 5971–5973, 2005.
71. Wu, B., Liang, J., Yang, J., Jia, C., Yang, X.-J., Zhang, H., Tang, N., Janiak, C., Sulfate ion encapsulation in caged supramolecular structures assembled by second-sphere coordination. *Chem. Commun.*, 15, 1762–1764, 2008.
72. Custelcean, R., Sellin, V., Moyer, B.A., Sulfate separation by selective crystallization of a urea-functionalized metal–organic framework. *Chem. Commun.*, 15, 1541–1543, 2007.
73. Tehrani, A.A., Abedi, S., Morsali, A., Wang, J., Junk, P.C., Urea-containing metal–organic frameworks as heterogeneous organocatalysts. *J. Mater. Chem. A*, 3, 20408–20415, 2015.
74. Zhu, C., Tang, H., Yang, K., Wu, X., Luo, Y., Wang, J., Li, Y., A urea-containing metal-organic framework as a multifunctional heterogeneous hydrogen bond-donating catalyst. *Catal. Commun.*, 135, 105837, 2020.
75. Afshariazar, F. and Morsali, A., Target-Architecture Engineering of a Novel Two-Dimensional Metal–Organic Framework for High Catalytic Performance. *Cryst. Growth Des.*, 19, 4239–4245, 2019.
76. Hall, E.A., Redfern, L.R., Wang, M.H., Scheidt, K.A., Lewis acid activation of a hydrogen bond donor metal–organic framework for catalysis. *ACS Catal.*, 6, 3248–3252, 2016.
77. Zhu, C., Mao, Q., Li, C., Zhou, Y., Wu, X., Luo, Y., Li, Y., A readily available urea based MOF that act as a highly active heterogeneous catalyst for Friedel–Crafts reaction of indoles and nitrostryenes. *Catal. Commun.*, 104, 123–127, 2018.
78. Dong, X.-W., Liu, T., Hu, Y.-Z., Liu, X.-Y., Che, C.-M., Urea postmodified in a metal–organic framework as a catalytically active hydrogen-bond-donating heterogeneous catalyst. *Chem. Commun.*, 49, 7681–7683, 2013.
79. Rao, P.C. and Mandal, S., Friedel–Crafts Alkylation of Indoles with Nitroalkenes through Hydrogen-Bond-Donating Metal–Organic Framework. *ChemCatChem*, 9, 1172–1176, 2017.
80. Ju, Z., Yan, S., Yuan, D., *De novo* tailoring pore morphologies and sizes for different substrates in a urea-containing MOFs catalytic platform. *Chem. Mater.*, 28, 2000–2010, 2016.
81. Roberts, J.M., Fini, B.M., Sarjeant, A.A., Farha, O.K., Hupp, J.T., Scheidt, K.A., Urea metal–organic frameworks as effective and size-selective hydrogen-bond catalysts. *J. Am. Chem. Soc.*, 134, 3334–3337, 2012.
82. Siu, P.W., Brown, Z.J., Farha, O.K., Hupp, J.T., Scheidt, K.A., A mixed dicarboxylate strut approach to enhancing catalytic activity of a *de novo* urea

derivative of metal–organic framework UiO-67. *Chem. Commun.*, 49, 10920–10922, 2013.
83. Zhang, H., Gao, X.-W., Wang, L., Zhao, X., Li, Q.-Y., Wang, X.-J., Microwave-assisted synthesis of urea-containing zirconium metal–organic frameworks for heterogeneous catalysis of Henry reactions. *CrystEngComm*, 21, 1358–1362, 2019.
84. Wang, X.-J., Li, P.-Z., Chen, Y., Zhang, Q., Zhang, H., Chan, X.X., Ganguly, R., Li, Y., Jiang, J., Zhao, Y., A Rationally Designed Nitrogen-Rich Metal–Organic Framework and Its Exceptionally High CO2 and H2 Uptake Capability. *Sci. Rep.*, 3, 1149, 2013.
85. Li, Q.-Y., Quan, Y., Wei, W., Li, J., Lu, H., Ni, R., Wang, X.-J., Synthesis, crystal structure and gas uptake properties of a urea-functionalized rht-type metal–organic framework. *Polyhedron*, 99, 1–6, 2015.
86. Marshall, R.J., McGuire, J., Wilson, C., Forgan, R.S., Crystallographic investigation into the self-assembly, guest binding, and flexibility of urea functionalised metal–organic frameworks. *Supramol. Chem.*, 30, 732–741, 2018.
87. Zhang, X., Zhang, Z., Boissonnault, J., Cohen, S.M., Design and synthesis of squaramide-based MOFs as efficient MOF-supported hydrogen-bonding organocatalysts. *Chem. Commun.*, 52, 8585–8588, 2016.
88. McGuirk, C.M., Katz, M.J., Stern, C.L., Sarjeant, A.A., Hupp, J.T., Farha, O.K., Mirkin, C.A., Turning On Catalysis: Incorporation of a Hydrogen-Bond-Donating Squaramide Moiety into a Zr Metal–Organic Framework. *J. Am. Chem. Soc.*, 137, 919–925, 2015.
89. Liu, Y., Ma, J., Wu, P., Zheng, J.-J., Tian, X., Jiang, M., He, Y., Dong, H., Wang, J., A nanoporous metal–organic framework as a renewable size-selective hydrogen-bonding catalyst in water. *Dalton Trans.*, 48, 11855–11861, 2019.

9

Carbonyl, Carboxy and Imide Functionalized Metal–Organic Frameworks

Abstract
Functional metal-organic frameworks based on carbonyl, carboxy and imide functions are discussed deeply in this section. Structural and practical aspect of each function is discussed to gain clear insight for designing functional metal organic frameworks with these carbonyl-based functions.

Keywords: Carbonyl, carboxy, imide, naphthalene diimide, functional metal-organic frameworks, metal-carboxylate solids, ion conductivity, donor-acceptor systems

9.1 Functionalized Metal–Organic Frameworks by Carbonyl Function

9.1.1 General Chemical Properties of Carbonyl Functional Group

In the structure of carbonyl function (C=O), owing to higher electronegativity of O-atom, there is a partial negative charge on the O-atom and carbon–oxygen double bond possesses strong dipole moment. In addition, oxygen atom is of two lone electron pairs. As results, O-atom is electron-rich, nucleophile and Lewis basic and host–guest chemistry of carbonyl atom is based on its strong dipole moment and high electron density on oxygen atom. Considering these chemical properties, carbonyl function can interact with hydrogen bond donor guests, Lewis acids like metal ions and molecules containing partially positive or electron-deficient sites. In structural view, carbonyl function is rigid because no possible rotation could be observed around carbon–oxygen double bond because of sp2 hybridization of carbonyl-C atom.

Ali Morsali and Sayed Ali Akbar Razavi. Functional Metal-Organic Frameworks: Structure, Properties and Applications, (201–230)

9.1.2 Function–Application Properties

As mentioned, carbonyl function is rigid and polar. So, this function is potentially ideal for development of rigid and polar frameworks. In combination of carbonyl and amide decorated MOFs we can say that both of them are of polar pore surface but amide decorated MOFs are relatively flexible and hydrogen bond donor/acceptor hosts while carbonyl decorated MOFs are rigid and just hydrogen bond acceptor sites.

Application of carbonyl decorated MOFs depended on chemical environment of carbonyl site. Two common strategy applied in development of carbonyl decorated MOFs. In on strategy, carbonyl function is place in the main chain of the ligand between phenyl rings. In second strategy, carbonyl function applied in the structure of MOFs by fluorenone motif. Depending on the each strategy, host–guest chemistry of carbonyl decorated MOFs is different. However, as a part of the amide function, the polar carbonyl group (C=O), is paid less attention.

Based on first strategy (introduction of carbonyl function between phenyl rings), carbonyl decorated MOFs applied in detection of H-bond donor guests [1, 2] and selective adsorption of CO_2 [3].

Banglin Chen and coworkers synthesized ketone decorated Tb-CBC framework (with formula [Tb(BCB)(DMF)]·(DMF)1.5$(H_2O)_2$ where H_3BCB = 4,4',4"-benzenetricarbonyltribenzoic acid) and applied it as fluorescent sensor for detection of small organic molecules (Figure 9.1) [1]. Up on excitation

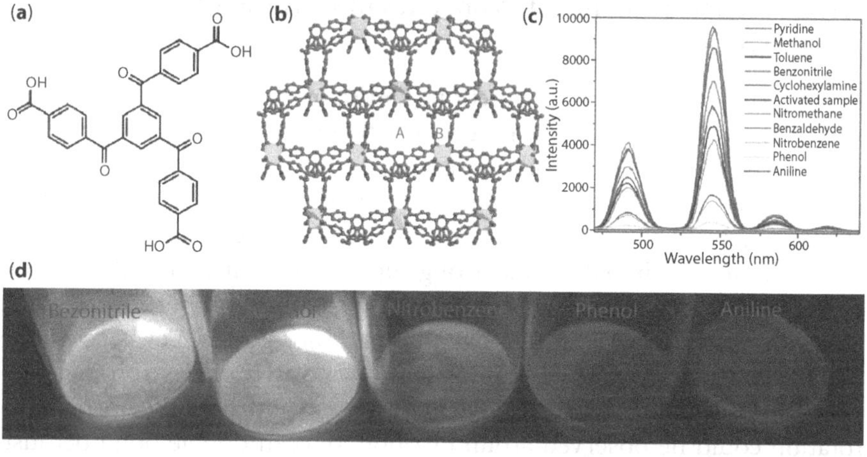

Figure 9.1 Application of Tb-BCB in detection of small organic molecules. (a) Structure of H3BCB ligand. (b) Crystal structure of Tb-BCB. (c) Change in fluorescent emission spectra of Tb-BCB in presence of different molecules. (d) The photo images of Tb-BCB in the presence of various guest molecules [1].

at 382 nm, the H_3BCB ligand emits fluorescence characteristic emission around 440 nm. Up on same excitation wavelength, the emission spectra of Tb-BCB is based on f–f transitions characteristic peaks of Tb(III) ions centered at 491, 545, 586, and 621 nm. This emission imply the energy transfer from BCB^{3-} linkers to Tb(III) centers. Addition of a small amount of different guest molecules to methanol solution of Tb-BCB leads to the completely different degrees of luminescence quenching effects so that hydrogen bond donor molecules, aniline and phenol, show the strongest quenching in emission peak of Tb-BCB. In fact, a small amount of aniline can even turn off the luminescence. The host-guest interaction between Tb-BCB framework and aniline through Tb-BCB(C=O)·(NH)aniline results in electron transfer from the framework to the analyte which leads to reduction in energy transfer from BCB^{3-} linkers to Tb(III) centers.

Wei Huang and coworkers developed a carbonyl functionalized MOF (Cu-cdip with formula $[Cu_2(cdip)(H_2O)_2]\cdot 2DMF\cdot 3H_2O$) based on carbonyl decorated 5,5'-carbonyldiisophthalic acid (H_4cdip) (Figure 9.2) and owing to its polarity applied for selective CO_2 separation [3]. This MOF adsorb 522.1 $cm^3\ g^{-1}$ for N_2 at 77 K and 1 bar with BET surface area of 1,539 $m^2\cdot g^{-1}$. Moreover that

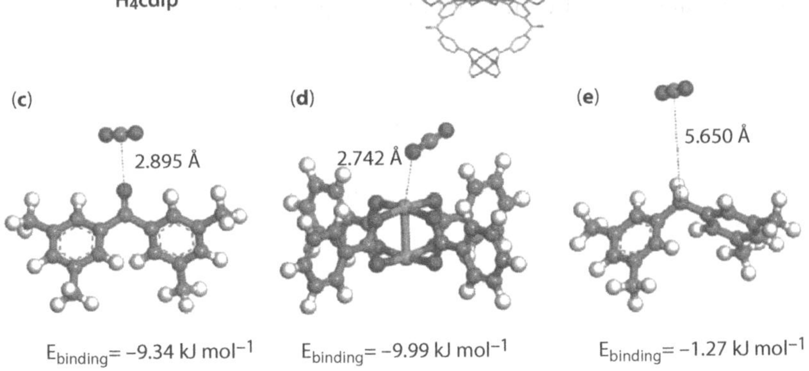

Figure 9.2 Application of Cu-cdip in gas adsorption. (a) Structure of the H_4cdip ligand. (b) Rhombicuboctahedral cage, with the Cu(II) ions pointing toward the center in the Cu-cdip. First-principles calculations for the binding energy of CO_2 molecule at different adsorption sites in Cu-cdip; (c) polar ketone site, (d) open Cu(II) sites and (e) non-functional ligand [3].

carbonyl sites, this MOF has open Cu(II) sites. This MOF applied in H_2 storage and could adsorb 3.05 wt% H_2 at 77 K and 1 bar which is higher than some of very well-known MOFs like PCN-14 (2.7 wt% H_2), NOTT-110 (2.6 wt% H_2), Mg-MOF-74 (2.2 wt% H_2) and ZIF-8 (1.27 wt% H_2). Also, Cu-cdip applied in CO_2 separation with exceptionally high CO_2 (46.7 wt % at 273 K and 1 bar) uptake which is higher than famous MOFs like Mg-MOF-74 (44.8 wt% CO_2) and NOTT-125 (40 wt% CO_2) at same conditions. Theoretical and experimental calculations show that polar carbonyl groups and open Cu(II) sites are most favorite adsorption sites. Based on these results, the binding energy of CO_2 adsorbed around the carbonyl site is −9.34 kJ·mol^{-1}, which is very comparable with that of an open Cu(II) site (−9.99 kJ·mol^{-1}). In contrast, if the ketone group is removed from the ligand, the CO_2 binding energy is calculated to be only −1.27 kJ·mol^{-1}. These comparisons imply that presence of carbonyl sites is effective as well as open Cu(II) sites.

As mentioned another strategy in synthesis of carbonyl based functions is synthesis of functional MOFs based on fluorene function. This aromatic carbonyl bearing functions is highly electroactive. Consequently, fluorene decorated MOFs applied as a platform for development of MOFs with special photophysical properties [4–6]. Also, owing to their polarity and rigidity, fluorene decorated MOFs applied in the field of gas adsorption [7, 8].

Kaskel and coworkers reported a carbonyl decorated MOF with fluorenone motif, $[Zr_6O_4(OH)_4(fdc)_6)\cdot 0.1BA\cdot 7.6H_2O]$ denoted as DUT-122 where BA = benzoic acid and H_2fdc = 9-fluorenone-2,7-dicarboxylic acid, which shows colorimetric guest-responsive fluorescent properties in presence of saturated vapors of different organic solvents like protic, aprotic, polar and non-polar, aromatic and aliphatic ones (Figure 9.3) [6]. This observation of solvent

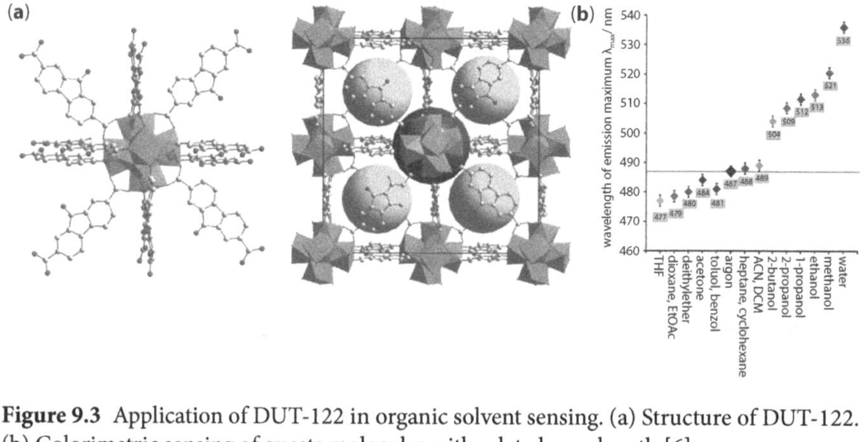

Figure 9.3 Application of DUT-122 in organic solvent sensing. (a) Structure of DUT-122. (b) Colorimetric sensing of guests molecules with related wavelength [6].

responsive from DUT-122 framework is based on sensitivity of carbonyl functionalized electroactive fluorenone motifs to the polarity different solvents.

Xinhui Zhou and coworkers reported the first single-lanthanide ratiometric luminescent thermometer MOF with formula $(Me_2NH_2)_3[Eu_3(fdc)_4(NO_3)_4]\cdot 4H_2O$ where H_2fdc is 9-fluorenone-2,7-dicarboxylic acid) (Figure 9.4) [4]. Time-resolved luminescence studies show that in Eu-fdc framework, energy difference between the H_2fdc triplet's excited state (17,794 cm^{-1}) and the $^5D_0$ of Eu^{3+} level (1,724 1 cm^{-1}) is small enough (less than 1,500 cm^{-1}) to allow for a strong thermally activated ion-to-ligand back energy transfer. As a consequence, both the ligand and Eu(III) emissions may be observed. Owing to hindered Eu(III)-to-fdc^{2-} back energy transfer at low temperature, the Eu(III) fluorescence emission peaks are strong. Up on increasing the temperature and so higher possibility of nonradiative-decay pathways, the fluorescence emission intensity of the ligand and Eu(III) centers will decrease. In more details, the emission of fdc^{2-} ligand decreases only ≈16% up to 150 K and 23% up to 230 K and the Eu(III) emission decreases ≈95% up to 150 K and 99% up to 230 K. This reveals that Eu(III) emission peak is strongly dependent to the temperature while the fdc^{2-} emission is essentially constant on the low temperature range and overally has dependence with higher temperature (65% decrease at 320 K). Finally, the ratio of the integrated areas of the $^5D_0 \rightarrow {}^7F_2$ transition (I_{Eu}) and the ligand's emission (I_{FDC}) are used to define the thermometric parameter $\Delta = I_{Eu}/I_{FDC}$ and, thus, to convert emission intensities into absolute temperatures. The Eu^{3+} to the ligand triplet emission intensity ratio allows measuring the temperature in the 12–320 K range, with a relative thermal sensitivity up to 2.7% K^{-1} at 170 K (0.33% K^{-1} at 300 K) and a repeatability of up to 96%.

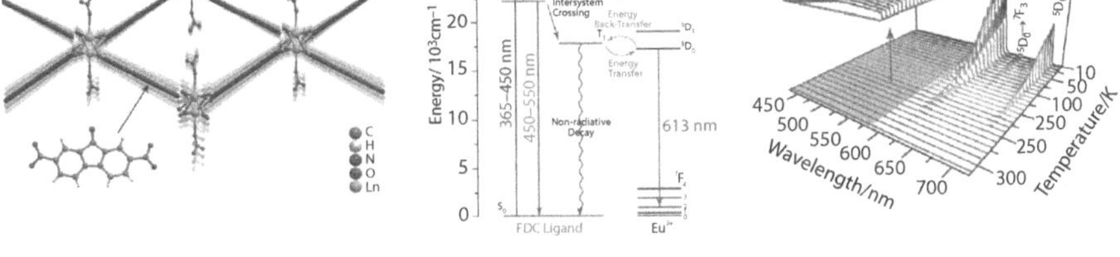

Figure 9.4 Application of Eu-fdc in ratiometric temperature sensing. (a) Schematic representation of the likely building units and crystal structure features of (Me2NH2)3[Ln3 (FDC)4(NO3)4]·4H2O. (b) Simplified energy diagram of the photoluminescence processes active in Eu-fdc (for simplicity only one Eu(III) local site is considered). (c) Emission spectra of Eu-fdc recorded between 12 and 423 K (excited at 365 nm). The inset shows a magnification of the fdc2– ligand emission in the 400–575 nm region [4].

9.1.3 Function–Structure Properties

As mentioned functionalization of pore walls of MOFs with carbonyl functions is an effective strategy for synthesis of polar, rigid and stable frameworks. Actually, the rigidity of the structure is because of rigidity of carbonyl functionalized ligands. In carbonylation strategy, sp^3 methylene moiety is replaced with polar sp^2 (-CO-) group. As a result of change in hybridization, the ligand and framework become rigid. Figure 9.5 depicted the carbonylation strategy for simple carbonyl based ligands (H_4mdip and H_2cdip) and fluorene based ligands (H_2fdc and H_2ofdc).

Matsuda and coworkers compared isostructural non-functionalized LMOF-201 framework (with formula [Zn_2(ofdc)$_2$(bipy)] where bipy is 4,4'-bipyridine and H_2fdc is fluorene-2,7-dicarboxylic acid) with carbonyl functionalized LMOF-202 (with formula ([Zn_2(fdc)$_2$(bipy)] where H_2fdc is fluorene-2,7-dicarboxylic acid). Conversion of flexible H_2fdc ligand to rigid H_2ofdc ligand through carbonylation strategy introduced slight differences but structural behavior of these MOFs up on guest-adsorption is completely different (Figure 9.5) [7]. CO_2 adsorption isotherm of rigid carbonyl decorated LMOF-202 framework is typical type-I at 195 K while for non-functionalized flexible LMOF-201 framework this isotherm is gate opening type indicating the large hysteresis because of structural change during adsorption measurements. *In-situ* X-ray powder diffraction shows that PXRD pattern of rigid carbonyl decorated LMOF-202 framework is identical during and after adsorption while for non-functionalized flexible LMOF-201 framework,

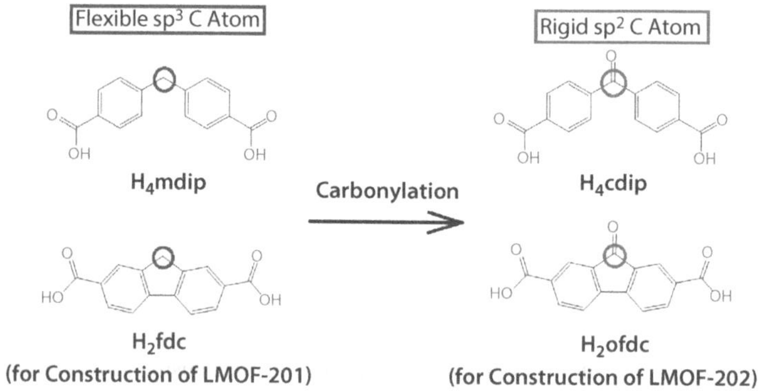

Figure 9.5 Carbonylation of sp^3 atoms for construction of rigid and polar carbonyl functionalized MOFs. H_2fdc is fluorene-2,7-dicarboxylic acid, H_4mdip is 5,5'-methylene-di-isophthalic acid, H_2ofdc is 9-oxofluorenone-2,7-dicarboxylic acid and H_4cdip is 5,5'-carbonyldiisophthalic acid [9].

PXRD pattern for guest free sample clearly differs from that of the as-synthesized sample.

9.2 Functionalized Metal–Organic Frameworks by Carboxy Function

9.2.1 General Chemical Properties of Carboxy Function

Functional behavior of carboxy (-CO-OH) can be described based on hydroxy and carbonyl functional groups. On one hand, owing to presence of carbonyl group(C=O), carboxy function can act as electron rich site with high dipole moment. In this regard, carboxy function can act as Lewis basic, H-bond acceptor, electron-rich or partially negative site to interact with Lewis acids, H-bond donor, electron-deficient or partially positive species. On the other hand, owing to presence of (-OH) group, carboxy function is Brønsted acid (with higher acidity than sole hydroxy) and H-bond donor site. These multiple chemical properties show that carboxy function is a suitable function to strengthen the host–guest chemistry of MOFs.

Owing to presence of two electron-rich O-atoms, carboxy functions can coordinate to metal ions as chelating site. Despite the non-complicated coordination modes of carboxylate and multi-carboxylate family of ligands, carboxylate $\left[-CO_2^-\right]$ (Figure 9.6), is one the functions that is widely applied in coordination chemistry to construct materials such as metal–carboxylate complexes, zero dimensional compounds, two and three-dimensional frameworks [10].

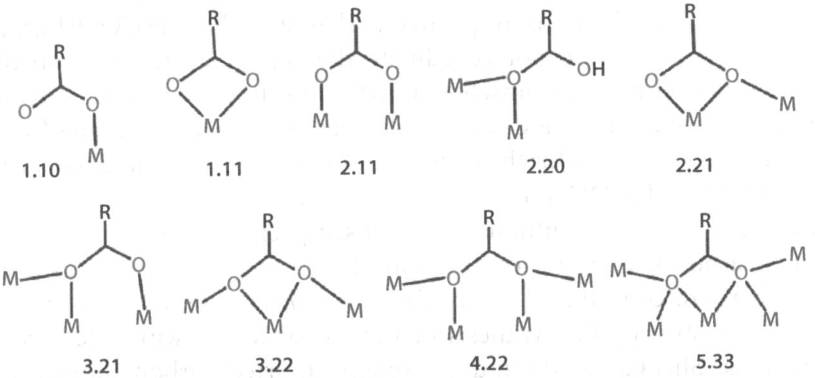

Figure 9.6 Coordination modes of the carboxylate ligands [10].

9.2.2 Synthesis of Functionalized Metal-Organic Frameworks with Free Carboxy Function

Owing to versatile host–guest chemistry, different coordination modes and strong coordination ability, carboxy function applied in the development of novel MOFs as guest-interactive and coordinating sites. It is completely known that carboxylate group extensively applied as coordinating functional group for development of novel MOFs. This is because of the fact that coordination polymers based carboxylate linkers are of regular structure and high crystallinity. For this purpose, carboxy function is place at terminal phenyl rings of the ligand with high ligand template effect and deprotonate during self-assembly process. However, for synthesis of functional MOFs bearing free carboxy functions, some strategies must be developed. Some applied strategies in this regard are (I) control over pH of the solution during self-assembly process [11] (II) control over synthesis condition when carboxylate ligands are places at phenyl ring with low ligand template effect [12] (III) immobilization of carboxy-bearing ligands on inorganic nodes through solvent assisted ligand incorporation [13] and (IV) post-synthesis modification [14].

Radu Custelcean and coworkers developed an effective strategy for synthesis of carboxy decorated MOFs based on control over pH of the solution (Figure 9.7) [11]. In this strategy, they synthesized $[CuCl(BNA)_2]{\cdot}Cl(H_2O)_4$ by slow diffusion of $CuCl_2$ in MeOH into an aqueous solution of binicotinic acid hydrochloride (BNA·HCl) as pyridine–carboxylate ligand with solution pH equal to 1.6. The results of single crystal determination shows that N atoms of pyridine are coordinated to the Cu(II) ions while carboxy functions remained free. The presence of COOH groups was supported by their characteristic infrared bands: a strong peak at 1,722 cm^{-1} and a broad absorbance in the range of 2,500–3000 cm^{-1} corresponding to the C=O and O–H stretching vibrations, respectively. They state that the COOH groups remain free, which do not coordinate the copper centers nor do they hydrogen bond among themselves or to the pyridine N atoms, because low pH and presence of Cl^- and water molecules, which are both better hydrogen bonding acceptors for the COOH group and thus inhibited the formation of COOH·HOOC interactions. In another work, it is established that through control over synthesis conditions, especially solvent, and location of the carboxylate groups on the central or terminal phenyl rings of the ligand, free or coordinated carboxylate functions can be achieved [12].

In next strategy for synthesis of functional MOFs with free carboxy functions, Shengqian Ma and coworkers inserted carbon dioxide into aryl C–H bonds of the backbone of a MOF to generate free carboxylate

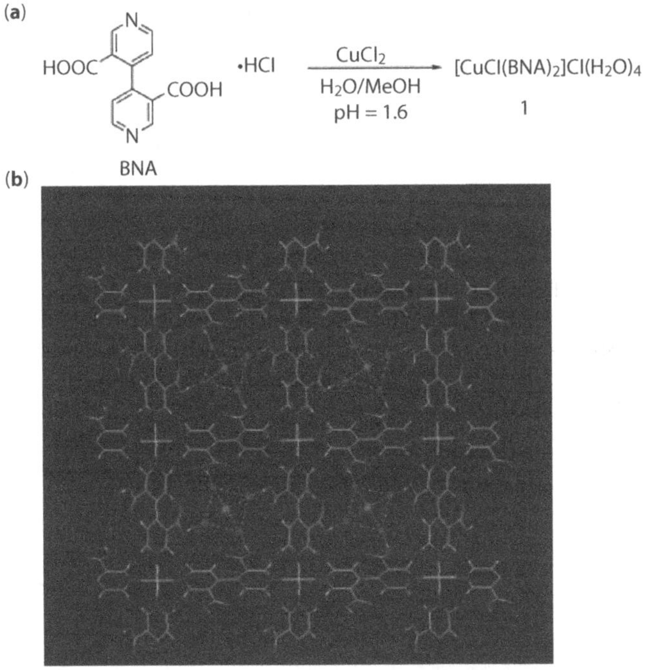

Figure 9.7 Synthesis of $[CuCl(BNA)_2]\cdot Cl(H_2O)_4$ through control over pH. (a) Structure of the ligand and synthesis process. (b) View down the c axis showing the square grid network containing channels decorated with COOH groups that bind $Cl^{-}(H_2O)_4$ clusters [11].

groups (Figure 9.8) [14]. For this aim, they selected UiO-67(dcppy) (2-phenylpyridine-5,4'-dicarboxylic acid) to generate the post-modified framework UiO-67(dcppy)-COOH with free carboxy group. This strategy based on utilizing CO_2 for heterogeneous C-H activation and carboxylation reactions on MOFs is a practical startegy for CO_2 chemical transformations under mild reaction conditions.

Based on solvent-assisted ligand incorporation strategy (SALI), Hupp and coworkers immobilized carboxylate ligand on labile Zr-nodes of NU-1000 and UiO-66 frameworks (Figure 9.9) [13]. They also show that this method works for incorporation of phosphonate-terminated ligands.

9.2.3 Function–Application Properties

As mentioned carboxy function has various chemical properties. So, different strategies applied to incorporate free carboxy function into the pores of functional MOFs as guest-interactive site.

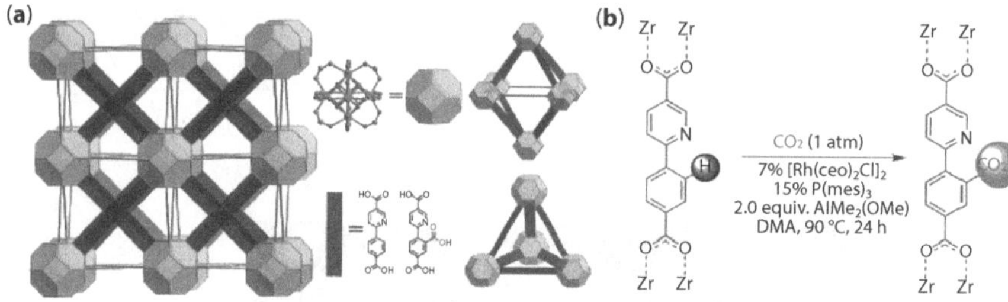

Figure 9.8 Post-synthesis CO_2 insertion into the structure of UiO-67(dcppy). (a) Structures of either UiO-67(dcppy) or UiO-67(dcppy)-COOH, composed of the cuboctahedral SBU, $[Zr_6(OH)_4O_4(CO_2)_{12}]$, and linear linkers, sustained by octahedral and tetrahedral cages. (b) Schematic representation of insertion of CO_2 into the aryl C-H bond within UiO-67(dcppy) [14].

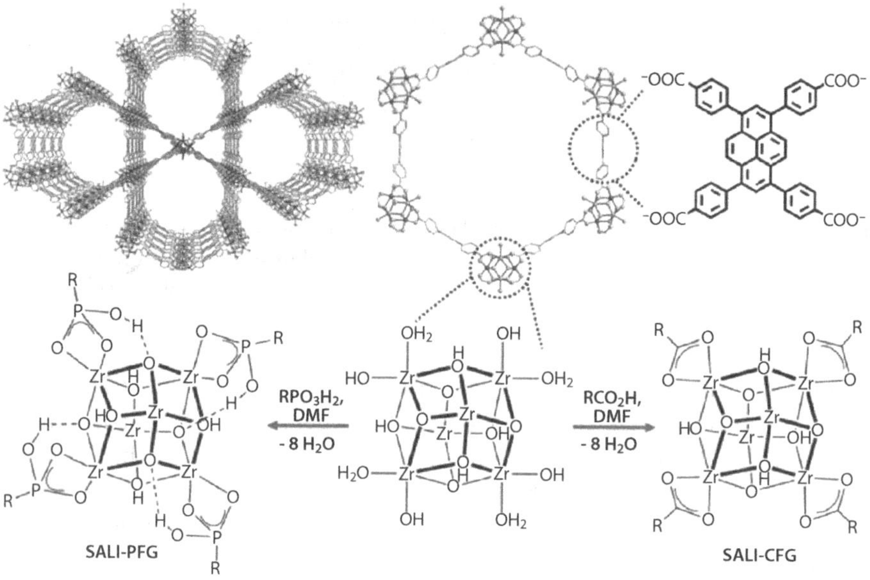

Figure 9.9 Molecular Representations of NU-1000 (top) and Depictions of Solvent-Assisted Ligand Incorporation (SALI) (bottom) [13].

Owing to their acidity, they can be applied as heterogeneous Brønsted acid catalyst in reactions such as methanolysis of epoxides [14]. Also, carboxy functionalized MOFs applied as platform to stabilize some species like proteins, nanoparticles and organocatalytic sites through hydrogen bonding or coordination interactions for different catalytic purposes [15–17]. Also, the Brønsted acidity of carboxy function and its ability to transfer H atoms and interact through H-bonding with different guest like water and imidazole makes it a suitable functional group for development of proton conductive MOFs [18–25].

Carboxy function is of strong dipole moment. So, it can interact with some gas molecules like CO_2 through different type of reactions like dipole(carboxy)·(CO_2)quadrupole interaction, carboxy(-OH)·(O)carbon dioxide hydrogen bond and carboxy(C=O)·(C)carbon dioxide donor–acceptor interactions [22, 26–28].

Multiple chemical properties of carboxy function make it an ideal guest-interactive site for interaction with different types of analytes such as metal ions through coordination and chelation interaction [29–38], hydrogen bond donor or acceptor small molecules [39–42].

UiO-MOFs applied as most favorable platform for development of carboxy decorated MOFs in different application like CO_2 adsorption, metal ion detection, proton conductivity and removal of toxic chemicals. Carboxy decorated UiO-MOF can be synthesized through post synthesis ligand incorporation of carboxy ligands on Zr-nodes or solvothermal procedure.

Hupp and coworkers introduced oxalic acid into the framework of UiO-66 through immobilization on Zr-nodes (Figure 9.10) [42]. They combine removal capacity of UiO-66-ox and UiO-66 in adsorption of NH_3, NO_2 and SO_2. UiO-66 can adsorb 2.0 mmol·g^{-1} NH_3, 0.1 mmol·g^{-1} SO_2 and 3.8 mmol·g^{-1} NO_2 while these values are improved for UiO-66-ox (2.5 mmol·g^{-1} NH_3, 0.8 mmol·g^{-1} SO_2 and 8.4 mmol·g^{-1} NO_2). NH_3 adsorb on UiO-66 by hydrogen bonding to Zr-nodes and subsequently to other NH_3 molecules. In case of NH_3 adsorption by UiO-66-ox, NH_3 react with the free carboxylic acid groups to form an ammonium-carboxylate acid-base species. Since there is not any strong possible interaction between UiO-66 and SO_2, this framework has very low SO_2 uptake while high capacity of UiO-66-ox to SO_2 is due to the chemisorption interaction between the free carboxylic acid group and SO_2 because of formation of $C(=O)OSO_2$ species that is strongly bond to UiO-66-ox through free carboxylate groups. UiO-66-ox capacity to NO_2 is more than double of UiO-66. This increase is consistent with ~6.3 NO_2 molecules per SBU or free carboxylic acid which is beyond adsorption by stoichiometric reaction. The ammonia increased capacity is due to the acid–base interaction while increased capacity for SO_2 and NO_2 is because of chemical properties of carboxy function. This work clearly shows that introduction of free carboxy function into the pore of MOFs is very effective strategy in removal of hazardous chemicals owing to its multiple chemical features.

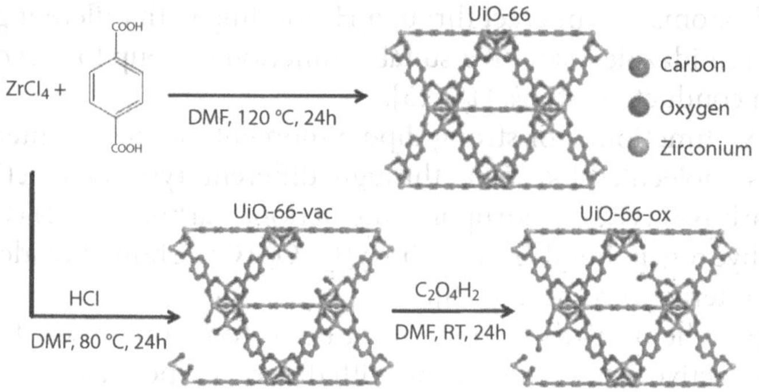

Figure 9.10 Synthesis of UiO-66-ox by Hupp and coworkers through solvent assisted ligand incorporation [42].

In other work by Christian Serre and coworkers, it is proved that owing to presence of carboxy group into the pores of UiO-66, carboxy decorated UiO-66, UiO-66-$(COOH)_2$, has CO_2/N_2 15:85 selectivity value of 56 with zero-coverage adsorption enthalpies equal to −34.8 kJ•mol^{-1} at 303 K (Figure 9.11) [27]. Based on preliminary Monte Carlo simulations they proved that CO_2 molecules are primarily distributed in the regions close to the μ_3-OH group of the $Zr_6O_4(OH)_4$ oxo-cluster node and the –COOH organic functions.

Devautour-Vinot and coworkers applied UiO-66-$(COOH)_2$ as a superprotonic platform for proton conductivity. This MOF bears high concentration of COOH groups [21]. So, the pores of UiO-66-$(COOH)_2$ are hydrophilic and Brønsted acidic. The oxygen atom and OH group of carboxylate unit inside the frameworks can construct complex H-bonded networks with each other or with water molecules. Up on temperature increasing from 25 to 90 °C at 40% humidity, the conductivity of UiO-66-$(COOH)_2$ increases from 7.47×10^{-6} S·cm^{-1} to 2.06×10^{-5} S·cm^{-1}. At 95% humidity, the conductivity of UiO-66-$(COOH)_2$ increases from 8.5×10^{-4} S·cm^{-1} at 25 °C to 2.30×10^{-3} S·cm^{-1} at 90 °C indicating the superprotonic behavior (conductivity more than 10^{-4} S·cm^{-1}) of UiO-66-$(COOH)_2$ at 95 % humidity and 90 °C. Studies show that reorganization of water molecules at higher temperature with equal distribution at tetrahedral and octahedral cages favors proton transfer over long distances which is assisted by a hydrogen-bonded water network.

In summary (I) designing uncoordinated carboxy functionalized MOFs by direct methods is challenge yet and some alternative methods like SALI and PSM should be replaced, (II) high affinity of carboxy

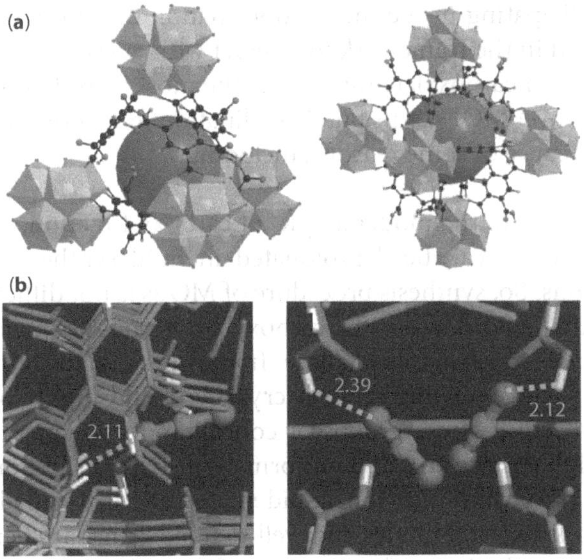

Figure 9.11 Application of UiO-66(COOH)$_2$ in CO_2 adsorption. (a) The UiO-66(Zr)-(COOH)$_2$ crystalline structure; tetrahedral cage (Left) and octahedral cage (Right). (b) Local views of the snapshots extracted from the GCMC simulations at 1.0 bar and 303 K, emphasizing the interactions between the CO_2 molecule and μ_3-OH group of the inorganic node (Left) and the -COOH organic functions (Right). Some framework atoms are deleted for clarity. Color codes: Zr: sky blue; O: red; C: gray, and H: white [27].

function to metal ions through chelation is of beneficial for construction of MOFs and sensing of metal ions, (III) The Brønsted acidity of carboxy function is practical for catalysis and proton conductivity and (IV) ability of carboxy function to incorporate in hydrogen bonding is very effective for enrichment of host-guest interactions and extending in dimensionality of coordination polymers through non-covalent supramolecular interactions.

9.2.4 Function–Structure Properties

Similar to other carbonyl-based functional groups, carboxy function applied in the structure of MOFs, but unlike them carboxy function usually applied as a coordinating site in constructing of structure of MOFs. The carboxyl group is one of the most widely occurring functional groups as coordinative site to metallic centers of MOFs. Because this functional group usually participate in the self-assembly process of synthesized MOFs through participating in the metal coordination or/and does not remain uncoordinated in the framework to interact with guests.

As mentioned carboxylate function is the most favorite function in the synthesis of novel MOFs. This public inclination to carboxylate coordinating sites is because of some reasons, including:

(I) Owing to the moderate pK_a of carboxy function, carboxy function can be deprotonated in mild synthesis conditions. So, synthesis procedure of MOFs is not difficult.
(II) After coordination of carboxylate groups to metal ions they construct crystalline frameworks with repeatable structure. Owing to such crystalline and highly ordered structure, these groups of coordination polymers are of high surface area and uniform pore size distribution.
(III) Carboxylate function is bidentate function with limited coordination modes and well-defined reticular chemistry. So, prediction of the structure of target MOF is much easier rather phosphonate and sulfonate based MOFs.
(IV) Owing to their chelation strength and their tendency to different kind of metal ions, they are able to develop novel framework with different types of metal ions.

Despite these advantageous of carboxylate based frameworks, they generally suffer from low stability in air and water. Sometimes, they suffer from low thermal stability, too. This observation is due to the fact that great numbers of carboxylate-based frameworks are constructed using non-oxophil metal centers late 3d late metal ions like Zn^{2+}, Cu^{2+}, Co^{2+}, and Ni^{2+}. Based on hard-soft acid-based theory, since the carboxylate function is hard and 3d late metal ions are soft, the coordination interaction between these building blocks are not much strong. Also, the negative charge of carboxylate function (–COO–) is delocalized on both oxygen atoms of the function which turns carboxylate into a less strongly coordinating site to the mentioned metal ions. MOF-5 with formula $[Zn_4O(bdc)_3]_n$ (H_2bdc is 1,4-benzenedicarboxylic acid) suffers from insufficient stability in presence of water and humid air because this MOF is developed based on coordination of hard carboxylate groups to non-oxophil soft Zn(II) ions [43]. So, in presence of water molecules and their attack to Zn_4O metal clusters, weak Zn(II)-carboxylate coordination bonds are interrupted and structure will collapse.

Different strategies applied to overcome such instability of MOFs. In first strategy and based on hard-soft chemistry, oxophil and hard metal ions (like

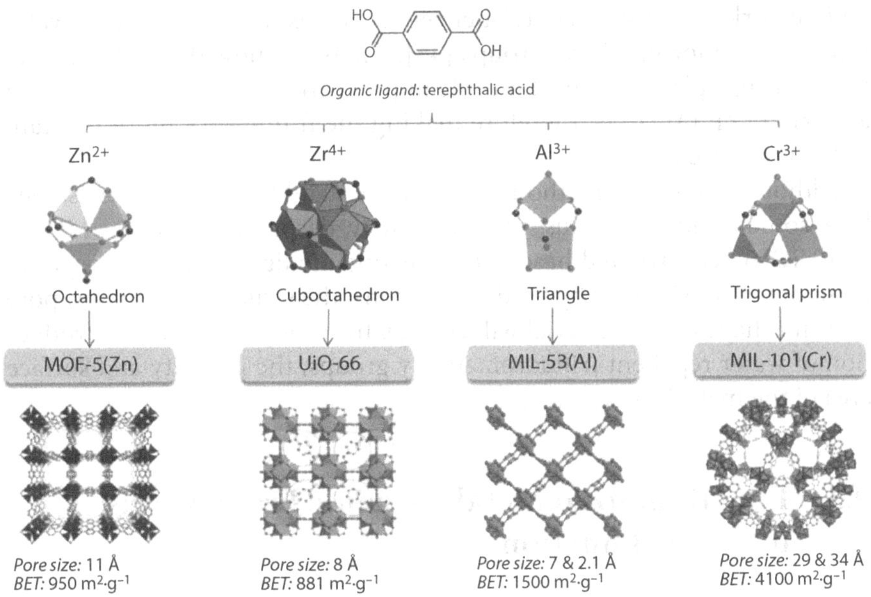

Figure 9.12 Inorganic nodes of some MOFs based on terphthalic acid linker. Combination of carboxylate linker with oxophil hard metal ions like Cr(III), Al(III) and Zr(IV) develop highly stable MOFs.

Zr(IV), Cr(III) and Al(III)) applied along with carboxylate donor ligands to construct highly stable MOFs (Figure 9.12) [44]. In this regard, UiO-66 (with formula $Zr_6O_4(OH)_4(bdc)_6$ where H_2bdc is benzenedicaroxylic acid) is synthesized which is highly stable organic solvents, water, humid air and even acidic solutions. Due to the high charge density and bond polarization, there is a strong affinity between Z(IV) and carboxylate O atoms in most carboxylate-based Zr-MOFs. As a highly stable Zr-based MOF in acidic solution, UMCM-309 framework, based on tritopic 1,3,5-(4-carboxylphenyl) benzene linker and $Zr_6(\mu_3\text{-}O)_4(\mu_3\text{-}OH)_4(RCO_2)_6(OH)_6(H_2O)_6$ cluster inorganic nodes, can save its crystallinity for over four months in highly acidic 1 M HCl solution [44]. However, although Zr-carboxylate based MOFs are highly stable in acidic media, but they suffer from insufficient instability in basic solution. Based on hard-soft chemistry we mentioned that synthesis of pyrazolate based framework using soft late 3d metal ins results in development of pyrazolate based frameworks with ultrahigh thermal and chemical stability in basic media. Another strategy to improve the stability of MOFs is decoration of pore walls of the frameworks with repellent functions. Based on this strategy and with the aim of improvement of stability of MOF-5 in aqueous media, Masel and coworkers substituted 1,4-benzenedicarboxylate ligand with water repellent trifluoromethoxy groups [45]. Analyses show that trifluoromethoxy groups play an important role in preventing water molecules from attacking (Zn_4O) cores, therefore making the framework more resistant to the moisture in the air.

Although these two methods improve the stability of the materials, they have some disadvantageous. In first strategy, despite the fact that MOF-5 and UiO-66 constructed based on same organic ligand but with replacement of soft Zn(II) ions by hard Zr(IV) ions, the structure, topology, pore size and shape of the material will change. In second strategy by introduction of water repellent trifluoromethoxy groups, the porosity and surface area of the material decreases.

9.3 Functionalized Metal–Organic Frameworks by Imide Function

9.3.1 General Chemical Properties of Imide Function

An imide (-CO-NR-CO-) is a functional group consisting of two acyl groups bound to nitrogen. As a useful function to design functional materials for supramolecular science, imide function particularly applied in the design of aromatic functionalized molecules with phenyl, naphthalene and perylene cores (Figure 9.13). Desirable electronic and spectroscopic

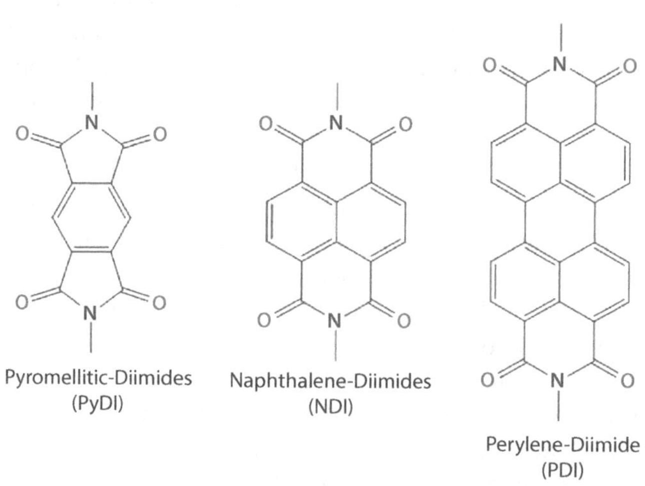

Figure 9.13 Structure of imide functionalized aromatic cores.

characters and better structure properties of naphthalene-diimides (NDI) over pyromellitic-diimides (PyDI) and perylene-diimide (PDI) helps NDI to be considered as ideal component for architecting functional materials for both covalent and non-covalent systems in supramolecular chemistry.

Naphthalene-diimide is neutral, conjugated planar, thermally stable and chemically robust with high melting point which is containing very various and fully practical chemical features like high electron deficiency, good charge carrier mobility and so high conductivity, special electron transfer and redox active properties. These characteristics of NDI make it suitable for different range of applications like electronics applications and electron transduction systems, photovoltaic devices, biological and medical applications, artificial photosynthesis and energy transfer, molecular chemosensors and biological sensors, photochromism and electrochromism [46].

Special electron deficiency character of NDI has very important role in its supramolecular chemistry and host–guest interactions. Such electron deficiency is the reason for NDI's high electron affinity, π–anion and π–π interactions, oxidative ability and donor–acceptor charge transfer (CT) complex systems formation. Moreover than control over host–guest interaction of NDI, this π-deficient character and π–π interactions of NDI can be utilized to implement strong directional interactions for crystal engineering of naphthalene-diimide functionalized materials [46].

A most important feature of the NDI chromophore is its mirror image fluorescence, which has a fluorescence quantum yield very close to unity. Indeed, high fluorescence quantum yields (~0.9) are observed for NDIs in all common organic solvents (aliphatic, aromatic, chlorinated, and dipolar solvents). Incorporation of aromatic functional groups on the diimide nitrogen sites will produce non-fluorescent or weakly fluorescent ligands, while introduction of alkyl groups often give rise to the typical white-blue fluorescence. Core substituted NDIs are more efficient in producing highly colorful and functional organic materials with many different photophysical properties. These observations clearly show NDIs are easily tunable for designing new efficient functional materials from biology to electronic [47].

It has been shown that designing NDI containing materials is of the most interdisciplinary fields of material and supramolecular science because of its diverse noncovalent interactions including hydrogen bonding, π–π stacking, π–ion interactions and π–halogen interactions. As results, NDI functionalized materials applied in pH sensing and molecular recognition like anion detection (e.g., basic anions like fluoride) through anion–π interactions and electron transfer, cation detection (e.g., colorimetric detection of Cu^{2+}, Hg^{2+} and Fe^{3+}) through intramolecular charge transfer and other important species (e.g., amino acids, explosives, and reactive oxygen/nitrogen species, etc.) [47].

NDIs are of electron accepting properties, capable of forming long-lived charge separated states and undergo chemically or electrochemically single reversible one-electron reduction at moderate potential (NDI reduction potential = −1.10V vs. Fc/Fc^+ in CH_2Cl_2) to form stable radical anions in high yield as redox-active units. Because of these characters NDIs applied in designing photochromic materials through photoinduced electron transfer, solar cells through light/solar energy into chemical energy, n-type semiconductors, catenane and rotaxane supramolecular switching machines, supramolecular self-assembled architectures and donor–acceptor systems.

9.3.2 Function–Application Properties

One of the most important goals of supramolecular chemistry is to assemble structural building blocks into arranged and ordered arrays, which do not suffer from structural defects and disorder, to create optimal properties and new applications through self-assembly. Given this attitude, it is rational to apply NDIs functional group into the highly regular structure of metal–organic frameworks. It is anticipated that the combination of different and unique properties of NDI such as (I) high electron deficiency or π-acidity (II) photo-redox active nature (III) ability to engage in a large variety of supramolecular noncovalent interactions (IV) forming stable radical anions and (V) charge mobility and conductivity, with porosity and regular structure of MOFs allow the design of new generation of redox and optically active materials with unique signal transduction. Owing to this versatile chemistry, NDI functionalized MOFs applied in different field of applications like hydroxide conductivity, Li storage, charge transfer and electrical conductivity, photocatalysis, optical chemosensing, gas separation and designing photochromic materials.

One of the most well-known applications of NDI functionalized MOFs is designing stimuli-responsive frameworks [48–55]. Up on exposure to external sources of light or electricity, NDI decorated MOFs undergo an electro/photo-induced redox activity and generate stable radical anions. Upon formation of stable NDI^- radical anion, the color of the MOFs changes because NDI/NDI^- redox couples have significantly different electroactive absorption bands in Vis region. So, photochromic and electrochromic behavior of NDI decorated MOFs accompanied by optical color change. When the external resources of light and electricity are switched off, the NDI^- radical anion transform into NDI core and the color of the framework returns to as-synthesized sample. This observation clearly shows that photochromic or electrochromic transformation between NDI^- radical anion and NDI core is a reversible process. Close crystal packing of NDI cores as well as their sufficient orbital overlap dominates the kinetic of electro/photo-induced redox activity and formation of NDI/NDI^- redox couple. It is found that orthogonal orientation of NDI cores in the crystal structure give rise in radical stability while π-stacked NDI cores with short distances results in short life time of NDI^- radical anion. Structural analysis by powder X-ray diffraction (PXRD) and electron resonance spectroscopy (ESR) clarify that stimuli-responsive formation of NDI/NDI^- redox couple occurs only based on chemical charge transfer and radical formation and structure of the framework is identical before and after color change. In situ ESR analysis displays that as-synthesized sample shows no ESR peak while stimulated sample shows an ESR peak.

As decelerated, NDI core with high charge mobility can generate NDI/NDI^- redox couple upon external electricity source which notify that NDI decorated MOFs potentially can be applied as n-type semiconductor. So, in an electrical charge transfer mechanism, NDI decorated chains of a framework can transport electrons thanks to high charge mobility and n-type characteristics of NDI cores [56].

Materials constructed based on donor–acceptor systems are strong tools for development of photoactive and conductive materials. Although charge transfer mechanism in donor–acceptor systems based on organic solids is highly studied, but researches on coordination solids in this topic is limited. MOFs are good platform for development of charge-transfer donor–acceptor systems because MOFs are crystalline and can be synthesized using organic ligands with electron donor and electron acceptor sites. Owing to its electron deficiency and n-type characters, NDI core can act as an ideal acceptor function in development of MOF-based charge transfer donor-acceptor systems [57]. When a MOF developed with suitable electron donor and electron acceptor functions, a possible charge transfer from electron donor/rich function to electron acceptor/deficient function can be detected. The mechanism of this charge transfer can be by "through-space mechanism" or "through-bond mechanism". Owing to the charge transfer, partial degree of charge separation could be observed so that electron acceptor site is partially negative and electron donor site is partially positive. In other word, donor function transforms into a radical cation while acceptor function transforms into a radical anion. In MOF-based donor acceptor systems, HOMO or conduction band in manly located on the donor function while LUMO is largely located on acceptor function. Using external sources of light of electricity, the level of charge transfer can be optimized proportional to the target application.

As mentioned aromatic-diimide cores are electronically active and are able to form radical anions upon light irradiation. This character is ideal for development of highly efficient heterogeneous photocatalysts by functionalized coordination polymers with aromatic-diimide cores. Duan and coworkers constructed a functional coordination polymer, named as Zn-PDI, with perylene-diimide (PDI) core based on bis(N-carboxymethyl) peryleneimide) ligand (H_2PDI) (Figure 9.14) [58]. Upon light irradiation PDI core behave in different ways. In first, PDI core transfers its excited photoinduced electrons to O_2 molecules to generate $O_2^{-\cdot}$ species for oxidation of alcohols and amines. In second way, excited PDI cores can adsorb one electron from sacrificial electron donor species (triethylamine) to generate PDI^- core. Excited Zn-PDI framework with $PDI^{-\cdot}$ core can transfer the electron to aryl halide molecules to convert C–X (X = halide) bonds to C–C or C–H bonds.

Functional MOFs with Diimide aromatic cores, especially NDI and pyridine-diimide cores (PyDI), applied in gas adsorption and separation based on two strategies. In first strategy, they applied without any modification treatment [59, 60]. Based on this strategy, diimide aromatic cores interact with gas molecules by π-deficient nature of the aromatic core (through charge transfer interactions) and high dipole moment of carbonyl sites (through electrostatic interactions). In second strategy, diimide

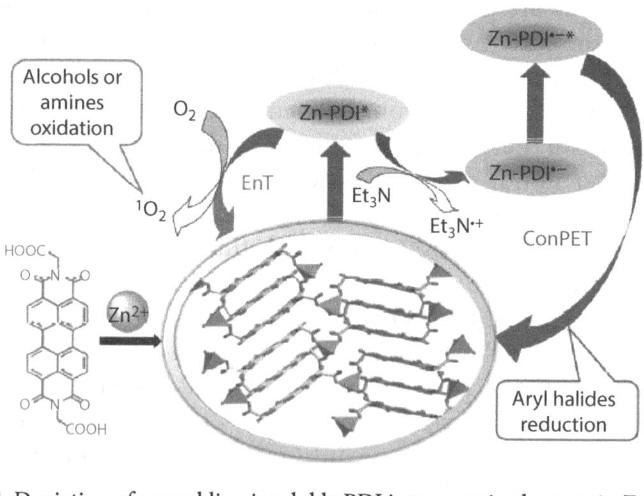

Figure 9.14 Depiction of assembling insoluble PDI into organized arrays in Zn-PDI for synthesis of an efficient photocatalyst to reduce aryl halides and oxidize alcohols and amines under visible light.

aromatic cores modify through redox reaction with alkaline metals to enhance the columbic host–guest interactions [61–64].

Based on first strategy, Banglin Chen and coworkers synthesized FJU-101 (with formula $[Cu_4(L)]$ where H_4L = N,N'-bis(5-isophthalicacid)naphthalenediimide) with open naphthalene diimide functional groups for room temperature high methane storage (Figure 9.15) [65]. FJU-101 has Brunauer–Emmett–Teller (BET) surface area and pore volume of 1,909 $m^2 \cdot g^{-1}$ and 0.762 $cm^3 \cdot g^{-1}$, respectively. The cylindrical channels of FJU-101 with the dimensions of 7.5 × 7.5 Å^2 (along the c-axis) are functionalized with open Cu(II) sites and carbonyl group sites of NDI function. This MOF can adsorb and deliver 212 cm^3(STP)$\cdot cm^{-3}$ and 144 cm^3(STP)$\cdot cm^{-3}$ methane at room temperature and 65 bar. Such enhanced methane uptake in FJU-101a in comparison with other MOFs like MFM-130, SNU-50 and UTSA-40 is attributed to the polar carbonyl sites of NDI cores, which can generate strong electrostatic interactions with CH_4 molecules. Although methane uptake of FJU-101 is smaller than some isostructure MOFs with a larger pore volume, but the CH_4 packing density in FJU-101 sample is higher than those of most widely studied MOFs like PCN-14, NOTT-101, NOTT-102, NOTT-103 and NJU-Bai-43. It indicates that the accessible carbonyl groups can serve as secondary functional sites to increase the CH_4 storage capacity.

Based on second strategy for application of diimide-cores in gas adsorption by MOFs, Hupp and coworkers synthesized $Zn_2(NDC)_2$(diPyNI) (denoted as **1**, NDC = 2,6-naphthalenedicarboxylate, diPyNI = *N,N'*-di-(4-pyridyl)-1,4,5,8-naphthalenetetracarboxydiimide) and reduced with lithium metal in dimethylformamide (DMF) as solvent to synthesis **1**$^-$Li$^+$ (Figure 9.16) [61]. The N_2 adsorption isotherm of **1** and **1**–Li$^+$ differs significantly so that at high nitrogen loading **1**$^-$Li$^+$ is able to accommodate more guests and the desorption curve has a distinct hysteresis loop that

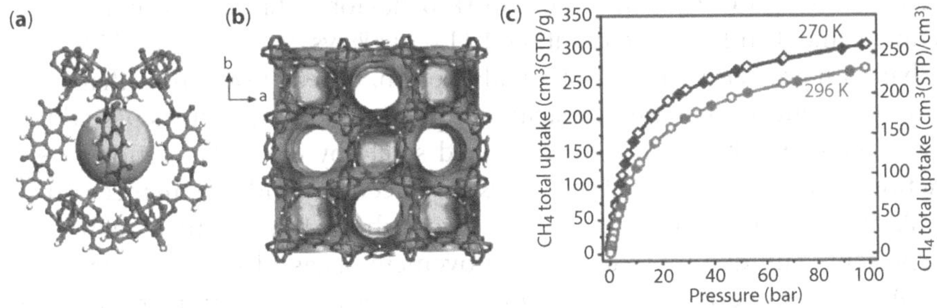

Figure 9.15 Application of FJU-101 in methane storage. Nanosized cage (a) and 3D framework of FJU-101 (b). (c) High-pressure methane adsorption isotherms of FJU-101a at 270 K and 296 K. Solid symbols: adsorption; open symbols: desorption [66].

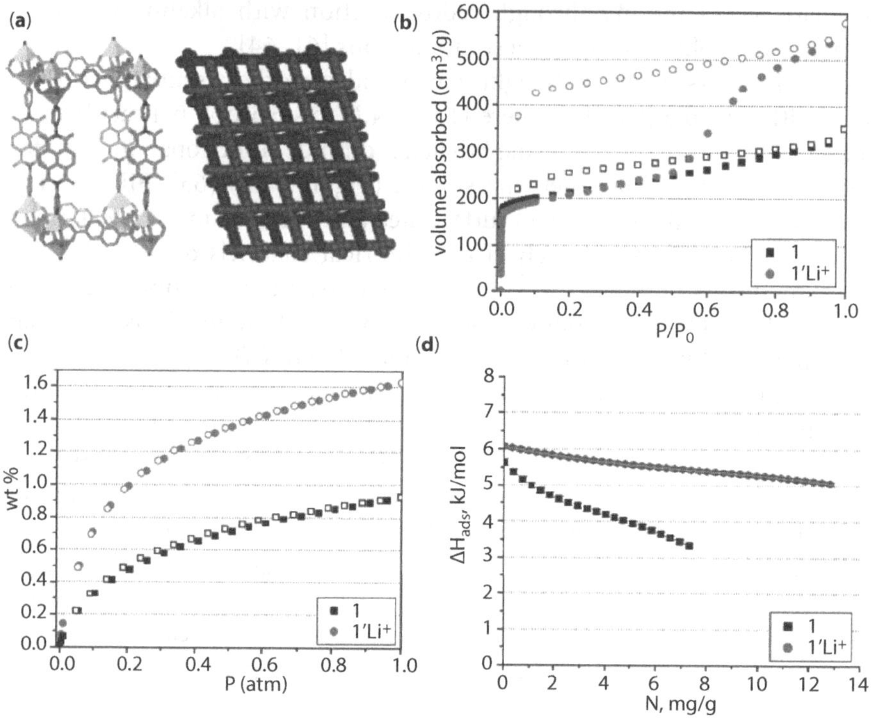

Figure 9.16 Application of 1 and 1^-Li^+ in H_2 storage. (a) Crystal structure of 1 omitting interwoven second network (Left) and 2-fold interpenetrated structure (Right). The yellow polyhedra represent the zinc ions: carbon, gray; oxygen, red; nitrogen, blue. (b) Nitrogen isotherms of 1 and 1^-Li^+: closed symbols, adsorption; open symbols, desorption. (c) H_2 adsorption isotherms of 1 and 1^-Li^+ at 77 K; (d) isosteric H_2 heat of adsorption of 1 and 1^-Li^+ [61].

does not rejoin the adsorption curve until $P/P_0 \approx 0.01$. This type of hysteresis is a sign of dynamic framework behavior. **1**$^-$Li$^+$ shows remarkable enhancement in H_2 capacity rather **1** (1.63 wt% vs. 0.93 wt%) at 77 K and 1 bar with greater isosteric heat of adsorption. Calculations show that such improvement in hydrogen capacity is equal to 60 H_2 molecules per added Li^+. Definitely, this cannot be explained solely by direct H_2–cation interactions. Other factors like framework displacement and/or enhanced strut polarizability are involved as well. This work shows that chemical reduction represents a unique approach for improvement of gas adsorption capacities.

As we saw, redox activity of diimide-cores is a tool for improvement of gas adsorption properties of MOFs. This redox activity can be useful in other application like Li-storage. Since, the process of Li-storage is based on a redox electrochemical procedure, NDI decorated MOFs can be of potential as cathode material in this field because owing to its redox activity, NDI core can progress the Li-Li^+ oxidation-reaction mechanism and accessible pore volume of the framework can store the lithium species. It is observed that C=O bond of NDI core reversibly reduced to C=C bond in lithiation process and C=C bond turn into C=O bond in delithiation process in the electrochemical process of Li-storage [67].

NDI core is an electroactive core in visible region with high π-deficiency which can involve in donor-acceptor interactions. NDI core can generally interacts electron rich species of those guests containing electron-rich especially electron-rich aromatic molecules through π(deficient)–π(rich) interaction and anions with anion–π interactions. In this regard NDI functionalized MOFs applied for chemosensing of electron-rich analytes like organic amines [68] and substituted phenyl ring with electron-rich methyl, methoxy and iodide groups [69]. Such color changes for NDI is due to the donor–acceptor interactions from electron-rich analytes to electron deficient NDI core. Since NDI core has low-lying HOMO and LUMO, is able to accept a density of electron from electron-rich analytes. As a result, electron density and electronic transitions of NDI can change which results in color change in visible region. Also, NDI decorated MOFs show solvatochromic behavior in presence of different organic solvents owing to sensitivity of NDI core to the solvent polarity [68]. Similarly, π-deficiency of NDI core is useful for designing ion conductive materials. Xiang and coworkers developed FJU-66 (with formula $[Cu_6(NDI)_3]\cdot 2DMF\cdot 6MeOH\cdot 2H_2O$ where H_2NDI is 2,7-bis(3,5-dimethyl)dipyrazol-1,4,5,8-naphthalene-tetracarboxydiimide) with high ion conductivity based on hydroxide ions [70]. For hydroxide conductivity, they applied [EVIm]OH ionic compound where $[EVIm]^+$ is 1-ethyl-3-vinylimidazolium. The cationic electron rich $[EVIm]^+$ ion interact with NDI core through various supramolecular interactions which create OH^- anions free from the resistance of the cation–anion electrostatic force. As a result, FJU-66 shows very high OH^- conductivity up to 0.1 $S\cdot cm^{-1}$ with the lowest activation energy E_a of 0.11 eV at low OH^- concentration of 0.34 $mmol\cdot g^{-1}$. In other work by Sourav Saha and coworkers, they developed a NDI functionalized MOF ($Cu_2(BPY)_2(NDIDS)$ where BPY is 4,4'-bipyridine and NDIDS is a functional ligand with NDI core and sulfonate coordinating sites) with noticeable lithium conductivity using $LiClO_4$ ionic compound [71]. Pristine MOF shows poor conductivity (4.65×10^{-10} $S\cdot m^{-1}$) at room temperature due to inadequate charge carrier density and electron delocalization pathway. After infiltration of $LiClO_4$ ion, the conductivity reaches up to 2.3×10^{-4} $S\cdot m^{-1}$. NDI core could bind to ClO_4^- ions, through anion–π and CH·anion interactions. In this case Li(I) ions remain free and conductivity increases. These two examples clearly show that selection of correct counter ion which is able to interact with NDI core, is a useful strategy for development of ion conductive MOF-based materials.

References

1. Guo, Z., Song, X., Lei, H., Wang, H., Su, S., Xu, H., Qian, G., Zhang, H., Chen, B., A ketone functionalized luminescent terbium metal–organic framework for sensing of small molecules. *Chem. Commun.*, 51, 376–379, 2015.
2. Wang, C.-C., Yang, C.-H., Tseng, S.-M., Lee, G.-H., Sheu, H.-S., Phyu, K., A new moisture-sensitive metal-coordination solids {[Cd(C4O4)(bipy)(H2O)2]·3H2O}∞ (bipy=4,4′-bipyridine). *Inorg. Chim. Acta*, 357, 3759–3764, 2004.
3. Feng, G., Peng, Y., Liu, W., Chang, F., Dai, Y., Huang, W., Polar Ketone-Functionalized Metal–Organic Framework Showing a High CO2 Adsorption Performance. *Inorg. Chem.*, 56, 2363–2366, 2017.
4. Li, L., Zhu, Y., Zhou, X., Brites, C.D.S., Ananias, D., Lin, Z., Paz, F.A.A., Rocha, J., Huang, W., Carlos, L.D., Visible-Light Excited Luminescent Thermometer Based on Single Lanthanide Organic Frameworks. *Adv. Funct. Mater.*, 26, 8677–8684, 2016.
5. Yan, D., Gao, R., Wei, M., Li, S., Lu, J., Evans, D.G., Duan, X., Mechanochemical synthesis of a fluorenone-based metal organic framework with polarized fluorescence: An experimental and computational study. *J. Mater. Chem. C*, 1, 997–1004, 2013.
6. Drache, F., Bon, V., Senkovska, I., Adam, M., Eychmüller, A., Kaskel, S., Vapochromic Luminescence of a Zirconium-Based Metal–Organic Framework for Sensing Applications. *Eur. J. Inorg. Chem.*, 2016, 4483–4489, 2016.
7. Ma, Y., Harada, Y., Hori, A., Hijikata, Y., Li, L., Matsuda, R., Tuning the flexibility of interpenetrated frameworks by a small difference in the fluorene moiety. *Dalton Trans.*, 46, 15200–15203, 2017.
8. Wang, B., Huang, H., Lv, X.-L., Xie, Y., Li, M., Li, J.-R., Tuning CO2 Selective Adsorption over N2 and CH4 in UiO-67 Analogues through Ligand Functionalization. *Inorg. Chem.*, 53, 9254–9259, 2014.
9. Ali Akbar Razavi, S. and Morsali, A., Linker functionalized metal-organic frameworks. *Coord. Chem. Rev.*, 399, 213023, 2019.
10. Goura, J. and Chandrasekhar, V.J.C.r., Molecular metal phosphonates. *Chem. Rev.*, 115, 6854–6965, 2015.
11. Custelcean, R. and Gorbunova, M.G., A Metal–Organic Framework Functionalized with Free Carboxylic Acid Sites and Its Selective Binding of a Cl(H2O)4-Cluster. *J. Am. Chem. Soc.*, 127, 16362–16363, 2005.
12. Wu, Y., Yang, G.-P., Zhao, Y., Wu, W.-P., Liu, B., Wang, Y.-Y., Three new solvent-directed Cd (ii)-based MOFs with unique luminescent properties

and highly selective sensors for Cu2+ cations and nitrobenzene. *Dalton Trans.*, 44, 3271–3277, 2015.
13. Deria, P., Bury, W., Hod, I., Kung, C.-W., Karagiaridi, O., Hupp, J.T., Farha, O.K., MOF Functionalization *via* Solvent-Assisted Ligand Incorporation: Phosphonates vs Carboxylates. *Inorg. Chem.*, 54, 2185–2192, 2015.
14. Gao, W.-Y., Wu, H., Leng, K., Sun, Y., Ma, S., Inserting CO2 into Aryl C-H Bonds of Metal–Organic Frameworks: CO2 Utilization for Direct Heterogeneous C-H Activation. *Angew. Chem.*, 128, 5562–5566, 2016.
15. Liu, Y., Xi, X., Ye, C., Gong, T., Yang, Z., Cui, Y., Chiral Metal–Organic Frameworks Bearing Free Carboxylic Acids for Organocatalyst Encapsulation. *Angew. Chem. Int. Ed.*, 53, 13821–13825, 2014.
16. Jayaramulu, K., Suresh, V.M., Maji, T.K., Stabilization of Cu2O nanoparticles on a 2D metal–organic framework for catalytic Huisgen 1,3-dipolar cycloaddition reaction. *Dalton Trans.*, 44, 83–86, 2015.
17. Shih, Y.-H., Lo, S.-H., Yang, N.-S., Singco, B., Cheng, Y.-J., Wu, C.-Y., Chang, I.-H., Huang, H.-Y., Lin, C.-H., Trypsin-Immobilized Metal–Organic Framework as a Biocatalyst In Proteomics Analysis. *ChemPlusChem*, 77, 982–986, 2012.
18. Rought, P., Marsh, C., Pili, S., Silverwood, I.P., Sakai, V.G., Li, M., Brown, M.S., Argent, S.P., Vitorica-Yrezabal, I., Whitehead, G., Modulating proton diffusion and conductivity in metal–organic frameworks by incorporation of accessible free carboxylic acid groups. *Chem. Sci.*, 10, 1492–1499, 2019.
19. Xie, X.-X., Yang, Y.-C., Dou, B.-H., Li, Z.-F., Li, G., Proton conductive carboxylate-based metal–organic frameworks. *Coord. Chem. Rev.*, 403, 213100, 2020.
20. Qin, Y., Xue, M.-H., Dou, B.-H., Sun, Z.-B., Li, G., High protonic conduction in two metal–organic frameworks containing high-density carboxylic groups. *New J. Chem.*, 44, 2741–2748, 2020.
21. Borges, D.D., Devautour-Vinot, S., Jobic, H., Ollivier, J., Nouar, F., Semino, R., Devic, T., Serre, C., Paesani, F., Maurin, G., Proton Transport in a Highly Conductive Porous Zirconium-Based Metal–Organic Framework: Molecular Insight. *Angew. Chem. Int. Ed.*, 55, 3919–3924, 2016.
22. He, T., Zhang, Y.Z., Wu, H., Kong, X.J., Liu, X.M., Xie, L.H., Dou, Y., Li, J.R., Functionalized Base-Stable Metal–Organic Frameworks for Selective CO2 Adsorption and Proton Conduction. *ChemPhysChem*, 18, 3245–3252, 2017.
23. Yang, F., Huang, H., Wang, X., Li, F., Gong, Y., Zhong, C., Li, J.-R., Proton Conductivities in Functionalized UiO-66: Tuned Properties, Thermogravimetry Mass, and Molecular Simulation Analyses. *Cryst. Growth Des.*, 15, 5827–5833, 2015.
24. Shigematsu, A., Yamada, T., Kitagawa, H., Wide Control of Proton Conductivity in Porous Coordination Polymers. *J. Am. Chem. Soc.*, 133, 2034–2036, 2011.
25. Wu, H., Yang, F., Lv, X.-L., Wang, B., Zhang, Y.-Z., Zhao, M.-J., Li, J.-R., A stable porphyrinic metal–organic framework pore-functionalized by

high-density carboxylic groups for proton conduction. *J. Mater. Chem. A*, 5, 14525–14529, 2017.

26. Wang, H.-H., Hou, L., Li, Y.-Z., Jiang, C.-Y., Wang, Y.-Y., Zhu, Z., Porous MOF with Highly Efficient Selectivity and Chemical Conversion for CO2. *ACS Appl. Mater. Interfaces*, 9, 17969–17976, 2017.
27. Yang, Q., Vaesen, S., Ragon, F., Wiersum, A.D., Wu, D., Lago, A., Devic, T., Martineau, C., Taulelle, F., Llewellyn, P.L., Jobic, H., Zhong, C., Serre, C., DeWeireld, G., Maurin, G., A Water Stable Metal–Organic Framework with Optimal Features for CO2 Capture. *Angew. Chem. Int. Ed.*, 52, 10316–10320, 2013.
28. Yang, Q., Wiersum, A.D., Llewellyn, P.L., Guillerm, V., Serre, C., Maurin, G., Functionalizing porous zirconium terephthalate UiO-66Zr. for natural gas upgrading: A computational exploration. *Chem. Commun.*, 47, 9603–9605, 2011.
29. Hao, J.-N. and Yan, B., A water-stable lanthanide-functionalized MOF as a highly selective and sensitive fluorescent probe for Cd2+. *Chem. Commun.*, 51, 7737–7740, 2015.
30. Yang, Y., Chen, L., Jiang, F., Wan, X., Yu, M., Cao, Z., Jing, T., Hong, M., Fabricating a super stable luminescent chemosensor with multi-stimuli-response to metal ions and small organic molecules through turn-on and turn-off effects. *J. Mater. Chem. C*, 5, 4511–4519, 2017.
31. Jia, X.-X., Yao, R.-X., Zhang, F.-Q., Zhang, X.-M., A Fluorescent Anionic MOF with Zn4(trz)2 Chain for Highly Selective Visual Sensing of Contaminants: CrIII. Ion and TNP. *Inorg. Chem.*, 56, 2690–2696, 2017.
32. Gadzikwa, T., Farha, O.K., Mulfort, K.L., Hupp, J.T., Nguyen, S.T., A Zn-based, pillared paddlewheel MOF containing free carboxylic acids *via* covalent post-synthesis elaboration. *Chem. Commun.*, 25, 3720–3722, 2009.
33. Wu, W., Li, B., Gu, C., Wang, J., Singh, A., Kumar, A., Luminescent sensing of Cu2+, CrO42- and photocatalytic degradation of methyl violet by ZnII. metal–organic framework MOF. having 5,5′-(1H-2,3,5-triazole-1,4-diyl) diisophthalic acid ligand. *J. Mol. Struct.*, 1148, 531–536, 2017.
34. Liang, Y.-T., Yang, G.-P., Liu, B., Yan, Y.-T., Xi, Z.-P., Wang, Y.-Y., Four super water-stable lanthanide–organic frameworks with active uncoordinated carboxylic and pyridyl groups for selective luminescence sensing of Fe3+. *Dalton Trans.*, 44, 13325–13330, 2015.
35. Sun, W., Wang, J., Zhang, G., Liu, Z., A luminescent terbium MOF containing uncoordinated carboxyl groups exhibits highly selective sensing for Fe3+ ions. *RSC Adv.*, 4, 55252–55255, 2014.
36. Fu, W.-Q., Liu, M., Gu, Z.-G., Chen, S.-M., Zhang, J., Liquid Phase Epitaxial Growth and Optical Properties of Photochromic Guest-Encapsulated MOF Thin Film. *Cryst. Growth Des.*, 16, 5487–5492, 2016.
37. Cao, J., Gao, Y., Wang, Y., Du, C., Liu, Z., A microporous metal–organic open framework containing uncoordinated carbonyl groups as postsynthetic

modification sites for cation exchange and Tb3+ sensing. *Chem. Commun.*, 49, 6897–6899, 2013.
38. Meng, X., Zhong, R.-L., Song, X.-Z., Song, S.-Y., Hao, Z.-M., Zhu, M., Zhao, S.-N., Zhang, H.-J., A stable, pillar-layer metal–organic framework containing uncoordinated carboxyl groups for separation of transition metal ions. *Chem. Commun.*, 50, 6406–6408, 2014.
39. Shi, B., Zhong, Y., Guo, L., Li, G., Two dimethylphenyl imidazole dicarboxylate-based lanthanide metal–organic frameworks for luminescence sensing of benzaldehyde. *Dalton Trans.*, 44, 4362–4369, 2015.
40. Arici, M., Luminescent 2D + 2D → 2D Interpenetrated Zn(II)-Coordination Polymer Based on Reduced Schiff Base Tricarboxylic Acid and Bisimidazole. Ligand for Detection of Picric Acid and Fe3+ Ions. *Cryst. Growth Des.*, 17, 5499–5505, 2017.
41. Yue, Z., Chen, Z., Yao, M., Wang, H., Li, G., Selective pyridine recognition by an imidazole dicarboxylate-based 3D cadmiumii. MOF. *RSC Adv.*, 4, 33537–33540, 2014.
42. DeCoste, J.B., Demasky, T.J., Katz, M.J., Farha, O.K., Hupp, J.T., A UiO-66 analogue with uncoordinated carboxylic acids for the broad-spectrum removal of toxic chemicals. *New J. Chem.*, 39, 2396–2399, 2015.
43. Greathouse, J.A. and Allendorf, M.D., The Interaction of Water with MOF-5 Simulated by Molecular Dynamics. *J. Am. Chem. Soc.*, 128, 10678–10679, 2006.
44. Bai, Y., Dou, Y., Xie, L.-H., Rutledge, W., Li, J.-R., Zhou, H.-C., Zr-based metal–organic frameworks: Design, synthesis, structure, and applications. *Chem. Soc. Rev.*, 45, 2327–2367, 2016.
45. Wu, T., Shen, L., Luebbers, M., Hu, C., Chen, Q., Ni, Z., Masel, R.I., Enhancing the stability of metal-organic frameworks in humid air by incorporating water repellent functional groups. *Chem. Commun.*, 46, 6120–6122, 2010.
46. Bhosale, S.V., Jani, C.H., Langford, S.J., Chemistry of naphthalene diimides. *Chem. Soc. Rev.*, 37, 331–342, 2008.
47. Al Kobaisi, M., Bhosale, S.V., Latham, K., Raynor, A.M., Bhosale, S.V., Functional naphthalene diimides: Synthesis, properties, and applications. *Chem. Rev.*, 116, 11685–11796, 2016.
48. Liu, J.-J., Xia, S.-B., Liu, D., Hou, J., Suo, H., Cheng, F.-X., Multifunctional naphthalene diimide-based coordination polymers: Photochromism and solventchromism. *Dyes Pigm.*, 177, 108269, 2020.
49. Liu, J.-J., Wang, Z.-J., Xia, S.-B., Liu, J., Shen, X., Photochromic and photocontrolled luminescence properties of two metal–organic frameworks constructed from a naphthalene diimide derivative. *Dyes Pigm.*, 172, 107856, 2020.
50. AlKaabi, K., Wade, C.R., Dincă, M., Transparent-to-dark electrochromic behavior in naphthalene-diimide-based mesoporous MOF-74 analogs. *Chem*, 1, 264–272, 2016.
51. Dietl, C., Hintz, H., Rühle, B., Schmedt auf der Günne, J., Langhals, H., Wuttke, S., Switch-On Fluorescence of a Perylene-Dye-Functionalized

Metal–Organic Framework through Postsynthetic Modification. *Chem. Eur. J.*, 21, 10714–10720, 2015.
52. Haldar, R., Mazel, A., Joseph, R., Adams, M., Howard, I.A., Richards, B.S., Tsotsalas, M., Redel, E., Diring, S., Odobel, F., Wöll, C., Excitonically Coupled States in Crystalline Coordination Networks. *Chem. Eur. J.*, 23, 14316–14322, 2017.
53. Han, L., Qin, L., Xu, L., Zhou, Y., Sun, J., Zou, X., A novel photochromic calcium-based metal–organic framework derived from a naphthalene diimide chromophore. *Chem. Commun.*, 49, 406–408, 2013.
54. Garai, B., Mallick, A., Banerjee, R., Photochromic metal–organic frameworks for inkless and erasable printing. *Chem. Sci.*, 7, 2195–2200, 2016.
55. Wei, F., Ye, Y., Huang, W., Lin, Q., Li, Z., Liu, L., Chen, S., Zhang, Z., Xiang, S., A naphthalene diimide-based MOF with mog net featuring photochromic behaviors and high stability. *Inorg. Chem. Commun.*, 93, 105–109, 2018.
56. Castaldelli, E., Imalka Jayawardena, K.D.G., Cox, D.C., Clarkson, G.J., Walton, R.I., Le-Quang, L., Chauvin, J., Silva, S.R.P., Demets, G.J.-F., Electrical semiconduction modulated by light in a cobalt and naphthalene diimide metal–organic framework. *Nat. Commun.*, 8, 2139, 2017.
57. Leong, C., Chan, B., Faust, T., D'Alessandro, D., Controlling charge separation in a novel donor–acceptor metal–organic framework *via* redox modulation. *Chem. Sci.*, 5, 4724–4728, 2014.
58. Zeng, L., Liu, T., He, C., Shi, D., Zhang, F., Duan, C., Organized Aggregation Makes Insoluble Perylene Diimide Efficient for the Reduction of Aryl Halides *via* Consecutive Visible Light-Induced Electron-Transfer Processes. *J. Am. Chem. Soc.*, 138, 3958–3961, 2016.
59. Mukherjee, S., Babarao, R., Desai, A.V., Manna, B., Ghosh, S.K., Polar Pore Surface Guided Selective CO2 Adsorption in a Prefunctionalized Metal–Organic Framework. *Cryst. Growth Des.*, 17, 3581–3587, 2017.
60. Prasad, T.K., Hong, D.H., Suh, M.P., High Gas Sorption and Metal-Ion Exchange of Microporous Metal–Organic Frameworks with Incorporated Imide Groups. *Chem.—Eur. J.*, 16, 14043–14050, 2010.
61. Mulfort, K.L. and Hupp, J.T., Chemical reduction of metal-organic framework materials as a method to enhance gas uptake and binding. *J. Am. Chem. Soc.*, 129, 9604–9605, 2007.
62. Bae, Y.-S., Hauser, B.G., Farha, O.K., Hupp, J.T., Snurr, R.Q., Enhancement of CO2/CH4 selectivity in metal–organic frameworks containing lithium cations. *Micropor. Mesopor. Mat.*, 141, 231–235, 2011.
63. Leong, C.F., Faust, T.B., Turner, P., Usov, P.M., Kepert, C.J., Babarao, R., Thornton, A.W., D'Alessandro, D.M., Enhancing selective CO2 adsorption *via* chemical reduction of a redox-active metal–organic framework. *Dalton Trans.*, 42, 9831–9839, 2013.
64. Mulfort, K.L. and Hupp, J.T., Alkali Metal Cation Effects on Hydrogen Uptake and Binding in Metal–Organic Frameworks. *Inorg. Chem.*, 47, 7936–7938, 2008.

65. Ye, Y., Lin, R.-B., Cui, H., Alsalme, A., Zhou, W., Yildirim, T., Zhang, Z., Xiang, S., Chen, B., A microporous metal–organic framework with naphthalene diimide groups for high methane storage. *Dalton Trans.*, 49, 3658–3661, 2020.
66. Ye, Y., Lin, R.-B., Cui, H., Alsalme, A., Zhou, W., Yildirim, T., Zhang, Z., Xiang, S., Chen, B., A microporous metal–organic framework with naphthalene diimide groups for high methane storage. *Dalton Trans.*, 49.12, 3658–3661, 2020.
67. Tian, B., Ning, G.-H., Gao, Q., Tan, L.-M., Tang, W., Chen, Z., Su, C., Loh, K.P., Crystal Engineering of Naphthalenediimide-Based Metal–Organic Frameworks: Structure-Dependent Lithium Storage. *ACS Appl. Mater. Interfaces*, 8, 31067–31075, 2016.
68. Mallick, A., Garai, B., Addicoat, M.A., Petkov, P.S., Heine, T., Banerjee, R., Solid state organic amine detection in a photochromic porous metal organic framework. *Chem. Sci.*, 6, 1420–1425, 2015.
69. Takashima, Y., Martínez, V.M., Furukawa, S., Kondo, M., Shimomura, S., Uehara, H., Nakahama, M., Sugimoto, K., Kitagawa, S., Molecular decoding using luminescence from an entangled porous framework. *Nat. Commun.*, 2, 168, 2011.
70. Li, Z., Zhang, Z., Ye, Y., Cai, K., Du, F., Zeng, H., Tao, J., Lin, Q., Zheng, Y., Xiang, S., Rationally tuning host–guest interactions to free hydroxide ions within intertrimerically cuprophilic metal–organic frameworks for high OH- conductivity. *J. Mater. Chem. A*, 5, 7816–7824, 2017.
71. Panda, D.K., Maity, K., Palukoshka, A., Ibrahim, F., Saha, S., Li+ Ion-Conducting Sulfonate-Based Neutral Metal–Organic Framework. *ACS Sustainable Chem. Eng.*, 7, 4619–4624, 2019.

10

Fluorine and Phosphonate Functional Metal–Organic Frameworks

Abstract
This chapter belongs to those functions which cannot be sorted in the previous groups of functional group. The content of this chapter focuses on phosphonate/phosphonic acid and fluorine functions. A brief summary about metal-phosphonate frameworks and the role of fluorine atom on the host-guest chemistry of MOFs is discussed in this section.

Keywords: Fluorine, trifluoromethyle, phosphonic acid, metal-phosphonate frameworks, fluorinated metal-organic frameworks, functional metal-organic frameworks

10.1 Functionalized Metal–Organic Frameworks by Phosphonic Acid/Phosphonate Functions

Similar to carboxy/carboxylate and sulfonic acid/sulfonate functions, phosphonic acid/phosphonate functions are applied as both coordinating and guest-interactive sites.

As guest-interactive site, phosphonic acid decorated MOFs are applied to catalyze some reactions like acetalization, Friedel–Crafts, and iso-Pictet–Spengler owing to the Brønsted acidity [6], to adsorb CO_2 molecules through phosphonic acid(P-OH)·(O)CO_2 interaction [7, 8] while phosphonate can act as a coordinating or chelating agent as well as an antenna to sensitize emissive f–f states of lanthanide metal ions [9]. Phosphonate/phosphonic acid decorated MOFs are applied as platform for development of proton conductive MOF-based materials because of their Brønsted acidity and ability to interact as hydrogen bond/donor sites [10–15].

As coordinating site, phosphonate based ligands can form a large variety of compounds including metal–phosphonate molecules and

---

Ali Morsali and Sayed Ali Akbar Razavi. Functional Metal-Organic Frameworks: Structure, Properties and Applications, (231–238) © 2021 Scrivener Publishing LLC

organic–inorganic hybrid materials. Phosphonate-based ligands are applied in the synthesis of coordination polymers more than sulfonate yet less than carboxylate. In this regard there are points that must be mentioned in comparison of phosphonate based ligands and carboxylate based ligands [16, 17].

(I) Phosphonate coordinating site is a strong chelation agent which can form stronger coordination bonds with metal ions rather carboxylate. This stronger metal–phosphonate interaction results in higher thermal and chemical stability (especially in air and humidity) of metal–phosphonate frameworks than metal–carboxylate frameworks.
(II) Phosphonate groups has strong tendency to coordinate to maximum number of metal ions so that one phosphonate group with three P–O bonds can coordinate up to eight metal ions which results in construction of packed structures with negligible porosity for coordination polymers based on phosphonate ligands. So, designing porous phosphonate coordination polymers is still a challenge. Also, such a high packing and strong bonds in metal–phosphonate frameworks make them less soluble which are serious problem in the way of synthesis of crystalline metal–phosphonate frameworks.
(III) In metal–phosphonate frameworks, the higher metal ion valence leads in lower solubility if the framework so that tetravalent metal–phosphonates are insoluble even in strong acids. Since, solubility of metal–phosphonate frameworks based on high valence metal ions (like Al^{3+}, Pb^{4+} and Zr^{4+}) is extremely low, it is so hard to synthesis novel crystalline metal–phosphonate frameworks based on high valence metal ions. However, these materials are water resistant and thermally stable.

In some case metal–phosphonate frameworks are similar with metal–sulfonate frameworks rather metal–carboxylate frameworks. For example, Structural predictability of metal–phosphonate frameworks is harder than metal–carboxylate frameworks because of higher number of P–O coordinating sites. Also, the polarity of the pore wall and porosity of phosphonate frameworks is larger than metal–carboxylate frameworks because of presence of uncoordinated polar P–O bonds in the pores.

In comparison of metal–phosphonate frameworks and metal–carboxylate frameworks we can denote that metal–phosphonate frameworks are of higher chemical and thermal stability, polarity as well as lower porosity, solubility, crystallinity and structural predictability. Newly, using carboxylate-phosphonate mixed ligands is a strategy to benefits from advantageous of both metal–phosphonate (stability and polarity) and metal–carboxylate (porosity and crystallinity) frameworks.

10.2 Functionalized Metal–Organic Frameworks by Fluorine Function

Fluorine most electronegative and low polarizable atom, so F–X (X is another atom except fluorine) bond is highly polar with high density of negative charge on fluorine atom. Since, chemical bonds based on F atoms are highly polar, synthesis of MOFs based on F containing ligands is a good strategy to develop MOFs with polar framework.

Fluorine based ligands applied extensively in the structure of MOFs as guest-interactive sites. Fluorine based ligands can be categorized in two groups; inorganic and organic fluorine based ligands. Organic fluorine-based ligands are based on highly polar C–F bonds mostly in fluoro (-F), trifluoromethyle (-CF_3), and perfluoroalkyle (-$(CF_2)_n$-) forms. Normally, inorganic fluorine-based ligands are SiF_6^{2-} [18–26] TiF_6^{2-} [27], BF_4^- [28, 29], PF_6^- [30–32], $AlF_5(H2O)_2^-$ [33], $NbOF_5^-$ [34, 35] and $CF_3SO_3^-$ [36, 37]. Both fluorine based organic and inorganic ligands applied extensively in field of gas adsorption because highly polar F-X bonds is favorite adsorption sites for different gas molecules like H_2 [38–42], CH_4 [43, 44], CO_2 [41, 42, 45–49], hydrocarbons [50, 51], fluorocarbons and chlorocarbons [52]. The host–guest interaction between gas molecules and F-X bonds is based on electrostatic interaction between F-X bond and gas molecules. Also, the improved polarity of the framework affects on the gas adsorption parameters like adsorption enthalpy and uptake capacity.

In structural view it is well-known that F containing functions like trifluoromethyle groups are hydrophobic. This character of fluorinated (F, CF_3...) groups applied for improvement of humidity and water stability of MOFs [53, 54].

The results of reviewing the published paper about fluorine decorated MOFs show that despite the simple chemistry of fluorine which centered at its polarity and hydrophobicity, fluorine decorated MOFs applied extensively in field of gas adsorption. This is due to the fact that gas molecules adsorbed on the polar surface. Such polar surface can be provided by fluorine decorated MOFs very well. Also, their hydrophobicity, is a tool to improve the stability of carboxylate based MOFs in aqueous media.

References

1. Dau, P.V. and Cohen, S.M., The influence of nitro groups on the topology and gas sorption property of extended Zn(ii)-paddlewheel MOFs. *CrystEngComm*, 15, 9304–9307, 2013.
2. Zhang, M., Wang, Q., Lu, Z., Liu, H., Liu, W., Bai, J., A nitro-decorated NbO-type metal–organic framework with a highly selective CO2 uptake and CH4 storage capacity. *CrystEngComm*, 16, 6287–6290, 2014.
3. Cmarik, G.E., Kim, M., Cohen, S.M., Walton, K.S., Tuning the Adsorption Properties of UiO-66 *via* Ligand Functionalization. *Langmuir*, 28, 15606–15613, 2012.
4. Biswas, S., Ahnfeldt, T., Stock, N., New Functionalized Flexible Al-MIL-53-X (X = –Cl, –Br, –CH3, –NO2, –(OH)2) Solids: Syntheses, Characterization, Sorption, and Breathing Behavior. *Inorg. Chem.*, 50, 9518–9526, 2011.
5. Wang, H., Wang, X., Liang, M., Chen, G., Kong, R.-M., Xia, L., Qu, F., A Boric Acid-Functionalized Lanthanide Metal–Organic Framework as a Fluorescence "Turn-on" Probe for Selective Monitoring of Hg2+ and CH3Hg+. *Anal. Chem.*, 92, 3366–3372, 2020.
6. Gong, W., Chen, X., Jiang, H., Chu, D., Cui, Y., Liu, Y., Highly stable Zr (IV)-based metal–organic frameworks with chiral phosphoric acids for catalytic asymmetric tandem reactions. *J. Am. Chem. Soc.*, 141, 7498–7508, 2019.
7. Zhai, F., Zheng, Q., Chen, Z., Ling, Y., Liu, X., Weng, L., Zhou, Y., Crystal transformation synthesis of a highly stable phosphonate MOF for selective adsorption of CO2. *CrystEngComm*, 15, 2040–2043, 2013.
8. Llewellyn, P.L., Garcia-Rates, M., Gaberová, L., Miller, S.R., Devic, T., Lavalley, J.-C., Bourrelly, S., Bloch, E., Filinchuk, Y., Wright, P.A., Serre, C., Vimont, A., Maurin, G., Structural Origin of Unusual CO2 Adsorption Behavior of a Small-Pore Aluminum Bisphosphonate MOF. *J. Phys. Chem. C*, 119, 4208–4216, 2015.
9. Fu, R., Hu, S., Wu, X., Rapid and sensitive detection of nitroaromatic explosives by using new 3D lanthanide phosphonates. *J. Mater. Chem. A*, 5, 1952–1956, 2017.
10. Begum, S., Wang, Z., Donnadio, A., Costantino, F., Casciola, M., Valiullin, R., Chmelik, C., Bertmer, M., Kärger, J., Haase, J., Krautscheid, H., Water-Mediated Proton Conduction in a Robust Triazolyl Phosphonate Metal–Organic Framework with Hydrophilic Nanochannels. *Chem. Eur. J.*, 20, 8862–8866, 2014.

11. Umeyama, D., Horike, S., Inukai, M., Itakura, T., Kitagawa, S., Inherent Proton Conduction in a 2D Coordination Framework. *J. Am. Chem. Soc.*, 134, 12780–12785, 2012.
12. Bao, S.-S., Otsubo, K., Taylor, J.M., Jiang, Z., Zheng, L.-M., Kitagawa, H., Enhancing Proton Conduction in 2D Co–La Coordination Frameworks by Solid-State Phase Transition. *J. Am. Chem. Soc.*, 136, 9292–9295, 2014.
13. Ortiz, A.U., Boutin, A., Gagnon, K.J., Clearfield, A., Coudert, F.-X., Remarkable Pressure Responses of Metal–Organic Frameworks: Proton Transfer and Linker Coiling in Zinc Alkyl Gates. *J. Am. Chem. Soc.*, 136, 11540–11545, 2014.
14. Ramaswamy, P., Wong, N.E., Gelfand, B.S., Shimizu, G.K.H., A Water Stable Magnesium MOF That Conducts Protons over 10^{-2}·S·cm^{-1}. *J. Am. Chem. Soc.*, 137, 7640–7643, 2015.
15. Pili, S., Argent, S.P., Morris, C.G., Rought, P., García-Sakai, V., Silverwood, I.P., Easun, T.L., Li, M., Warren, M.R., Murray, C.A., Proton conduction in a phosphonate-based metal–organic framework mediated by intrinsic "free diffusion inside a sphere. *J. Am. Chem. Soc.*, 138, 6352–6355, 2016.
16. Gagnon, K.J., Perry, H.P., Clearfield, A.J.C.r., Conventional and unconventional metal–organic frameworks based on phosphonate ligands: MOFs and UMOFs. *Chem. Rev.*, 112, 1034–1054, 2011.
17. Goura, J. and Chandrasekhar, V.J.C.r., Molecular metal phosphonates. *Chem. Rev.*, 115, 6854–6965, 2015.
18. Cui, X., Chen, K., Xing, H., Yang, Q., Krishna, R., Bao, Z., Wu, H., Zhou, W., Dong, X., Han, Y., Li, B., Ren, Q., Zaworotko, M.J., Chen, B., Pore chemistry and size control in hybrid porous materials for acetylene capture from ethylene. *Science*, 353, 141–144, 2016.
19. Noro, S.i., Kitagawa, S., Kondo, M., Seki, K., A new, methane adsorbent, porous coordination polymer [{CuSiF6 (4, 4′-bipyridine) 2} n]. *Angew. Chem. Int. Ed.*, 39, 2081–2084, 2000.
20. Kanoo, P., Reddy, S.K., Kumari, G., Haldar, R., Narayana, C., Balasubramanian, S., Maji, T.K., Unusual room temperature CO2 uptake in a fluoro-functionalized MOF: Insight from Raman spectroscopy and theoretical studies. *Chem. Commun.*, 48, 8487–8489, 2012.
21. Nugent, P., Rhodus, V., Pham, T., Tudor, B., Forrest, K., Wojtas, L., Space, B., Zaworotko, M., Enhancement of CO2 selectivity in a pillared pcu MOM platform through pillar substitution. *Chem. Commun.*, 49, 1606–1608, 2013.
22. Elsaidi, S.K., Mohamed, M.H., Schaef, H.T., Kumar, A., Lusi, M., Pham, T., Forrest, K.A., Space, B., Xu, W., Halder, G.J., Liu, J., Zaworotko, M.J., Thallapally, P.K., Hydrophobic pillared square grids for selective removal of CO2 from simulated flue gas. *Chem. Commun.*, 51, 15530–15533, 2015.
23. Elsaidi, S.K., Mohamed, M.H., Pham, T., Hussein, T., Wojtas, L., Zaworotko, M.J., Space, B., Crystal Engineering of a 4,6-c fsc Platform That Can Serve as a Carbon Dioxide Single-Molecule Trap. *Cryst. Growth Des.*, 16, 1071–1080, 2016.

24. Burd, S.D., Ma, S., Perman, J.A., Sikora, B.J., Snurr, R.Q., Thallapally, P.K., Tian, J., Wojtas, L., Zaworotko, M.J., Highly Selective Carbon Dioxide Uptake by [Cu(bpy-n)2(SiF6)] (bpy-1 = 4,4′-Bipyridine; bpy-2 = 1,2-Bis(4-pyridyl) ethene). *J. Am. Chem. Soc.*, 134, 3663–3666, 2012.
25. Nugent, P., Belmabkhout, Y., Burd, S.D., Cairns, A.J., Luebke, R., Forrest, K., Pham, T., Ma, S., Space, B., Wojtas, L., Eddaoudi, M., Zaworotko, M.J., Porous materials with optimal adsorption thermodynamics and kinetics for CO2 separation. *Nature*, 495, 80, 2013.
26. Elsaidi, S.K., Mohamed, M.H., Simon, C.M., Braun, E., Pham, T., Forrest, K.A., Xu, W., Banerjee, D., Space, B., Zaworotko, M.J., Thallapally, P.K., Effect of ring rotation upon gas adsorption in SIFSIX-3-M (M = Fe, Ni) pillared square grid networks. *Chem. Sci.*, 8, 2373–2380, 2017.
27. Chen, K.-J., Scott, H.S., Madden, D.G., Pham, T., Kumar, A., Bajpai, A., Lusi, M., Forrest, K.A., Space, B., Perry, J.J., Zaworotko, M.J., Benchmark C2H2/CO2 and CO2/C2H2 Separation by Two Closely Related Hybrid Ultramicroporous Materials. *Chem*, 1, 753–765, 2016.
28. Kotani, R., Kondo, A., Maeda, K., Gate adsorption of CO2 on a flexible one-dimensional copper-based coordination polymer crystal. *Chem. Commun.*, 48, 11316–11318, 2012.
29. Kondo, A., Noguchi, H., Ohnishi, S., Kajiro, H., Tohdoh, A., Hattori, Y., Xu, W.-C., Tanaka, H., Kanoh, H., Kaneko, K., Novel expansion/shrinkage modulation of 2D layered MOF triggered by clathrate formation with CO2 molecules. *Nano Lett.*, 6, 2581–2584, 2006.
30. Noro, S.-i., Tanaka, D., Sakamoto, H., Shimomura, S., Kitagawa, S., Takeda, S., Uemura, K., Kita, H., Akutagawa, T., Nakamura, T., Selective Gas Adsorption in One-Dimensional, Flexible CuII Coordination Polymers with Polar Units. *Chem. Mater.*, 21, 3346–3355, 2009.
31. Noro, S.-i., Matsuda, R., Hijikata, Y., Inubushi, Y., Takeda, S., Kitagawa, S., Takahashi, Y., Yoshitake, M., Kubo, K., Nakamura, T., High CO2/CH4 Selectivity of a Flexible Copper(II) Porous Coordination Polymer under Humid Conditions. *ChemPlusChem*, 80, 1517–1524, 2015.
32. Noro, S.-i., Hijikata, Y., Inukai, M., Fukushima, T., Horike, S., Higuchi, M., Kitagawa, S., Akutagawa, T., Nakamura, T., Highly Selective CO2 Adsorption Accompanied with Low-Energy Regeneration in a Two-Dimensional Cu(II) Porous Coordination Polymer with Inorganic Fluorinated PF6– Anions. *Inorg. Chem.*, 52, 280–285, 2013.
33. Cadiau, A., Belmabkhout, Y., Adil, K., Bhatt, P.M., Pillai, R.S., Shkurenko, A., Martineau-Corcos, C., Maurin, G., Eddaoudi, M., Hydrolytically stable fluorinated metal–organic frameworks for energy-efficient dehydration. *Science*, 356, 731–735, 2017.
34. Bhatt, P.M., Belmabkhout, Y., Cadiau, A., Adil, K., Shekhah, O., Shkurenko, A., Barbour, L.J., Eddaoudi, M., A Fine-Tuned Fluorinated MOF Addresses the Needs for Trace CO2 Removal and Air Capture Using Physisorption. *J. Am. Chem. Soc.*, 138, 9301–9307, 2016.

35. Cadiau, A., Adil, K., Bhatt, P., Belmabkhout, Y., Eddaoudi, M., A metal-organic framework-based splitter for separating propylene from propane. *Science*, 353, 137–140, 2016.
36. Noro, S.-i., Fukuhara, K., Hijikata, Y., Kubo, K., Nakamura, T., Rational Synthesis of a Porous Copper(II) Coordination Polymer Bridged by Weak Lewis-Base Inorganic Monoanions Using an Anion-Mixing Method. *Inorg. Chem.*, 52, 5630–5632, 2013.
37. Kondo, A., Chinen, A., Kajiro, H., Nakagawa, T., Kato, K., Takata, M., Hattori, Y., Okino, F., Ohba, T., Kaneko, K., Kanoh, H., Metal-Ion-Dependent Gas Sorptivity of Elastic Layer-Structured MOFs. *Chem. Eur. J.*, 15, 7549–7553, 2009.
38. Huang, Y. and Ke, S.H., Hydrogen Storage in MOF-5 with Fluorine Substitution: A van der Waals Density Functional Theory Study. *Adv. Mat. Res.*, 716, 244–247, 2013.
39. Hulvey, Z., Falcao, E.H.L., Eckert, J., Cheetham, A.K., Enhanced H2 adsorption enthalpy in the low-surface area, partially fluorinated coordination polymer Zn5(triazole)6(tetrafluoroterephthalate)2(H2O)2•4H2O. *J. Mater. Chem.*, 19, 4307–4309, 2009.
40. Forrest, K.A., Pham, T., Georgiev, P.A., Pinzan, F., Cioce, C.R., Unruh, T., Eckert, J., Space, B., Investigating H2 Sorption in a Fluorinated Metal–Organic Framework with Small Pores Through Molecular Simulation and Inelastic Neutron Scattering. *Langmuir*, 31, 7328–7336, 2015.
41. Pachfule, P., Chen, Y., Jiang, J., Banerjee, R., Fluorinated Metal–Organic Frameworks: Advantageous for Higher H2 and CO2 Adsorption or Not? *Chem. Eur. J.*, 18, 688–694, 2011.
42. Gupta, M., Chatterjee, N., De, D., Saha, R., Chattaraj, P.K., Oliver, C.L., Bharadwaj, P.K., Metal–Organic Frameworks of Cu (II) Constructed from Functionalized Ligands for High Capacity H2 and CO2 Gas Adsorption and Catalytic Studies. *Inorg. Chem.*, 59, 1810–1822, 2020.
43. Chang, G., Wen, H., Li, B., Zhou, W., Wang, H., Alfooty, K., Bao, Z., Chen, B., A Fluorinated Metal–Organic Framework for High Methane Storage at Room Temperature. *Cryst. Growth Des.*, 16, 3395–3399, 2016.
44. Zhao, D., Yu, C., Jiang, J., Duan, X., Zhang, L., Jiang, K., Qian, G., A fluorinated Zr-based MOF of high porosity for high CH4 storage. *J. Solid State Chem.*, 277, 139–142, 2019.
45. Pal, T.K., De, D., Senthilkumar, S., Neogi, S., Bharadwaj, P.K., A Partially Fluorinated, Water-Stable Cu(II)–MOF Derived *via* Transmetalation: Significant Gas Adsorption with High CO2 Selectivity and Catalysis of Biginelli Reactions. *Inorg. Chem.*, 55, 7835–7842, 2016.
46. Zhang, D.-S., Chang, Z., Li, Y.-F., Jiang, Z.-Y., Xuan, Z.-H., Zhang, Y.-H., Li, J.-R., Chen, Q., Hu, T.-L., Bu, X.-H., Fluorous Metal-Organic Frameworks with Enhanced Stability and High H2/CO2 Storage Capacities. *Sci. Rep.*, 3, 3312, 2013.
47. Liang, W., Bhatt, P.M., Shkurenko, A., Adil, K., Mouchaham, G., Aggarwal, H., Mallick, A., Jamal, A., Belmabkhout, Y., Eddaoudi, M., A tailor-made

interpenetrated MOF with exceptional carbon-capture performance from flue gas. *Chem*, 5, 950–963, 2019.

48. Belmabkhout, Y., Zhang, Z., Adil, K., Bhatt, P.M., Cadiau, A., Solovyeva, V., Xing, H., Eddaoudi, M., Hydrocarbon recovery using ultra-microporous fluorinated MOF platform with and without uncoordinated metal sites: I—Structure properties relationships for C2H2/C2H4 and CO2/C2H2 separation. *Chem. Eng. J.*, 359, 32–36, 2019.
49. Chen, S., Wang, D.-W., Wang, S.-J., Jiang, J.-J., Su, C.-Y., Stable fluorinated 3D isoreticular nanotubular triazole MOFs: Synthesis, characterization and CO2 separation. *J. Porous Mater.*, 26, 1573–1579, 2019.
50. Fan, W., Liu, X., Wang, X., Li, Y., Xing, C., Wang, Y., Guo, W., Zhang, L., Sun, D., A fluorine-functionalized microporous In-MOF with high physicochemical stability for light hydrocarbon storage and separation. *Inorg. Chem. Front.*, 5.10, 2445–2449, 2018.
51. Belmabkhout, Y., Zhang, Z., Adil, K., Bhatt, P.M., Cadiau, A., Solovyeva, V., Xing, H., Eddaoudi, M., Hydrocarbon recovery using ultra-microporous fluorinated MOF platform with and without uncoordinated metal sites: I—structure properties relationships for C2H2/C2H4 and CO2/C2H2 separation. *Chem. Eng. J.*, 359, 32–36, 2019.
52. Chen, T.-H., Popov, I., Kaveevivitchai, W., Chuang, Y.-C., Chen, Y.-S., Jacobson, A.J., Miljanić, O.Š., Mesoporous Fluorinated Metal–Organic Frameworks with Exceptional Adsorption of Fluorocarbons and CFCs. *Angew. Chem. Int. Ed.*, 54, 13902–13906, 2015.
53. Joharian, M. and Morsali, A., Ultrasound-assisted synthesis of two new fluorinated metal–organic frameworks (F-MOFs) with the high surface area to improve the catalytic activity. *J. Solid State Chem.*, 270, 135–146, 2019.
54. Wu, T., Shen, L., Luebbers, M., Hu, C., Chen, Q., Ni, Z., Masel, R.I., Enhancing the stability of metal–organic frameworks in humid air by incorporating water repellent functional groups. *Chem. Commun.*, 46, 6120–6122, 2010.

Index

Printed in the USA/Agawam, MA
September 16, 2026

910661.007